INQUIRIES

Israel Scheffler
INQUIRIES

Philosophical Studies of Language, Science, & Learning

WIPF & STOCK · Eugene, Oregon

Wipf and Stock Publishers
199 W 8th Ave, Suite 3
Eugene, OR 97401

Inquiries
Philosophical Studies of Language, Science and Literature
By Scheffler, Israel
Copyright©1986 by Scheffler, Israel
ISBN 13: 978-1-62564-083-3
Publication date 5/31/2013
Previously published by Hackett Publishing Co., 1986

To Nelson Goodman

CONTENTS

Preface ix
Introduction xi

Part One.   *Language & Symbol*   1
1. An Inscriptional Approach to Indirect Quotation   7
2. On Synonymy and Indirect Discourse   14
3. Inscriptionalism and Indirect Quotation   21
4. Postscript on Inscriptionalism   28
5. Explanations, Desires, and Inscriptions   31
6. Symbolic Aspects of Ritual I: Ritual, Myth, and Feeling: Cassirer and Langer   41
7. Symbolic Aspects of Ritual II: Ritual and Reference: Five Modes   52
8. Reply to Gareth Matthews on Ritual and Reference   70
9. Four Questions of Fiction   74

Part Two.   *Science & Reality*   81
1. Explanation, Prediction and Abstraction   87
2. Thoughts on Teleology   103
3. Prospects of a Modest Empiricism   122
4. Inductive Inference: A New Approach   154
5. Reflections on the Justification of Induction   165
6. The Paradoxes of Confirmation   175
7. Selective Confirmation and The Ravens (with Nelson Goodman)   201
8. An Improvement in the Theory of Projectibility (with Robert Schwartz and Nelson Goodman)   207
9. Projectibility: A Postscript   212
10. What is Said to Be (with Noam Chomsky)   215
11. Reflections on the Ramsey Method   225
12. Epistemology of Objectivity   232
13. Vision and Revolution: A Postscript on Kuhn   259
14. The Wonderful Worlds of Goodman   271

Part Three. *Learning & Acting* 279
1. Anti-Naturalist Restrictions in Ethics 283
2. On Justification and Commitment 293
3. Is the Dewey-like Notion of Desirability Absurd? 303
4. Philosophical Models of Teaching 309
5. On Ryle's Theory of Propositional Knowledge 322
6. Moral Education and the Democratic Ideal 330
7. Basic Mathematical Skills: Some Philosophical and Practical Remarks 338
8. In Praise of the Cognitive Emotions 347
9. Dewey's Social and Educational Theory 363
10. Pragmatism as a Philosophy 375

Writings of Israel Scheffler 393
Index 397

PREFACE

THIS BOOK consists for the most part of papers that have appeared since 1953, reporting various inquiries into language, science, and learning. Several contain first formulations of ideas more fully developed in one or another of the books I published during this period. Some respond to criticisms of these ideas or of others presented in my books, while the remainder elaborate such ideas or treat of independent themes. Certain extracts from the books themselves are included to round out the presentation of my views. Introductions to the Parts offer recent reflections that carry further the original inquiries, and an extended study of ritual appears here for the first time in its present form.

I have learned from so many people during the period spanned by this book that I cannot possibly thank them all. Specific acknowledgments, along with information on sources, are given in notes to the selections themselves. Noam Chomsky, Nelson Goodman, and Robert Schwartz granted permission to include papers jointly written with me, for which I am grateful. Catherine Z. Elgin provided helpful criticism of initial drafts of my introductions, as well as acute discussion of various points of substance. I am indebted to her and to Samuel Scheffler for encouraging me to prepare this book and for invaluable assistance in thinking through its nature and organization. I must finally mention with thanks the secretarial help of JoAnne Sorabella, who worked valiantly, efficiently, and cheerfully in readying the manuscript, proofreading, and preliminary preparation of the index.

The book is dedicated to Nelson Goodman, teacher, colleague, and friend, for his eightieth birthday.

ISRAEL SCHEFFLER

Newton, Massachusetts
December 1985

INTRODUCTION

INQUIRY, in every case, reaches out to the new; the quest for certainty is the quest for an end to inquiry. Rejecting this quest shapes a distinctive epistemological attitude – one which is prominent in the pages to follow.

In nominalistic spirit, I reject reliance on the modalities, whether logical, natural, or linguistic, as fundamentally unclear. Necessity in all its guises is thus abjured and the notion of analyticity as well. Accordingly, the quest for a certainty based upon meaning is wholly abandoned.

Nor is there any room for certainty in my conception of scientific observation. Evidence, argument, test, and probability are obvious touchstones in science, but none harbors a hard core of immutable sensory judgment. Though observation may indeed dislodge theory, theory may also overrule observation, and one observation may conflict with another. Observation relates to theory not as the certain to the probable but rather as the particular to the general. Clashing with theory, an independently credible observation report may provoke a reequilibration of belief, but the resulting system may as well exclude as include it.

The foundational view of knowledge is thus opposed, along with the obverse doctrine of pure coherence. If science is not an airy fantasy, neither is it an edifice resting upon basic beliefs beyond the threat of change. A better image of science is that suggested by Peirce, of a cable made of many interwoven fibers and stronger than any of these, none of which is indispensable. Alternatively, there is the more recent suggestion of Popper, comparing science to a house built upon a swamp, and Neurath's metaphor of science as a ship being continually rebuilt upon the open sea. To capture the anti-foundational spirit of science while affirming its credibility is the task of the view of justification proposed below.

Because the pattern of justification that I propose applies not only to science but also to ethics, there is no haven for certainty in the latter realm either. No act or belief is an island. Each has systematic ramifications. That is to say, each can survive only within a changing community of surrounding acts or beliefs, whose claims demand equal consideration even if they do not uniformly carry equal weight. Such a picture does not, of course, imply that no act or belief is durable or stable. It implies only that stability is not a self-evident privilege or entitlement but rather an achievement, to be reconciled with the demands of continuing systematization.

This epistemological attitude extends also to philosophy, which, no more than the realms of meaning, science, or ethics, gives sanctuary to certainty. Philosophy has no direct access to higher realities, firmer principles, or keener insights than are available elsewhere. As I view it, philosophy is systematic interpretation and reflection. Reflecting upon prior belief and practice, it analyzes, questions, criticizes, and systematizes, thus modifying the initial objects of its attention.

It begins, not at the beginning, with a clean slate, but in the middle and after the fact. As Peirce put it in his critique of Descartes's method, "We cannot begin with complete doubt. We must begin with all the prejudices which we actually have when we enter upon the study of philosophy." Whatever illumination philosophy may yield is an outcome of its work upon such prejudices rather than a consequence of its transcendent certainties projected upon a vacant field.

Philosophy stands, not outside the sphere of common thought and experience but squarely within it. It carries no epistemological immunity to the common ills and distempers of inquiry. Logic, evidence, clarity, system, truth – all make their demands of philosophy no less than of other realms of the intellect and, where philosophers fall short, they can offer no special excuse. What is distinctive about philosophy is not its certainty but its persistent curiosity, not its infallibility but its interest in understanding every sphere of thought and life.

If philosophy is not self-sufficient, however, neither is it powerless. It presupposes, but it also reworks the pre-philosophical matter from which it sets out. Like science, from which it cannot be sharply distinguished, it may yield novelty and reveal the unseen.

Like science too, it is pluralistic, and in three main senses. First, its problems are plural, drawn from any region whatever of pre-philosophical interest or conviction. Second, its concepts are not preordained; they need to be designed or chosen out of an infinite array of alternatives. Third, its solutions are not uniquely related to the problems they address. For any systematic interpretation preserving the preferred truths of some pre-philosophical domain, there will be others that do the same. Yet, to provide *any* adequate interpretation, in philosophy no less than in science, is a significant feat; pluralism is not nihilism. An adequate interpretation represents a triumph of insight and order available to all – hence progress.

Such progress of the understanding is not limited in its import to the understanding. It colors our feelings, memories, perceptions, anticipations,

Introduction

and actions. Philosophy is no more a spectator sport than is science or art. Its analyses modify habits, its techniques channel visions, its visions organize sentiments and orient actions. Such connectivity of theory and practice, affirmed by pragmatism, rings true to me and inclines me toward that philosophy – in its broad outlines at least, and despite my criticisms, elsewhere, of its specific formulations.

There are, of course, variant readings of pragmatism, as of every philosophical movement, and some are too soft for my taste. That philosophy is social does not mean it is merely social; that it comments on culture does not imply it is only cultural commentary. Philosophy converses, but is not swallowed up in conversation. If its starting points are not fixed and its paths and destinations are plural, yet it traverses a hard terrain imposing a severe discipline. No more than science can it simply will its conclusions; no less than science does it strive for objectivity relative to independent constraints, variable as these may be. Objectivity without certainty, relativity but not subjectivism, truth consistent with pluralism – these are the pragmatic emphases I admire.

It follows that I see no rift between pragmatism and analytic methods in philosophy. Without argument and analysis, pragmatism is mere attitude, not philosophy proper. As I read the pragmatists, they came not to bury philosophical analysis but to apply it to novel effect over a wide range of contemporary problems. Nor can current analyses, of whatever school, be enclosed within a small circle of technical concerns. There can in fact be no conflict between range and argument, between vision and technique. This conviction, at any rate, has helped sustain the variety of analytical efforts represented in the selections to follow.

These selections address specific problems and comprise no overarching system of thought. But they are related by common themes and connecting orientations. My preference for meager means in logic and ontology is evident throughout. It is motivated, not by miserliness but by three considerations: that philosophical obscurity increases with extravagance, that triviality of result varies inversely with economy, and that a solution gained under stringent constraints is likely to be preserved under relaxation, but not vice versa.

The logical attitude pervading my work is nominalistic, eschewing abstract entities and the modalities while recognizing a variety of individuals. In the study of language, such nominalism takes the form of inscriptionalism – that is, a recognition of linguistic tokens alone, to the exclusion of types and other universals. The resulting analysis of indirect quotation, in terms of that-clause predicates of individuals, is described in Part 1 and discussed there and in Part 2 as well, in its later extension to contexts of belief, desire, teleology, and explanation.

The concern for economy in ontological matters is evident in papers in Parts 1 and 2 that range from the analysis of fiction to Ramsey interpretations of theory. Such concern is shown as well by the discussion in Part 2 of criteria of ontological commitment.

Now, the ontological commitment of a theory may be thought of as a type of putative reference and, when the theory is true, as a type of reference. "There are unicorns" purports to refer to unicorns but fails, for there are none; its ontological commitment falls short of reference. On the other hand, "There are cats" not only commits itself to the existence of cats but refers to them as well. Both Part 1 and Part 2 deal with reference – in the context of ritual and of science, respectively. The semantic view here presupposed is that reference ranges beyond the merely denotative and is restricted neither to science nor even to verbal expression.

It is, however, in any of its modes, cognitive in import, modifying awareness and informing belief. But cognition is itself not wholly isolated within the realm of belief, sharply separated from action, affect, and learning. Various links between cognition and these latter domains are articulated in the selections to follow. One link is expressed in the view, already noted, that reference is involved in ritual as well as in science – more generally, in the nonverbal as well as the verbal. Another is developed in the treatment of teleology, which joins belief with desire in the effort to explain action. A third link is outlined in the treatment of emotion (in Part 3) as not merely presupposing cognition and affecting it generally in turn but as encompassing such varieties as surprise, with specific cognitive functions.

A fourth link is forged by the discussions of ethical topics in Part 3. Naturalism in ethics is upheld and the emotive theory of ethics denied; the justification of actions is portrayed as similar in pattern to the justification of beliefs; and pragmatic analogies between the scientific and the democratic ethos are brought out. A fifth link, finally, is comprised by the treatment of learning (in Part 3) as acquisition of skill, insight, and habit as well as belief, as not only epistemic but also moral, emotional, and institutional in its significance.

The material of the book, edited only very lightly, has been organized into three parts. Part 1, "Language and Symbol," begins with my inscriptional interpretation of indirect discourse and treats also related problems of belief, desire, and fiction from an inscriptional point of view. Further, it offers an approach to the symbolism of ritual, commenting on the ideas of Cassirer and Langer and presenting an independent account of the multiple modes of ritual reference.

Part 2, "Science and Reality," deals with explanation, prediction, and induction, addressing as well related questions of empirical significance,

Introduction

teleology, and paradoxes of confirmation. It concerns itself, in addition, with the relations between science and reality – with the ontology of theories, credibility and truth, paradigms and objectivity, invented versions, and worlds not made but affirmed.

Part 3, "Acting and Learning," treats human action in its relations to valuing, feeling, and learning. Interpreting justification as bridging the realms of science and conduct, it further relates cognition and emotion, skill and understanding, doing and knowing. Having expressed such characteristic pragmatic emphases in discussions of teaching as well as of mathematical and moral education, it concludes with an explicit account of pragmatism as a philosophy.

PART ONE
Language & Symbol

THE FIRST PAPER here treats indirect quotation from an inscriptional point of view. Thereafter, ascriptions of belief and desire are similarly treated. Finally, interpretations of ritual and of fiction are undertaken, against a background of inscriptional assumptions.

Behind this expansion of scope as exhibited in the papers stand my two books, *The Anatomy of Inquiry* (1963) and *Beyond the Letter* (1979).[1] The former considerably elaborates the inscriptional analysis offered in the first paper, extending it to problems in the ontology of explanation and the interpretation of teleology. The latter develops inscriptionalism in independent directions, through analyses of ambiguity, vagueness, and metaphor in language. Partly because of such newer preoccupations and partly through study of Goodman's *Languages of Art* (1968),[2] I was led to attempt an account of ritual as a form of symbolism. The result is reported in the papers comprising the last segment of Part 1.

As I reflect now on the development of this work, I see a strong thread of interest not always evident to me earlier – an interest in taming the more wayward growths of philosophy. I was thus, early on, impelled to try to make sense of indirect discourse, that wild patch of semantics – to make sense of it rather than simply reject it or pretend that it was already transparent. Later, I was drawn to deal with belief and desire – the tangled roots of teleology – and, following that attempt, to inquire into the most undisciplined aspects of language: ambiguity, vagueness, and metaphor. This sequence of projects has culminated, not unnaturally, with my effort to understand the symbolic functioning of ritual and the semantics of fiction.

In each case, I have wanted neither to deny the phenomenon addressed nor to presuppose or duplicate it in my own explanatory discourse; I have tried, rather, to provide a clear interpretation that would satisfy philosophically.

I expressed my general point of view in *Beyond the Letter* where, defending the project of providing a theoretical analysis of ambiguity, vagueness, and metaphor, I wrote, "To admit [these features] as objects of our descriptive efforts does not commit us to incorporating them into our descriptive language. Conversely, to bar them, as far as possible, from our descriptive apparatus in no way commits us to denying their existence in actual instances of language use. It is possible, in short, to strive for a clear, literal, and precise account of such phenomena as vagueness, ambiguity, and metaphor."[3]

Now I should like to comment on two critiques of my theory of indirect discourse, those of Alonzo Church and of Donald Davidson. The first paper below proposed a nominalistic analysis of indirect quotation capable of withstanding Church's strictures against Carnap, which were intended to show "an insuperable objection" to "analyses that undertake to do away with propositions".[4] My proposal was to take the that-clause, in an indirect quoting context of the form "... writes that ———", as a single predicate denoting just those concrete inscriptions comprising rephrasals of its that-content. "Thales writes that all is water" was thus to be interpreted, according to my proposal, as "Some inscription i is such that it is both a that-all-is-water, and inscribed by Thales."

Church responded by arguing against nominalism in general that it is "too complicated in application."[5] I reply to this argument from complexity in the third paper below. He also raised the question how to interpret a new group of sentences nominalistically, e.g., "Church and Goodman have contradicted each other." I replied by acknowledging that further problems assuredly remain for nominalists (as they do for non-nominalists) and sketched a possible analysis for the above sentence.

In a later paper, Church contends that his new problem sentences are not "just a further problem to be solved" but rather "the crux of the matter."[6] He also finds fault with the analytic sketch I proposed. Now these problem sentences (unlike his original examples, "Seneca said that man is a rational animal" and "Columbus believed the world to be round") are indeterminately ascriptive, none specifying a sentence as content of the ascribed assertion. And such indeterminately ascriptive sentences are, as Quine has suggested, dispensable. Illustrating his point with the two sentences, "Paul believes something that Elmer does not" and "Eisenhower and Stevenson agree on something," Quine writes, "Such quantifications tend anyway to be pretty trivial in what they affirm, and useful only in heralding more tangible information."[7] Once such information is supplied, in the way of ascribing particular sentences, my theory of indirect discourse again acquires direct purchase on the problem.

Thus, even if Church's objections to my analytic sketch are assumed to be decisive, they suggest to the nominalist only the abandonment of the

Introduction

isolated quantifications in question, leaving intact his fundamental understanding of indirect discourse. And these objections are, in any case, powerless to demonstrate that an adequate nominalistic analysis of such sentences cannot be given.

Donald Davidson criticizes my theory on different grounds. He argues that my postulated that-clause predicates are infinite in number and unstructured, hence incapable of supporting a truth predicate characterizable "in Tarski's style." Any language containing such predicates is therefore, he maintains, unlearnable. Davidson's preferred theory is to take the "that" of indirect quotation as a demonstrative referring to its ensuing, but unasserted, utterance.[8]

Now, that the learnability of language is in general dependent upon rule and structure I regard as a wholly implausible doctrine. The learning of languages with metaphors, demonstratives, and indexicals, not to mention the acquisition of symbol systems of nonlinguistic sorts, would be difficult to understand, in any case, as rule-governed processes, and impossible to understand were Davidson's restriction to finite languages supporting recursive definitions in Tarski's style to be taken seriously. Relevant arguments to this effect have been offered in recent papers of Robert Schwartz and Catherine Elgin.[9]

But waiving this general point, what is Davidson's notion of structure and how does it serve to show my language unlearnable? Just after calling my that-clause predicates "unstructured," he says, "Quine does not put the matter quite this way, and he may resist my appropriation of the terms 'logical form' and 'structure' for purposes that exclude application to Scheffler's predicate. Quine calls the predicate 'compound' and describes it as composed of an operator and a sentence."[10] Though not "logically analyzable," such predicates are, in fact, structured and rule-governed; it it is easy to see how the rule guides learning.[11] In a discussion of related points, R. J. Haack remarks, "The contention that a language which contains Scheffler's predicates is unlearnable in principle is false simply because the language is learnable. Scheffler has learnt it, Quine has learnt it, Davidson has learnt it and I have learnt it."[12]

Haack further points out several ways in which structural relations are properly imputable to that-clause predicates. That such predicates are, in any case, not simply unanalyzable is, he says, "clear from the fact that the rule for forming the single predicates is also one from which sentences can be recovered from the predicates." He argues further that Tarski-type truth definitions are not precluded for languages with such predicates and shows how a class of sentences containing such predicates can be provided truth conditions.

Davidson's own theory seems to me to be vulnerable, by his lights, to the criticisms he makes of mine. To begin with, his theory depends on the demonstrative "that," which changes its reference with each ensuing utterance in contexts of indirect discourse. This means that his language contains "an infinite number of predicates," each "unstructured" and each "in the eyes of semantic theory, unrelated to the rest." These quoted phrases, used in criticizing the that-clause predicates of my theory, thus apply to the demonstratives required by his own and should lead him to suppose his own language unlearnable.

Davidson maintains that "an adequate account of the logical form of a sentence" must exhibit its semantic character as "owed to how it is composed, by a finite number of applications of some of a finite number of devices that suffice for the language as a whole, out of elements drawn from a finite stock (the vocabulary) that suffices for the language as a whole."[13] The demonstrative "that," upon which his theory rests, is not, however, in the semantic sense, an element of vocabulary, nor, a fortiori, is it a compositional device. Rather, it is a syntactic means for forming an infinite number of *predicates,* each semantically "unstructured" and "unrelated to the rest."

Scott Weinstein has proposed a truth definition for demonstrative utterances but concludes with the caution: "According to our theory, in order to determine the condition under which an utterance is true, we must determine what sentence it is an utterance of and what the referents of any demonstratives which occur in that utterance are.... We have made no attempt to say anything informative about these relations in our theory."[14] In other words, to apply Weinstein's definition, the reference and, hence, the predicative status of the particular utterance of "that" must first be established. And this must presumably be established afresh for each utterance in every new context altering the demonstrative reference, that is, for a theoretically infinite number of cases. How is this procedure itself learned, on Davidson's account?

Davidson claims that his theory provides the structure that mine lacks. But the content of the that-clause is, on his view, equally inert. It is displayed, not asserted. What follows the demonstrative "that" gives, as he says, "the content of the subject's saying, but has no logical or semantic connection with the original attribution of a saying. This last point is no doubt the novel one, and upon it everything depends: from a semantic point of view the content-sentence in indirect discourse is not contained in the sentence whose truth counts."[15] Hardly a novel point in my opinion, this feature defines the problem and is incorporated in my treatment no less than in Davidson's.

On the other hand, he claims, "The familiar words coming in the train of the performative of indirect discourse do, on my account, have structure, but it is familiar structure and poses no problem for theory of truth not there before indirect discourse was the theme."[16] The structure presumably consists in the fact that the that-content, though unasserted, can be processed as a sentence in the speaker's language, even though "not contained in the sentence whose truth counts." But this much structure is certainly available on my theory, for which the "that" operator forms a predicate in the first instance by application to a sentence of the speaker's language. Though not "contained" logically within the quoting sentence, and unasserted by the quoting speaker, it is certainly recognizable and understandable as a sentence of that speaker's language, posing no more problem for the theory of truth on my account than on Davidson's.

Finally, a curiosity: Davidson quotes the Oxford English Dictionary as holding that "The use of *that* is generally held to have arisen out of the demonstrative pronoun pointing to the clause which it introduces."[17] Whatever may be said of Davidson's theory, it will not transpose generally to other languages, e.g., to French, for which the needed demonstrative is lacking. The point is of some interest since, in considering the proposed analysis of "Galileo said that the earth moves" as "Galileo spoke a sentence that meant in his language what 'The earth moves' means in English," Davidson dismisses the analysis in the following words, "To see how odd this is, however, it is only necessary to reflect that the English words 'said that,' with their built-in reference to English, would no longer translate (by even the roughest extensional standards) the French 'dit que.' "[18] Apparently the same incapacity of the theory he himself adopts later in the same paper is no longer seen as an impediment.

NOTES

1. I. Scheffler, *The Anatomy of Inquiry* (New York: Alfred A. Knopf, 1963; now Indianapolis: Hackett Publishing, 1981); I. Scheffler, *Beyond the Letter* (London: Routledge & Kegan Paul, 1979).

2. Nelson Goodman, *Languages of Art* (Indianapolis: Hackett Publishing, 1968).

3. *Beyond the Letter,* 6.

4. A. Church, "On Carnap's Analysis of Statements of Assertion and Belief," *Analysis* 10 (1950): 97-99.

5. A. Church, "Propositions and Sentences," in *The Problem of Universals,* ed. I. M. Bochenski, A. Church, and N. Goodman (Notre Dame, Indiana: University of Notre Dame Press, 1956).

6. A. Church, "Sobre el análisis del discurso indirecto propuesto por Scheffler," in *Semántica Filosófica: Problemas y Discusiones,* ed. Thomas Moro Simpson (Buenos Aires: Siglo Veintiuno, 1973), 363-69.

7. W. V. Quine, *Word and Object* (Cambridge, Mass: Technology Press of M.I.T. and John Wiley & Sons, 1960), 215. I have noted my agreement with Quine that such isolated quantifications are indeed expendable. See *Anatomy of Inquiry*, 108-10.

8. Donald Davidson, "Theories of Meaning and Learnable Languages," in *Logic, Methodology and Philosophy of Science*, ed. Y. Bar-Hillel (Amsterdam, 1965), 383-393, and "On Saying That," *Synthèse* 19 (1968-69): 130-46.

9. Robert Schwartz, "Infinite Sets, Unbounded Competences, and Models of Mind," in *Perception and Cognition,* ed. C. W. Savage, vol. 9, *Minnesota Studies in the Philosophy of Science* (1978), 183-200; and Catherine Z. Elgin, "Representation, Comprehension, and Competence," *Social Research* 51 (No. 4, Winter, 1984): 906-25.

10. Davidson, "On Saying That," op. cit. See also Quine's reply to Kaplan, in *Synthèse* 19 (1968-69): 314.

11. On related points, see p. 48 and especially n. 5 on p. 135 of my *Beyond the Letter*. On structure of that-clause predicates, see also Christopher S. Hill, "Toward a Theory of Meaning for Belief Sentences," *Philosophical Studies* (1976), pp. 209-26, esp. p. 218 and n. 10, p. 225.

12. R. J. Haack, "Davidson on Learnable Languages," *Mind* 87 (1978): 230-49. (See also R. J. Haack, "On Davidson's Paratactic Theory of Oblique Contexts," *Noûs* 5 (No. 4, 1971): 351-61).

13. "On Saying That," op. cit.

14. Scott Weinstein, "Truth and Demonstratives," *Noûs* 8 (1974): 179-84.

15. "On Saying That," op. cit.

16. Ibid.

17. Ibid.

18. Ibid.

1. AN INSCRIPTIONAL APPROACH TO INDIRECT QUOTATION

FOLLOWING Goodman[1] in treating inscriptions framed by quotes as concrete general rather than abstract singular terms,[2] and considering every inscription denoted by a given quotes-inscription to be a *replica*[3] of every other so denoted (including the framed content of the quotes-inscription), we understand from any statement that John writes "P", that John inscribes some replica of a given quotes-content, i.e. that of the quoting statement. A direct-quoting sentence such as "J writes 'P' " may, then, be analyzed as asserting: "(Ex)(Ey)(x = J · 'P' y · Inscribes xy)", where " 'P' " is construed as a single predicate of certain concrete inscriptions.

For indirect quotation, i.e. sentences of the form "... writes that _____", a structurally analogous treatment is here proposed, construing that-clauses in such contexts as single predicates of concrete inscriptions, and taking every inscription denoted by a given that-clause as a *rephrasal* of every other so denoted (including the that-content of the that-clause, i.e. all of the latter exclusive of the "that"). From any statement that John writes that P we understand that John inscribes some rephrasal of the appropriate that-content of the quoting statement. An indirect-quoting sentence such as "J writes that P" may now be analyzed as asserting: "(Ex)(Ey)(x = J · That-Py · Inscribes xy)", where "That-P" is construed as a single predicate of certain concrete inscriptions.

It is, of course, clear that not every rephrasal-pair is a replica-pair, and conversely, that not every replica-pair is a rephrasal-pair. Hence, not every replica of a given inscription denoted by some "that-P" is also so denoted, since some such replica may, for example, differ crucially in context, form part of a different language, or contain indicators. Nevertheless, though having replicas which are not rephrasals, no unique inscription-event is itself embedded in more than one appropriate context, or is part of more than one language. The shift from words and propositions to inscription-events enables us, then, to speak with determinacy of simply a that-P, just as we speak of a "P", without further specifications to avoid ambiguities of language, context, or intent.

This paper appeared in *Analysis* 14 (No. 4, 1954): 83-90.

Professor A. Church[4] in recent comments on Professor Carnap's analysis of belief-statements,[5] has offered what he thinks "may be an insuperable objection" against "analyses that undertake to do away with propositions in favour of such more concrete things as sentences". The proposal suggested above, for a purely inscriptional interpretation of indirect quotation, must hence be confronted with Professor Church's arguments. The attempt will be made to show that our nominalistic analysis completely escapes the force of Professor Church's strictures.

Church uses two sentences as illustrative throughout his paper: (1) *Seneca said that man is a rational animal,* and (A) *Columbus believed the world to be round.* He cites five different proposals for the analysis of (1) in terms of sentences. These proposals are as follows:

(2) Seneca wrote the words 'Man is a rational animal'.
(3) Seneca wrote the words, 'Rationale enim animal est homo'.
(4) Seneca wrote words whose translation from Latin into English is: 'Man is a rational animal'.
(5) Seneca wrote words whose translation from some language S' into English is: 'Man is a rational animal'.
(6) There is a language S' such that Seneca wrote as sentence of S' words whose translation from S' into English is 'Man is a rational animal'.

Church criticizes each of the five proposed analyses as follows:

(2) is false, while (1) is true.

(3) reproduces Seneca's words without saying what meaning was attached to them, while (1) conveys the content of what Seneca said without revealing his words.

(4) omits the information that Seneca intended his words as a Latin sentence, rather than as a sentence of some other language in which conceivably the identical words might have some quite different meaning.

(5) must be rejected for the same reason. Moreover if 'language' means 'semantical system' and does not necessarily refer to a historical phenomenon, then (5) is true of anything Seneca wrote.

(6) does not in itself allow us to infer (1), but only together with an additional factual premise, to the effect that 'Man is a rational animal' means in English that man is a rational animal. (6) can be shown to be inadequate as an analysis of (1) in this way, moreover.[6] Translate (6) and (1) into some other language (say German) and observe that the translated statements have different meanings (suppose the German reader of the two translated sentences to understand no English). The German translation of (1) is "Seneca hat gesagt, dass der Mensch ein vernünftiges Tier sei". In translating

(6) "English" becomes "Englisch", and " 'Man is a rational animal' " becomes " 'Man is a rational animal' ".

Since (6) is similar to Carnap's analysis, and the same criticism applies to it, Church takes his objection to (6) to be equally an objection to Carnap's theory. But he analyzes his second illustrative sentence (A) strictly in Carnap's terms to show how the objection to (6) applies here. Thus, following Carnap's analysis,[7] Church analyses (A) as:

(B) There is a sentence S'_i in a semantical system S' such that
(a) S'_i is intensionally isomorphic to "the world is round" and
(b) Columbus was disposed to an affirmative response to S'_i.

This formulation must, however, be improved, as Church points out, since "intensional isomorphism...is a relation between ordered pairs consisting each of a sentence and a semantical system". Thus (B) must be changed to

(C) There is a sentence S'_i in a semantical system S' such that
(a) S'_i as sentence of S' is intensionally isomorphic to "the world is round" as English sentence and (b) Columbus was disposed to an affirmative response to S'_i as sentence of S'.

And for (1) the analogue of (C) is given as:

(7) there is a sentence S'_i in a semantical system S' such that
(a) S'_i as sentence of S' is intensionally isomorphic to 'Man is a rational animal' as English sentence and (b) Seneca wrote S'_i as sentence of S'.

In the case of (C) as well as (7) the same objection is offered as was offered to (6), and the device of translation is said to point up, in each of these cases, the inadequacy of the proposed analysis.

According to Church, the foregoing criticisms assume that "English" and "Englisch" have pragmatic senses in English and German respectively, i.e., "the language current in Great Britain at time t". If, on the other hand, the sense of these words were taken to be "the language for which such and such semantical rules hold", with a sufficient list supplied to ensure that there is only one language referred to, then the objection that (1) is not a consequence of (6) or (7), "would...be less immediate, and it is possible that it would disappear".

In order to meet this contingency, Church offers the following objection to (7) as an analysis of (1). Suppose we have the German translation of (1), "Seneca hat gesagt, dass der Mensch ein vernünftiges Tier sei", which we call "(1')". As (1) is analyzed by (7), so (1') will be analyzed by (7") according to Carnap's proposal, i.e. "Es gibt einen Satz S'_i auf einem semantischen System S', so dass (*a*) S'$_i$ als Satz von S' intensional isomorph zu 'Der Mensch ist ein vernünftiges Tier' also deutscher Satz ist, und (*b*) Seneca S'$_i$ als Satz von S' geschrieben hat." Now, though (1) and (1') are translations of each other, (7) and (7"), in English and German respectively, are not translations of each other, and, in particular, they are not intensionally isomorphic, and "because of the exact parallelism between them, the two proposals stand or fall together". Furthermore, "if we consider the English sentence (*a*) "John believes that Seneca said that man is a rational animal" and its German translation (*a'*), we see that the sentences to which we are led as supposed analyses of (*a*) and (*a'*) may even have opposite truth-values in their respective languages; for John, though knowing the semantical rules of both English and German, may nevertheless fail to draw certain of their logical (or other) consequences." (p. 99).

In order to see whether or not any of Church's criticisms applies to the analysis which was here offered of indirect discourse, let us restate our analysis in terms of Church's examples. Church's sentence (1) becomes: "(Ex)(Ey)(x = Seneca · that-man-is-a-rational-animal y · Inscribes xy)". In English, we would say "Seneca wrote a that-man-is-a-rational-animal", or "There is a that-man-is-a-rational-animal S, such that Seneca wrote S".

Let us call our latter English analysis "E", and see whether or not any of Church's remarks in relation to his cited proposals 2-7 are applicable to E. E and (1) must always have the same truth value, so Church's criticism of (2) as differing in truth-value is inapplicable here. (3) is criticized on the grounds that "it reproduces Seneca's words without saying what meaning was attached to them", in contrast to (1) which conveys the content of what Seneca said without revealing his actual words. E, however, just like (1), fails to reveal Seneca's actual words, but conveys the content of what he said nonetheless, so that it escapes the criticism of (3). (4) is criticized on the grounds that, though it attributes to Seneca words whose translation from Latin into English is 'Man is a rational animal', it fails to say that Seneca intended his words as a Latin sentence. But, obviously, this objection holds only for an analysis, which like (4), mentions Latin. It does not hold for an analysis, which, like E, makes no mention of any language whatsoever. As a matter of fact, if E were to mention Latin, it would be doing more than (1), which also makes no mention of any particular language. Analogous remarks would suffice to show that Church's criticism of (5) is also inapplicable, since E mentions no particular language. As pointed out above, the utility of inscriptionalism in this connection is just that it enables avoidance of problems of multiple context and language.

For the same reason, *and since E mentions no particular directly-quoted expression,* the criticism of (6) pointing out the necessity of a factual premise giving the meaning of the mentioned expression in (6), is obviously inapplicable to E. That this criticism of (6) is inapplicable to E is, moreover, especially clear from the fact that the translation-device employed by Church to point up the inadequacy of (6) fails to undermine E. This translation-device rests upon the circumstance that (6) contains a directly-quoted sentence, which is translated as itself, while (1) contains no such sentence and hence is translatable completely, in its quote-part, into other words, in another language.[8] Thus, one who understood this other language and no other, would understand (1) and (6) differently and would clearly require a factual premise giving the meaning of the quoted expression in (6), in terms of this other language, in order to produce equivalent sentences. It should now be obvious, from this description, why the translation-device fails to distinguish between (1) and E in the same sense that it distinguishes between (1) and (6). For E, like (1), contains no directly-quoted sentence, and, hence, like (1), it is completely translatable in its quote-part, into other words, in another language. Thus, one who understood this other language, and no other, would understand both (1) and E, and would require no additional premise of the kind necessary for (6). Since E fails to analyze (1) in terms of a directly-quoted expression, but rather construes its that-clause as a predicate of rephrasals, it is completely translatable into other words, including this predicate. E thus becomes E' in German, "Es gibt einen Dass-der-Mensch-ein-vernüntriges-Tier-sei S, so dass Seneca S geschrieben hat."

Church's criticisms of (7) and (C) are the same as his criticism of (6) and rest upon the same circumstance that a directly-quoted expression appears in the analysans but not in the analysandum, thus rendering translation uneven. It should be clear from our remarks above that these criticisms of (7) and (C) are also inapplicable to E, for the same reason that the objection to (6) was found to be inapplicable to E, i.e., the latter contains no directly-quoted expression and its construction of the English that-clause of (1) as a predicate of rephrasals makes it wholly translatable into other words.

It remains now to examine Church's modified objection to (7). It rests on the fact that (1) and its German counterpart (1') are analyzed, according to Carnap's proposal, by (7) and (7") respectively, and that the latter are not translations of each other, nor are they intensionally isomorphic. Furthermore, considering the sentence "John believes that Seneca said that man is a rational animal" (a) and its German translation (a'), their analyses in accordance with Carnap's proposal may even produce sentences with differing truth-values, for John is asserted in (a), according to Carnap, to affirm that Seneca wrote a sentence intensionally isomorphic to an English sentence S, while in (a'), he is asserted to affirm that Seneca wrote a sentence intensionally isomorphic to a German sentence S'.

Let us now examine our proposed analyses of (1) and (1'). Our analysis of (1) is E, while our analysis of (1') is E', "Es gibt einen Dass-der-Mensch-ein-vernünftiges-Tier-sei S, so dass Seneca S geschrieben hat". E and E' are certainly translations of each other, E' having been originally introduced by us as a translation of E. Furthermore, E and E' are intensionally isomorphic. According to our theory then, *the analysis of a given indirect Q is a translation of the analysis of the translation of Q.* This is precisely what, as Church points out, Carnap's proposal fails to achieve.

Finally, regarding the sentence (*a*) and its German translation (*a*'), the difficulty pointed out by Church does not arise in the case of our analysis, for both (*a*) and (*a*') assert that John affirms that Seneca inscribed a that-man-is-a-rational-animal. Whereas, then, Carnap's analysis has the unlikely consequence that John is affirming something about the semantical rules of every language into which (*a*) may happen to be translated, our analysis has no such consequence, for no semantic term, like "is intensionally isomorphic to" or "is a rephrasal of" appears in our analysans. This fact is of crucial importance.

If our survey of Church's arguments is indeed correct, then we may conclude that, no matter how telling they may be in relation to an analysis similar to Carnap's, they are completely inapplicable to an analysis such as ours. If this is true, then Church's criticisms fail to reveal an insuperable objection to every analysis which rejects abstract entities in favour of concrete individuals. Indeed, our analysis has treated neither of sentence-types, of word-types, nor of propositions, and yet has been shown to withstand the force of Church's arguments. We conclude that the analysis of indirect quotation does not necessarily require an ontology which includes abstract entities.

NOTES

I gratefully acknowledge the help of Professor Nelson Goodman, who commented upon an expanded version of this paper, and with whom I have frequently discussed related issues. I wish also to thank Professor W. V. Quine for valuable suggestions concerning the present version.

1. Goodman, N., *The Structure of Appearance,* ch. 9. For general references to nominalism, see Quine's "A theory of classes pre-supposing no canons of type", Nat. Acad. of Sci. Proc. 1936; his paper "On Universals", J. Symb. Log. 12:74-84; and Goodman and Quine's "Steps towards a Constructive Nominalism", Ibid, 105-22.

2. See Quine's *Methods of Logic,* Holt, 1950, section 34.

3. This usage is Goodman's not Pierce's.

4. "On Carnap's Analysis of Statements of Assertion and Belief," *ANALYSIS*, 10 (No. 5): 97-99.

5. In *Meaning and Necessity*, Univ. of Chicago Press, 1947.

6. Church attributes this method to a suggestion of Langford, in the *Journal of Symbolic Logic*, 2:53.

7. Op cit. p. 61.

8. This explanation is here (and throughout) couched, for brevity, in familiar platonistic terms; the transition to nominalism is readily made, however, e.g., replicas of (6)'s quotes-inscription in every translation of (6) denote replicas whose rephrasal-values in each of these translations are indeterminate, while (i) is not thus handicapped.

2. ON SYNONYMY AND INDIRECT DISCOURSE

THE NOTION of synonymy has recently been severely criticized, and its replacement by graded, continuous notions of one or another sort urged on general grounds.[1] At the same time, it has usually been assumed both by critics and defenders of the notion, that synonymy and indirect discourse are in the same boat, that analyzing the latter, for instance, requires no more than an acceptable decision on the former while it requires at least that. Defenders of synonymy have thus thought it sufficient to apply their interpretations of this idea to indirect discourse, while opponents have not thought to attack such application save by way of an attack on synonymy.

Contrary to this common assumption, I shall here urge the fruitfulness of considering indirect discourse independently,[2] arguing that the case for a graded conception of such discourse holds even granted a generally adequate intensional explication of synonymy. Since Professor Carnap's explication in terms of intensional isomorphism[3] seems to me the clearest and most explicit recent attempt at a joint analysis of synonymy and indirect discourse, my discussion will center on his treatment, though directed toward the general point just stated.

That Carnap's "intensional isomorphism" is inapplicable in a direct way to natural languages seems clear,[4] and it is this which has given much point to the general critique of synonymy. Bypassing this issue in line with our present thesis, however, suppose we do have languages enabling application of Carnap's concept. Imagine, e.g., some sentence N, indirectly referred to by a sentence Q ("John believes [or says] that W"), such that N and "W" (hereafter called the that-content of Q) are parts of languages whose semantical rules are completely stated. Yet Carnap's analysis of Q is not intended as merely nominal, but rather as an explication of presystematic usage.[5] Its adequacy may therefore be tested by reference to such usage. The following discussion will attempt to show that intensional isomorphism is too rigid to reflect the presystematic variability of the relation between N's and their corresponding that-contents in true Q's, i.e., that this relation in indirect discourse is often narrower and often wider than intensional isomorphism.

This paper appeared in *Philosophy of Science* 22 (No. 1, 1955): 39-44.

On Synonymy and Indirect Discourse · 15

Carnap's motive for using "intensional isomorphism" in analyzing indirect discourse is the need for a narrower relation than L-equivalence. His reasoning here may be indicated by a consideration of the following sentences:

(i) "John believes that D."
(ii) "John believes that D'."
(iii) "John is disposed to an affirmative response to some sentence in some language, which is L-equivalent to 'D'."

With an appropriate list of L-true sentences, we proceed to examine John. Says Carnap (p. 53), "We ask John, for every sentence or for its negation, whether he believes what it says or not. Since we know him to be truthful, we take his affirmative or negative answer as evidence for his belief or nonbelief." Assume 'D' to be a sentence which John affirms under questioning, and 'D'' to be one which he fails to affirm. Though L-equivalent, since both L-true, their interchange alters the truth of (i) to the falsity of (ii). This shows (iii) to be a faulty analysis of (i), since it remains true under the very same interchange, i.e., "John is disposed to an affirmative response to some sentence in some language, which is L-equivalent to 'D''" remains as true as (iii), John being affirmatively disposed to 'D', which is L-equivalent to 'D''.

The nub of (iii)'s inadequacy is that it compels us to say that John believes both that D and that D', provided he affirms 'D', while presystematically (as determined by our questioning procedure) John may very well be said to believe that D but not that D'. Hence Carnap suggests using the narrower relation of intensional isomorphism, in which (p. 55) "The two sentences must, so to speak, be understood in the same way; they must not only be L-equivalent in the whole but consist of L-equivalent parts, and both must be built up out of these parts in the same way." The suggested alternative to (iii) we write here in similar form to facilitate comparison:

(iv) "John is disposed to an affirmative response to some sentence in some language, which is intensionally isomorphic to 'D'."

The advantage of (iv) over (iii) is that it no longer compels us always to say that John believes that D and that D', when he affirms 'D'. For when they are L-equivalent (but not intensionally isomorphic), their interchange not only turns the true (i) into the false (ii), but also the true (iv) into the false "John is disposed to an affirmative response to some sentence in some language, which is intensionally isomorphic to 'D''."

But how can we rule out the possibility that the latter sentence may be true after all because John affirms some third sentence which is intensionally isomorphic to 'D'', even though he fails to affirm the latter? It seems generally curious that Carnap's treatment omits consideration of the possibility that 'D' and 'D'' in his example may be intensionally isomorphic as well as L-equivalent, and that his argument may be extended to disqualify (iv) just as (iii) was disqualified. Suppose, e.g., that 'T' is intensionally isomorphic to 'D', but that John fails to affirm 'T' while affirming 'D' under questioning. Thus (i) is true, but the following sentence is false:

(v) "John believes that T."

Though 'D' and 'T' are intensionally isomorphic, their interchange alters (i)'s truth to (v)'s falsity. However, it also exposes the inadequacy of (iv) which remains true under the identical interchange. That is, "John is disposed to an affirmative response to some sentence in some language, which is intensionally isomorphic to 'T' " is just as true as (iv), since John affirms 'D' which is intensionally isomorphic to 'T'.

The nub of (iv)'s inadequacy is that it forces us to say that John believes both that D and that T, provided he affirms 'D', while presystematically (as determined by our questioning procedure) John may perfectly well be said to believe that D but not that T. Now it might be argued that our example of John's believing one sentence and yet not believing another sentence intensionally isomorphic to it is impossible or at least implausible. For, so it might be said, such sentences are, as Carnap puts it, "understood in the same way", since built in the same way out of L-equivalent parts. But "understood" is here crucially ambiguous. Carnap's treatment of "intensional isomorphism" refers throughout to purely linguistic entities and not at all to pragmatics, i.e., to the psychological reactions of persons to sentences. Hence, "being understood in the same way" refers to some semantic characteristic, and no inference may be drawn from the intensional isomorphism of two sentences to the nature of the psychological reactions to them. To exclude the possibility that John may truly be said to believe one and not the other is to express a *psychological* theory as well as a *semantical* one, and a highly improbable one at that. For the same limitations which prevent John from seeing that a sentence is L-equivalent to another may prevent his seeing that one sentence is intensionally isomorphic to another.[6] Psychological abilities are continuous and not likely to follow sharp semantical boundaries. Generalizing now, if our notion of the relation between an indirectly mentioned N and some appropriate that-content is to be faithful to presystematic use and psychologically plausible, then no matter how semantically narrow this relation is conceived, short of identity, it will be too wide for some cases.[7]

Psychological considerations support an analogous conclusion in the other direction as well, i.e., human abilities are graded upward as well as downward from the ability to respond equivalently to intensionally isomorphic sentences. Hence, on general grounds, it seems implausible to hold that only a relation as narrow as intensional isomorphism will do for all contexts.

Suppose, e.g., we question not John but Russell, using *L*-true sentences from *Principia*. Secondly, there are cases where the pragmatic context does not require so narrow a relation but only some obvious extensional connection. A witness, let us suppose,[8] is asked to identify the culprit (Lightfingered Larry) among seven suspects on the police lineup and, not knowing his proper name, declares truly, "The third man from the left is the culprit". Reporting in indirect mode, the attendant states, "The witness says that Lightfingered Larry is the culprit." Though the witness' *N* and the attendant's that-content are hardly construable systematically as intensionally isomorphic, the latter's *Q* is surely true, even if the witness fails to affirm its that-content under questioning. Generally, where our purpose is to pick out or identify, we may be satisfied with less than intensional isomorphism.

Finally, the use of indicator phrases[9] cannot be generally formalized and systematically referred to in terms of intensional isomorphism. Since such phrases are nevertheless employed within that-contents of indirect discourse, intensional isomorphism cannot reflect such presystematic discourse with indicators. The point about indicators is that their designata must be determined by reference to the contexts of the inscriptions themselves, e.g., we do not know the designatum of a particular 'I'-inscription without knowledge of some feature of its context, in this case, its producer. Since a replica of a particular indicator-inscription may differ in context, it may differ as to designata.

Now Carnap's specification of "intensional isomorphism" depends on "*L*-equivalent", explained (P. 14) as follows: "Let '*P*' and '*Q*' be two predicators of degree one in S_1... They are equivalent if and only if '$P \equiv Q$' is true, hence, ... if and only if '$(x)(Px \equiv Qx)$' is true, hence if '*P*' holds for the same individuals as '*Q*'." If, in addition, the truth of '$(x)(Px \equiv Qx)$' can be established without referring to facts by merely using the semantical rules of S_1, then this sentence is *L*-true. If so, '*P*' and '*Q*' are *L*-equivalent in S_I. We cannot, however, test indicators thus, i.e., by putting them into sentences of the form '$(x)(Px \equiv Qx)$'; we cannot literally re-employ a given indicator-inscription with spatio-temporal boundaries, while to use a replica of it is not necessarily to discourse about the same designata. Thus, we can establish neither the *L*-equivalence nor the intensional isomorphism of indicators. This is not to deny that an appropriate nonindicator may often replace an indicator. But such a replacing term cannot itself be intensionally isomorphic to the original; it can at most agree with it by some extensional criteria. Also, such replacement, often undeniably convenient, neither alters nor explains the fact that indicators are employed in presystematic cases of indirect discourse where intensional isomorphism is out of the question for systematic representation.[10]

If the considerations adduced are relevant, then "intensional isomorphism" is too inflexible to serve adequately for the analysis of indirect discourse,[11] even granted its general adequacy in explicating synonymy, and independently of the issue of its direct applicability to natural languages. An analogous judgment seems plausible for other explications of synonymy.[12] For defenders of synonymy such considerations would seem to indicate a revised, independent analysis of indirect discourse, while for critics, who favor graded, and perhaps frankly psychological conceptions of likeness of meaning, an independent reinforcement of their position seems available here.

NOTES

I am deeply indebted to Prof. N. Goodman for comments on an earlier form of this paper, and for discussion of related topics. Thanks are also due Prof. C. G. Hempel and Mr. A. N. Chomsky for helpful comments and suggestions.

1. See for example Goodman's "On Likeness of Meaning", *Analysis,* 10: 1-7, White's "The Analytic and the Synthetic: An Untenable Dualism" in *John Dewey: Philosopher of Science and Freedom,* ed. S. Hook, Dial, 1950, and Quine's *From A Logical Point of View,* Harvard, 1953.

2. In a different connection, some aspects of such a consideration are discussed in the writer's "An Inscriptional Approach to Indirect Quotation", *Analysis,* 14: 83-90.

3. In Carnap, R. *Meaning and Necessity,* Univ. of Chicago Press, 1947.

4. See his remark, Ibid., p. 58, "The present definition makes no claim to exactness; an exact definition would have to refer to one or two semantical systems whose rules are stated completely."

5. See his statement, Ibid., p. 53, "Although sentences of this kind ... are, indeed, used and understood in everyday life without any difficulty, they have proved very puzzling to logicians who have tried to analyze them. Let us see whether we can throw some light upon them with the help of our semantical concepts."

6. See Goodman's argument in "On a Pseudo-Test of Translation", *Phil. Studies,* 3: 81-82, that *knowing* of a specified kind can be only a pseudo-test of translational equivalence. Perhaps one motivation for offering this pseudo-test is precisely the notion disputed in the present paper, that synonymy must be explicated so as to apply to indirect discourse, e.g., "John *knows* that ———", so that if interchangeability fails here, synonymy is precluded.

See also White's suggestion in "Ontological Clarity and Semantic Obscurity", *Jour. Phil.,* 48: 379, of the need for a psychological analysis of even synonymy. For our limited consideration of indirect discourse directly referring to reactions like stating, believing, etc., the point seems even clearer, in favor of some pragmatic requirement.

7. This criticism of (iv) seems more general than Mates' point in

"Synonymity", *Univ. of Calif. Pub. in Phil.*, 25, 1950, "that, for any pair of intensionally isomorphic sentences – let them be abbreviated by '*D*' and '*D*' ', – if anybody even doubts that whoever believes that *D* believes that *D'*, then Carnap's explication is incorrect." His argument assumes that the sentences:

(1) Whoever believes that *D* believes that *D*.
(2) Whoever believes that *D* believes that *D'*.

are intensionally isomorphic, and since nobody doubts (1), if anybody doubts (2), Carnap's analysis is faulty. Putnam, in "Synonymity and the Analysis of Belief Sentences", *Analysis*, Vol. 14, pp. 114-122, has tried to meet this argument by revising the notion "intensional isomorphism" to include identical logical structure, and hence to render (1) and (2) nonsynonymous. But, since Putnam's criterion *does* allow some different sentences to be intensionally isomorphic in his revised sense, (say '*D*' and '*T*' in our treatment above), his proposal, even if it meets Mates' objection, fails to upset our criticism of (iv) above, which does not depend on Carnap's omission of logical structure from his explanation of intensional isomorphism. Further, as Chomsky suggests, this point can be used to formulate an example analogous to that of Mates which seems difficult even for Putnam's revision: Suppose '*A*', '*B*', and '*C*' all intensionally isomorphic. Then the sentences:

(3) Whoever believes that *A* believes that *B*.
(4) Whoever believes that *A* believes that *C*.

are intensionally isomorphic even for Putnam's revision. Yet how is the factual possibility ruled out that nobody doubts that (3) but someone doubts that (4), or that John believes that (3) but not that (4)?

8. This example was suggested to me by Dr. Goodman, in conversation.

9. For a general, inscriptional treatment of indicators, see Ch. 11 of Goodman's *The Structure of Appearance*, Harvard, 1951.

10. The point made here seems applicable generally to theories of indirect discourse depending on senses or rules of terms. Frege, e.g., in "Ueber Sinn und Bedeutung", translated in Feigl and Sellars *Readings in Philosophical Analysis*, says that in such discourse, "words have their indirect nominata which coincide with what are ordinarily their senses." What is the ordinary sense of an indicator term?

11. In "An Inscriptional Approach to Indirect Quotation" (see Note 2. above), concerned with the structure rather than the basic terms for analyzing indirect discourse, I used the term "rephrasal" rather than "intensional isomorphism" for the relation between the that-content of a true *Q* and its appropriate *N*. There, the issue of structure is independent of choice of term. Here, in criticizing Carnap's term, I make no claim to be able to *define* any alternative like "rephrasal". But if my arguments in the present paper are well-taken, any alternative will need to be conceived as graded and contextually variable with purpose, perhaps with the particular verb introducing the indirect mode, etc., such variable relation being at times tighter than synonymy or translation, at times considerably weaker. To suggest these characteristics which appear to me necessary, I use the term "rephrasal".

12. The present paper was accepted for publication before the appearance of Professor A. Church's "Intensional Isomorphism and Identity of Belief"

(Phil. Stud. 5 (October 1954): 65-73), but the editors have been kind enough to allow me space for the following comments:

(A) Church proposes to replace intensional isomorphism by synonymous isomorphism as a "criterion of identity of belief," illustrating his rejection of the former by appeal to the historical difficulty of finding a proof of Fermat's Last Theorem (and hence the possibility of believing its denial while disbelieving an obviously false counterpart which, assuming Fermat's claim correct, is intensionally isomorphic to this denial).

But if the history of belief-responses to sentences is held relevant in controlling proposals for identity-criteria, must we not reject synonymous isomorphism too? If someone (apparently confused by the terms "optician," "oculist," "optometrist," and "ophthalmologist") says "I believe that eye-doctors are eye-doctors but not that eye-doctors are oculists," ought we not, in a perfectly good sense of "belief," take him literally at his word, especially in view of the patent history of confusion of these terms in the popular mind?

(B) Church seeks to rebut Mates' conclusion with respect to synonymous isomorphism by denying that one of the latter's illustrative sentences (see fn. 7 above) is really doubted while the other is not. Instead, Church construes the purported doubt as applying to an equivalent (but not synonymously isomorphic) metalinguistic sentence.

But this strategy can (with slight modification) be applied equally to save intensional isomorphism as well, i.e. to reconstrue ostensible divergence in belief-response to intensionally isomorphic pairs as holding actually of equivalent but not intensionally isomorphic sentences. Thus, e.g. suppose 'A' and 'B' to be intensionally isomorphic. Then " 'B' is true" is intensionally isomorphic to neither, and can be taken to be the real object of doubt in every case where it is claimed that 'A' is believed and 'B' doubted. Why, in sum, does Church draw the line just between intensional and synonymous isomorphism, when his argument from history can be applied *against* the latter and his reconstruction of doubt can be used *for* the former?

(C) As support for his rebuttal of Mates, Church cites a case where translation of Mates' pair results in identical sentences in a different language, while translation of the associated metalinguistic sentence preserves its distinctness. Now if intensionally as well as synonymously isomorphic sentences may ever be said to translate each other, then analogous support could be found for a resolution to save intensional isomorphism. If translation is, however, synonymous isomorphism, then Church's argument restates simply the original case of divergent beliefs about synonymous sentences, which Mates began with. To one who is willing to accept such a case, the translation argument offers no new deterrent. If, finally, Church's point concerns not the differing truth-values of "Mates doubts that (15) but not that (14)" and its German translation, but depends rather on the assumption that both represent the same reasoning while only the second is self-contradictory, then is not the assumption that translatability implies identity of reasoning itself a (partial) criterion of identity of belief, which is precisely what is at issue?

3. INSCRIPTIONALISM AND INDIRECT QUOTATION

SOME years ago, I proposed an inscriptional analysis of indirect quotation[1] and tried to show that it escaped the objections offered by Church[2] to Carnap's theory of belief-statements,[3] concluding that these objections therefore provide no conclusive argument against nominalistic interpretations of indirect discourse. Briefly put, my proposal was to construe the that-clauses, in indirect quoting contexts of the form "... writes that ———", as single indivisible predicates denoting concrete inscriptions forming rephrasals of one another; "Seneca writes that man is a rational animal" was thus to be taken as "There is an inscription i, such that i is both a that-man-is-a-rational-animal and written by Seneca". My paper has since been discussed by Professor Church and others, and some general issues raised that call for further comment.

As remarked above, my examination of Church's arguments against Carnap's theory was intended to show that my inscriptional proposal could withstand their force, thus rebutting Church's suggestion that they might constitute "an insuperable objection" to "analyses that undertake to do away with propositions". It is, therefore, somewhat misleading to say, as Professor Church now says, after adducing *new* arguments, "Scheffler has not established his claim to have provided a workable substitute for propositions".[4] My claim was to have shown Church's original arguments ineffective in one sample inscriptional case, and hence no insuperable bar to a nominalistic interpretation. I had no illusions of having rebutted in advance all other anti-nominalistic arguments that might be constructed. Professor Church's new considerations are of independent interest and will be discussed below, but they do not show that my original claim has not been established.

Professor Hempel states,[5] indeed, that my argument in support of this claim "is quite convincing", but doubts that the "ultimate purpose of nominalistic analysis, namely, replacement of relatively unclear notions by clearer and more precise ones", is served by my "technically nominalistic" proposal. The that-clause predicates required by my proposal are, suggests Hempel, "hardly more intelligible than the notion of the proposition expressed

---

This paper appeared in *Analysis* 19 (No. 1, 1958): 12–18.

by the corresponding that-clause", for "the task of stating criteria of application for any particular that-clause predicate would seem to face the same difficulties as the task of specifying the propositional meaning of a particular sentence belonging to a given language and occurring in a definite context".

This argument, however, seems to me to rest on a confusion of "clarity" in the sense of *philosophical intelligibility* with "clarity" in the sense of *determinacy of application*. Nominalistic philosophy, as I understand it, strives to increase the intelligibility of theories, not the determinacy or precision with which they can be applied. In showing how to avoid appeal to suspect entities such as entelechies, forces, properties, propositions, etc., nominalistic analyses are thought by nominalists *thereby* to increase intelligibility, though overall determinacy of application is unaffected. By meeting the requirements of technical nominalism with respect to eliminating commitment to propositions, my proposal is thus fully in accord with nominalistic goals. To suppose differently, on the basis of Professor Hempel's argument, we should have to rule that the interpretation of "is wise" in "Socrates is wise" as an indivisible, non-designative, predicate of individuals is no better from a nominalistic standpoint than a construal of the sentence as "Socrates is a member of the class Wise", in view of the fact that to state criteria of application for the predicate is no easier than to specify rules of admission to the class. Analogously, no nominalist would need to balk at thinking of Good Fortune as an entity and describing Jones as having had this entity, since the alternative description of Jones as having been very fortunate is no more easily tied to explicit and general criteria of application.[6] In sum, to the nominalist eager to devise ways of avoiding theoretical commitment to pseudo-entities, Professor Hempel seems to say, "Why go to all the trouble, when you can get along just as well *with* such commitments in applying your theory to cases?"[7]

If this question, far from representing the "ultimate purpose" of nominalism, sounds rather pragmatic in spirit, Professor Church's emphasis on the simplification effected by an abstract ontology sounds even more so. Arguing generally against finitistic nominalism in the sense of Goodman and Quine,[8] he suggests that a major difficulty is that "The theory is too complicated in application".[9] As is the case in mathematics and the natural sciences, he says, "a theory may be greatly simplified by incorporating into it additional entities beyond those which had originally to be dealt with, and I believe it to be a false economy which would forego simplification of a theory by such means. The notion of a concrete physical object, extended in space and persistent through time, is a case in point, as what had originally to be dealt with by the physical theories in which such objects appear was not these objects themselves but rather certain observations and physical experiences. Indeed the justification would seem to be basically the same for extended physical objects in macrophysical theory and for ideal sentences in logical syntax: both are postulated entities – some may prefer to say inferred entities – without which the theory would be intolerably complex if not impossible".[10]

This argument might, it seems to me, also be put forward in defence of such postulated and undeniably simplifying entities as the élan vital, the spirit of the age, the Will, the Devil, luck, the superego, the id, and the manifest destiny of nations. Surely, *without* these entities, the theory of man (besides losing a number of interesting characterizations and distinctions otherwise possible) becomes intolerably complex. Unfortunately, however, *with* these entities it seems to many people to become just intolerable. That is, it becomes unintelligible to them without further, complicating explanation. It is, for such people, simply beside the point that a theory they cannot understand has a surpassing simplicity as well as a demonstrated usefulness. It is beside the point, that is to say, so far as intelligibility is in question, though possibly relevant from other standpoints, e.g., aesthetic elegance, or convenience in certain practical applications such as pedagogy, psychotherapy, propaganda, or prediction.

That intelligibility is not an automatic concomitant of theoretical simplicity and utility is amply indicated by the fact that Professor Church's chief example – the case of concrete physical objects justified by their simplifying of our physical knowledge – is also an example of a perennial philosophical problem: how to understand the existence of "external" physical reality. Philosophers who take this problem seriously are not simply bad physicists, nor are they therefore to be construed as Super-physicists, legislating to their scientific subordinates. Rather, they are concerned to render intelligible to themselves theories they gladly admit are useful, elegant, and of great importance. Their enterprise is no more to be judged by its scientific fertility than are scientific theories to be judged by their philosophical interest.

It follows that to insist on philosophical clarification of the ontology of a given theory is not generally to interfere with the normal scientific career of the theory. The language of "renunciation", "rejection", "giving up", and "doing without", in ontological discussions is thus dangerously ambiguous.[11] To state that a given scientific assumption is philosophically unintelligible is not to counsel giving it up in science, but rather to ask for further, philosophical explanation. When Professor Church, therefore, says, "I believe it to be a false economy which would forego simplification of a theory by such means", he apparently thinks of the nominalist as motivated by the desire to effect a scientific economy, or as somehow opposed to the development of platonistic theories as a scientific pursuit. On the contrary, as I see it, the nominalist's desire is to put into intelligible language (from his point of view) and so to *understand* what it is the scientist or the platonist is saying. To suppose that, because he tries to do this, he is somehow opposed to legitimate scientific simplification is like saying that, because I don't understand Sanskrit, I am somehow opposed to it.[12]

Nor is anything so far said to be taken as conceding that the case of ideal sentences in logical syntax is indeed a scientific simplification analogous to simplifications in physical science. Thus far, I have argued that the philosopher's explanatory concern is independent of the scientist's or the practitioner's. But this does not imply that all postulations made in the name of science and simplicity are on a par. To revert to our earlier example, there are many who would distinguish sharply between the superego and the spirit of the age on purely scientific or pragmatic grounds: the latter, unlike the former, (it is said by these people) does no "causal" or "empirical" explaining, despite the fact that it helps to build neat historical theories. Like the dormitive virtue, it simplifies, but does no real scientific work. It is at least questionable whether ideal sentences are not more like the spirit of the age and the dormitive virtue than like the superego and the electron, since they are not tied to observation in any clear sense nor do they account for phenomena in the way acknowledged scientific theories do.[13]

Aside from considerations of relative simplicity, Professor Church now raises the question how to interpret a *new* group of sentences on a nominalistic basis:

(a) Church and Goodman have contradicted each other.
(b) Goodman will speak about individuals.
(c) Some assertions of Velikovsky are improbable.
(d) All assertions of Aristotle are falsehoods.

Church expresses doubt that either extensive analysis of the semantic predicates here or an axiomatic treatment of their relations to each other, to the that-clause predicates of my proposal, and to the syntax of inscriptions denoted by the latter, can be accomplished without reintroducing commitment to propositions. He does not offer a proof in support of his doubt, nor even a specific argument of the sort he offered previously against Carnap's analysis. In fact, he states, "The possibility remains that the claim [to have provided a workable substitute for propositions] might be substantiated by a longer and more detailed development, including solutions of the difficulties just discussed and treatment of a compatible finitistic syntax".[14] Professor Church, in effect, is pointing out that further problems remain for the nominalist. This is certainly true; as a matter of fact, some problems perhaps more naturally related to those of indirect quotation remain. Meckler, for example, cites as a limitation of my paper its failure to suggest an analysis for belief-sentences.[15] The existence of further problems as yet unsolved can, however, hardly qualify as an argument against any philosophical programme. Much less can it serve to establish the validity of a rival programme, and in particular, one that faces much the same or analogous problems, e.g., the task, mentioned by Hempel, "of specifying the propositional meaning of a particular sentence belonging to a given language and occurring in a definite context"[16] and that of stating conditions under which two *inscriptions* express or name the same or contradictory propositions.

Finally, it is not a part of my proposal (as Professor Church appears to think it is) that accidentally occurring inscriptions may not be considered for any purpose.[17] I excluded consideration of them just in cases of explicit indirect quotation where, if the quoting statements are true, there exist appropriate nonaccidentally occurring inscriptions. It does not follow that my peculiar cross-linguistic that-clause predicates must be used elsewhere, to analyse other statements than those of the form "... writes that ———". It is conceivable that some problems could be handled by relating inscriptions generally through rephrasal-relationships to statements in an interpreted base language, characterized by a nominalistic syntax. With contradiction syntactically described for the base language, for example, Church's problem-sentence (a) might be interpreted as:

(a') $(\exists x)(\exists y)(\exists z)(\exists w)$(Goodman inscribes x · Church inscribes y · Reph xz · Reph yw · Contradicts zw)

The same strategy seems feasible also for (b) and (c), the latter's analysis depending on characterization of a syntactic (or syntactic-pragmatic) relation K of confirmation in the base language as well as a description of accepted evidence-statements in this language. I am somewhat doubtful about this strategy in the case of (d), however, because of the general difficulty of defining truth and falsehood.[18] And still other problems, including that of belief-sentences, remain for the nominalist. No proof has been offered showing that they are insoluble. But even if they continue to resist the nominalist's best efforts, this will no more establish platonism than the continuing difficulties of physics establish the superiority of mysticism.

NOTES

1. Scheffler, I., "An Inscriptional Approach to Indirect Quotation," *Analysis* 14 (No. 4, 1954): 83-90.

2. Church, A., "On Carnap's Analysis of Statements of Assertion and Belief," *Analysis* 10 (No. 5, 1950): 97-99.

3. Carnap, R., *Meaning and Necessity,* University of Chicago Press, 1947

4. Church, A., "Propositions and Sentences", in *The Problem of Universals,*

a symposium by I. M. Bochenski, A. Church, and N. Goodman. The University of Notre Dame Press, 1956, 11.

5. Hempel, C. G., "Review of Scheffler, 'An Inscriptional Approach to Indirect Quotation,' " *Journal of Symbolic Logic* 22 (No. 1, 1957): 86.

6. That a term is applied with great precision does not mean that it needs no explanation, but rather that it is deserving of explanation; e.g., it is because we *do* know how to apply numerical terms, counterfactual constructions, inductive notions, that explaining them becomes important. To hold that everyone who applies them proficiently understands them theoretically is clearly wrong. It follows that, for example, to render primitive causal or counterfactual notions explicit and precise through formalization is not to provide an explanation. Critics of logical methods in philosophy are, I think, (sometimes) rightly objecting to such formalizing of obscure notions as a substitute for explaining them.

7. Professor Hempel objects also that to state conditions for the application of my term "rephrasal" is no easier than to do so for "synonymy." But my reason for using "rephrasal" is not that it is in any way clearer than "synonymy." Rather my argument is that synonymy is not adequate for the analysis of indirect discourse. See Scheffler, I., "On Synonymy and Indirect Discourse," *Philosophy of Science,* 22 (No. 1, 1955): 39-44, and especially footnote 11, and Hempel's review of this paper in *Journal of Symbolic Logic* 22 (No. 2, 1957): 208.

8. The reference is to Goodman, N. and W. V. Quine, "Steps Toward a Constructive Nominalism," *Journal of Symbolic Logic,* 12 (No. 4, 1947): 105-22.

9. "Propositions and Sentences," 11.

10. Ibid., 9.

11. See Goodman and Quine, op. cit., first paragraph.

12. Like explanatory discourses generally (say in the classroom), the nominalist's explanatory language is different from and, in one sense, more complicated than, the discourse explained. But no explanatory discourse is intended to supplant the original generally and there is another sense in which the explanatory discourse is simpler, i.e., it is more understandable. Thus an explanatory *definiens* may be both more complicated structurally and also more understandable than a given *definiendum.* The reader will by now have discerned some of the many different uses to which "simplicity" may be put, rendering obscure a variety of current philosophical arguments resting on some unspecified use, or confusing one with another.

13. For this argument I am indebted to N. Chomsky.

14. "Propositions and Sentences," 11.

15. Meckler, L., "An Analysis of Belief-Sentences," *Philosophy and Phenomenological Research* 16 (No. 3, 1956): 317-30, especially point (4) on p. 318.

16. *Journal of Symbolic Logic* 22 (No. 1, 1957): 86.

17. "Propositions and Sentences," 10.

18. A nominalized version of Tarski's definition for a sufficiently rich base language to incorporate rephrasals of all actual inscriptions (or those we care about) would seem to do the trick, however. This possibility is suggested by the paper of R. M. Martin and J. H. Woodger, "Toward an Inscriptional Semantics," *Journal of Symbolic Logic* 16 (No. 3, 1951): 191-203.

4. POSTSCRIPT ON INSCRIPTIONALISM

IN 1954, I published a nominalistic account of indirect quotation, construing it to express a relation between persons and inscriptions alone.[1] This account, first proposed in my dissertation in 1952,[2] had been prompted in good part by Church's anti-Carnapian arguments, suggesting the indispensability of propositions in analyzing indirect discourse.[3] My alternative idea was to take the that-clause, in a quoting context of the form '... writes that ———', as a single predicate denoting those inscriptions constituting rephrasals of its own sentential content, the quoting sentence as a whole to be understood as affirming some such rephrasal to have been produced by the person quoted. 'Seneca wrote that man is a rational animal' was, as I put it, to be interpreted as asserting that some rephrasal of the constituent 'Man is a rational animal' (i.e., some that-man-is-a-rational-animal) had been inscribed (written) by Seneca.

Miss Marilyn P. Frye, in an interesting discussion,[4] now offers three counterexamples to my proposal. First, a printing press may inscribe a rephrasal of 'P' but cannot write that P. Second, someone dictating a message, to the effect that P, may well be said to have written that P without having inscribed (or literally written) any that-P. Third, a person (e.g., a secretary) may inscribe (literally write) some that-P without being properly said to have written that P.

Now, in specifying *inscribing* (or writing, literally taken) as the appropriate producing relation between persons and inscriptions, my formulation did indeed require a simplifying assumption (as Miss Frye's examples show), to wit, that each rephrasal is the work of exactly one person. Collapsing the process of rephrasal production, this assumption identifies the inscriber with the (single) producer, leaving no room for any of Miss Frye's counterexamples to arise.[5]

The denial of the assumption means, however, that the person writing that P, can no longer be supposed literally to inscribe some that-P, nor can the inscriber of a that-P be assumed to have written that P. Here I believe Miss Frye to be quite right. But I disagree with her view that the issue is thereby fundamentally changed. What is involved, it seems to me, is only a

---

This paper appeared in the *Journal of Philosophy* 62 (No. 6, 1965): 158-60.

more flexible understanding of the relevant relation between person quoted and inscription produced. Whoever writes that P, must still be supposed to sustain some special producing relation R to some that-P, which holds when he inscribes this that-P himself, in fitting contexts and with assertive force (or, to use Miss Frye's notion, in "the performance of a language act") or when he brings about its inscription through the use of certain instrumentalities, in fitting contexts and with assertive force (or in "the performance of a language act"). Such relation, however it may be ultimately described, may be presumed again to eliminate all of Miss Frye's counterexamples.

Miss Frye notes, indeed, that 'wrote,' in one version of my original account, may have "the sense in virtue of which it is associated with language acts." To such reading she objects, however, on the ground that it would be "very odd," and she argues, moreover, that it must turn out to be circular, since 'X wrote S' would need to be taken as "X wrote such-and-such, employing the inscription S." Oddness is, however, remediable by stipulation, and a noncircular stipulation is, ironically, suggested repeatedly by Miss Frye herself in her "language act" formula. Understand 'writes', therefore, if you like, as "writes, in the sense associated with performance of a language act," or substitute your own favorite formula to similar effect. What you will have is still a two-place predicate, interpretable as representing a relation between a person and an inscription.

Miss Frye correctly points out that, in the discussions of Church and myself, there was a shifting from 'saying' to 'writing' to 'inscribing,' which did not worry us. Such elasticity had, I believe, the same motivation as that of the simplifying assumption above discussed: to abstract from analysis of the relevant producing relationship R, in order to focus more sharply on questions of logical form and ontology.[6] The substantive analysis of R is, of course, welcome. But an independent consideration of the structure of indirect quotation and its associated ontological commitments is also possible. Assuming that for every true "writing that" statement there are a suitable person and a suitable inscription standing in some producing relationship R, which itself always authorizes an appropriate "writing that" statement, my proposal is simply a device for analyzing the statement so as to affirm such relationship explicitly, while eluding anti-nominalistic strictures.

NOTES

1. "An Inscriptional Approach to Indirect Quotation," *Analysis* 14 (March, 1954): 83–90. See also my "On Synonymy and Indirect Discourse," *Philosophy of Science* 22 (January, 1955): 39–44; my "Inscriptionalism and Indirect

Quotation," *Analysis* (October, 1958): 12-18; and related discussions in my *The Anatomy of Inquiry* (New York: Knopf, 1963).

2. *On Quotation* (typescript), University of Pennsylvania Library, 1952.

3. Alonzo Church, "On Carnap's Analysis of Statements of Assertion and Belief," *Analysis,* 10 (April, 1950): 97-99. See also R. Carnap, *Meaning and Necessity* (Chicago: University of Chicago Press, 1947).

4. "Inscriptions and Indirect Discourse," Journal of Philosophy, 61 (Dec. 24, 1964): 767-72.

5. Such an assumption was never stated explicitly in any of my publications. It seems, however, to have been indicated (at least indirectly) in my dissertation, where, in describing general features of the quoting situation, I refer to "some specific *inscription* N, *inscribed* by some *author* J," and assume every quoting sentence to refer to "some quoted linguistic entity *as the work of some author*" (p. 10, italics added).

6. Church writes (*op. cit.,* p. 97), "In each case 'wrote' is to be understood in the sense, 'wrote with assertive intent.' And to simplify the discussion, we ignore the existence of spoken languages, and treat all languages as written."

5. EXPLANATIONS, DESIRES, AND INSCRIPTIONS

1. INTRODUCTION

Do THE NOTIONS of explanation, belief, and desire compel us to postulate facts explained, propositions believed, and states desired? In philosophical reconstruction of our talk of explanation, belief, and desire, are we driven not merely to abstraction but to intensionalism? I argued in my [1963], sections 6 and 8 that the answer is no, and suggested an inscriptional interpretation presupposing an exclusively concrete ontology.

Mr Gorovitz now raises several objections to my account (Gorovitz [1970]). He does not deny the difficulties of intensionalism but argues rather that my nominalistic construction is an inadequate alternative, concluding, in particular, that "the problems of ontological commitment associated with talk about explanation remain" (p. 255). I should like here to respond to his major lines of criticism, clarifying certain aspects of my proposal in the process.

2. SCOPE OF THE ANALYSIS

The verb "explain" is used in a variety of contexts and serves a multiplicity of purposes. The particular focus of the discussion in my [1963] was the use of the word as it emerged from reflection on the deductive pattern of explanation. For the sentence:

(1) k is blue

constituting a description of the event k, an appropriate explanans g deductively yielding (1) may, under this pattern, presumably be said to explain why k is blue. The question now arises as to how such an assertion (which I shall call an "explanation-description"):

(2) g explains why k is blue

is itself to be interpreted. The discussion in my [1963] is directed to this question.

This paper appeared in the *British Journal for the Philosophy of Science* 22 (1971): 362-69.

Mr Gorovitz says he wishes "to consider the question of what sort of thing it is that an explanation explains" ([1970], p. 247). But when he further says that "What Scheffler seeks, clearly, is an account of what it is that gets explained" (*ibid.* p. 248), he overstates the case. My discussion does not presuppose at the outset that a satisfactory analysis must yield "explains" as a relative predicate, nor is it supposed that, if it does yield a relative counterpart of "explains", say "E", the converse domain of the latter will coincide with that of the pre-analytic "explains". The analysis of (2) may, so far as *initial* preconceptions are concerned, accordingly not result in perpetuating such an expression as "what it is that gets explained", much less give an account of its use. The point is especially important because it renders ambiguous the critical phrase "objects of explanation". In one sense of the phrase, it purports to refer to putative elements in the converse domain of the ordinary "explains" (or of some co-extensive analysans). In another, and quite different, sense it purports to refer to those objects presupposed by whatever statements philosophical reflection offers to supplant (2) and its kin. The first sense cannot, without begging significant questions, be assumed to underlie the ontological issues as I conceive them. The question I address is how to interpret, i.e. provide a philosophically adequate substitute for (2), taken as a useful expression emerging from the deductive theory of explanation: philosophically adequate in that its logical form is explicit, its ontology concrete, and its capacity to serve the main purposes of the original reasonably evident.

3. ABSTRACTNESS AND INTENSIONALITY

Mr Gorovitz rightly states that "intensionality and abstractness are different characteristics" and remarks that "elimination of intensional objects from the ontology represents a more conservative asceticism than the elimination of abstract entities" (*ibid.* p. 249). I regret that he finds my [1963] unclear on these points and misses any indication, in my discussion, of the meaning of abstractness and of intensionality, as well as any comment as to whether facts are to be avoided because of their abstractness or their logical intensionality.

I agree that my treatment was insufficiently explicit on some of the general points mentioned, where I relied on accounts of others, notably Quine and Goodman. I cannot agree, however, that "Scheffler nowhere indicates... what it means to say of an entity that... it is intensional" (Gorovitz [1970], p. 248) for I did write of facts that they are "abstract ('logically intensional') entities, intermediate between chunk and descriptions, each such entity corresponding to some class of logically equivalent (true) descriptions uniquely" ([1963], p. 60).[1] As to my failure to comment on the possibility of a "conservative asceticism" that rejects intensions but accepts classes, I plead guilty. I suggest, however, that this failure was natural within the framework of my discussion, dealing as it did with *facts*, which violate even Gorovitz's conservative asceticism, whereas my inscriptional remedy (averse not only to intensions but also to classes) satisfies even the more radical type.

This remedy, proposed after detailed consideration of various alternatives, is to supplant (2) with:

$$(3) \quad (\exists x)(g\text{Ex} \cdot W(Bk)x)$$

where "$W(Bk)$" is a predicate denoting every logical equivalent of the sentence within its parentheses (inclusive of this sentence itself), the whole statement (3) to be taken as asserting that something is a why-k-is-blue sentence and stands to g as explanandum to associated explanans. Explanation is here, in effect, taken as relating sentences to other sentences. That is to say, (3) supplants (2) and offers us an analysis in terms of "E", which relates sentences to one another. In the open sentence corresponding to (3), i.e.:

$$(4) \quad x\text{Ey} \cdot W(Bk)y$$

the variables 'x' and 'y' are bindable, but range solely over sentences. Throughout, sentences are to be construed as concrete sentence-inscriptions. Thus neither intensional objects nor classes are presupposed, nor, certainly, relations. Mr Gorovitz is thus simply in error in saying that "Scheffler is committed to the existence of the relation that he claims holds between g and 'Bk'" ([1970], p. 249) "E" is not, within (4), either a name or a variable, nor does it occupy a position accessible to variables. It is a constant general relative term, and carries no commitment to any relation as an entity.

4. EXPLANATIONS AND EXPLANATION-DESCRIPTIONS

In commenting on (3), I remarked in my [1963] that some suitable inscription x can be presumed to exist since putting the question what explains why k is blue provides such an inscription, and, certainly, to produce the explanatory argument is itself to provide another such.[2] This comment was, I am afraid, misleading; it should rather have stressed, in accord with the suggested interpretation of "$W(Bk)$" that the explanation-description itself produces an inscription of appropriate sort. To be sure, the existence of a suitable question or explanatory argument may perhaps be additionally assumed as likely in certain (or possibly even typical) contexts, but failing such an assumption, the explanation-description alone guarantees the existence of a relevant explanandum-inscription.

Now Mr Gorovitz ([1970], p. 250) offers a case in which no question is asked but an explanation is offered, and this certainly provides no counterexample of the desired sort, for it does not rule out the production of a relevant inscription either in the explanation itself or in the explanation-description of the form (2) or (3). A second case suggested (*ibid.* p. 250) concerns a person who explains a puzzling event to himself in thought without producing an utterance or inscription at all. Mr Gorovitz takes this as a counterexample for he supposes me to hold "that to provide explanation is necessarily to include, in what one provides, an inscription of the required sort" (*ibid.*). But what I say in my [1963] is rather that producing "the explanatory argument" provides a requisite inscription. The explicit analysis of my [1963] is, moreover, directed to the interpretation of explanation-descriptions of the form (2), which themselves contain and hence guarantee the existence of appropriate explanandum-inscriptions.

The analysis in other words, seeks to construe such sentences as (2), and not such sentences as:

(5)   Jones explains (something) to himself.

It does not offer, or pretend to offer, an interpretation of all contexts of the ordinary verb "explain" but only those in which we purport to describe the explanatory force of given statements with respect to certain others taken as explananda. In particular, it offers no (pragmatic) theory as to the conditions under which some person Jones may properly be said to have *provided an explanation*. It is perfectly compatible with my [1963] proposal that Jones says nothing but is considered by us to be appropriately related to g: perhaps he suggests it or leads us to suppose he believes it. In any case, when *we* say that g explains why k is blue, our own explanation-description produces the requisite why-k-is-blue inscription.[3]

In sum, the purported counterexample fails since Mr Gorovitz errs in ascribing to me the "claim that there will always be, in the case of explanation, an appropriate inscription since we must necessarily provide one in the process of providing an explanation, at least, to someone else" ([1970], p. 252). The analysis of my [1963], to repeat, is *not* concerned with the process of providing an explanation but with the attribution of explanatory force to given statements, relative to some explanandum.

5. EXPLANATION AND DESIRE

In the course of analysing teleological explanation, I proposed in my [1963] to interpret:

(6)   John desires that John qualify for entrance to medical school

as:

(7) $(\exists x)(\text{That}(QJ)x \cdot \text{DTr}J x)$

to be read as:

Some x is such that it is a that-John-qualifies-for entrance-to-medical-school inscription, and is desired-true by John.

The effect is to supplant the ordinary "desires that" of (6) with the two-place predicate "DTr" relating persons and inscriptions. The variable "x" ranges over concrete inscriptions, as before. However, unlike "W(Bk)", the predicate "That(QJ)" does not denote every logical equivalent of its parenthesised insert, since the point, for desire, is to gain the effect, not of *facts*, but rather of *states*, which are more finely individuated, there being a different state for each differing sentence. The proposal of my [1963] is thus to take "That(QJ)" as a general term denoting all and only those inscriptions which are rephrasals of its insert, i.e. which are replicas of this insert, carry the same language affiliation, and lack indicator terms. Since the insert is itself included in the denotation of the predicate-inscription, the existence of the latter guarantees the existence of a suitable inscription denoted by this predicate. Thus, the existential quantification "$(\exists x)(\text{That}(QJ)x)$" of (7) is, in particular, true if (7) exists. The point of the whole analysis is to reconstruct teleological arguments which purport to explain action as a result of belief and desire, and the effort is to show that, given (7) and suitable additional premises which likewise presuppose nothing but persons and inscriptions, arguments can be formulated which serve the main purposes served by the unreconstructed teleological originals.

To this analysis, Mr Gorovitz objects that desire requires a prior, or at least a simultaneous object. If, as he suggests, John desires to meet a certain woman, but says or writes nothing to that effect, "John has a desire, although no appropriate inscription exists at the time of the desire. Surely the reality of his desire is not dependent on our subsequent description of it ... the possibility of a man having a desire – and hence of there being an object of that desire – in the absence of any appropriate inscription, shows that it is not, in general, inscriptions that are the objects of desire" ([1970], p. 252). Now this objection clearly rests on the assumption that desire presupposes a prior or simultaneous object. While this assumption may seem to be natural for ordinary contexts of "desires", it is certainly no part of the analysis of my [1963], which supplants the ordinary "desires" by "DTr", relating persons and inscriptions, with no restrictions on their relative temporal or spatial characteristics. The analysis is intended to interpret, not desire in general, but statements of the form of (6), and, as in the case of explanation, there is no antecedent assumption that the analysis must yield a relative predicate whose converse domain coincides in special ways with that of the ordinary "desires", taken as a relation between persons and intensional objects of one or another sort. Indeed, the latter relation is just what is suspect and occasions the fundamental difficulties motivating the analysis. The analysis thus does not purport to duplicate ordinary conceptions of desire but to replace them, offering, in (7), a way of gaining the effect of (6) as a premise in teleological explanation, without attendant intensional difficulties, indeed without abstract ontology altogether.

As in the case of "object of explanation", the phrase "object of desire" is thus ambiguous in a critical way, as between "object in the putative converse domain of the ordinary 'desires' ", and "object presupposed in the analysans offered to replace (6) and its kin". A simultaneity requirement, no matter how natural for the former, cannot be automatically transferred to the latter.

Given (6), (7) interprets it as describing John by associating him with a certain sort of inscription, one of which is displayed within (7) itself. Such description of John in effect consists in putting him into relation with inscriptions which replicate one of our own. The process is logically sound, despite its seeming strange, – as Quine suggests, "a little like describing a prehistoric ocean current as clockwise" (Quine [1966], p. 192). If, nevertheless, it is thought to be inherently paradoxical for John to sustain any relation to an object located at a later time, the thought must itself be rejected as unfounded. Thales stands in a relation of temporal precedence to Hegel, tenselessly stands, that is, for it makes little sense to ask when he so stands – in particular, whether he stood in that relation before Hegel was born, or to suppose that his preceding of Hegel did not begin until Hegel's birthday. Analogously, it is perfectly plain that John may stand in some relation (tenselessly) to some inscription that succeeds him in time, just as he may be related to a contemporary object remote from him in space. To be sure, our grounds for *asserting* any particular temporal precedence themselves vary over time, but that is another story altogether. Similarly, our grounds for asserting John's relation to some inscription themselves vary with our own circumstances in several respects, but that is an independent matter. Whatever conditions may ordinarily be supposed to warrant us in asserting (6) at any time, they warrant us in asserting (7) at that time, thus at one stroke producing the inscription which serves us as a parameter for simultaneously describing John's desiring behaviour, no matter what *its* temporal location.

The substantive conditions which provide the requisite warrant in either case are not specified by the analysis of my [1963], which restricts itself to interpreting only the logical form and ontological character of (6), presuming (7) true just under those conditions in which (6) is true. "Any further, substantive analysis of 'desire', specifying the operative conditions for the truth" of (6) and hence of (7), "is theoretically welcome, however, as an independent step... the 'desires-true' formulation is to be taken as true under just those conditions in which its ordinary 'desires that' counterpart is considered true. In particular, for an agent to desire-true some given inscription does *not* imply that he produce, possess, wish to possess, be aware of, or even understand the inscription in question" (Scheffler [1963], pp. 101-2). The upshot is that John's behaviour may be such as to render true our associating him with a subsequent inscription. So long as we have (6), we can interpret it after the manner of (7). In Gorovitz's case earlier cited, where John "desires that he meet the woman. But no inscription to that effect is produced" (Gorovitz [1970], p. 251), our own very statement that John desires that he meet the woman constitutes the production of the requisite inscription, and it matters not that John himself is reticent.

In analysing (6), my [1963], it must be said, offers no help with such statements as:

(8)   John desires something.

"Such quantifications", as Quine remarks in considering comparable cases of belief, "tend anyway to be pretty trivial in what they affirm, and useful only in heralding more tangible information". (Quine [1960], p. 215). In offering no interpretation of such sentences as (8), my [1963] is certainly limited, but it is not at all clear that this limitation is serious if, as I argued in my [1963], teleological arguments can be reconstructed without them. Mr. Gorovitz insists that the limitation of my analysis constitutes an objection to it. "If beliefs and desires have objects, surely we must somehow be able to account for such sentences as assert that two people share a common belief" (Gorovitz [1970], p. 253). But it is not at all clear to me why, if an analysis accomplishes a given purpose, it is faulty in not accomplishing some other. Every analysis is limited, and to judge the limitation *per se* to be a defect is to express a value judgment about the significance of what is omitted as well as a strategic judgment about how the omission is to be remedied. I agree with Quine, however, rather than with Gorovitz, that even the surrendering of (8), at least from the point of view of reconstructing teleological explanation, is not a significant price to pay in order to avoid the obscurities of intension and abstraction. That some price needs to be paid is clear, for classes and intensions give us genuinely increased powers of expression, not otherwise available. The point cannot therefore be to duplicate them nominalistically but only to gain their effect in limited but important contexts, in order to avoid the metaphysical difficulties that many of us find repugnant.

6. SENTENCES AND AMBIGUITY

Mr Gorovitz thinks to find a special difficulty in the wedding of Quine's conception of sentences as sequences of phonemes or characters to a relative construction of belief. The point is that the same such sequence may carry different semantic interpretations, so that taking it as object of a given belief will render the belief-attribution ambiguous. It is not at all clear that we have here, as Gorovitz thinks, "conclusive objections" to such a course, that "they show the difficulties of trying to account in an extensional, non-semantic way, for issues of the sort we have been considering" ([1970], p. 255). Certainly the conception of sentences as sequences is syntactic, but it does not follow that semantic determination is lacking. For the purposes of belief-attribution or explanation, we normally suppose some relativisation to an interpreted language, either our own, or some other. Quine has himself discussed this matter, and even expressed uneasiness about an explicit relativity to language, on the ground of obscurity of individuation (Quine [1966], pp. 193-4). Where we take our own language as an implicit parameter, we will certainly run into trouble in belief-attribution if such language is ambiguous, but there is no general reason for supposing that such ambiguities must always, and in principle, dog our steps. A nonambiguous language of restricted and regimented sort is at any rate possible, and in general desirable for scientific purposes. Ambiguity is no more of a difficulty for its uses in belief-attribution than for any other of its descriptive employments.

7. EXPLANATION AGAIN

A final counterexample is proposed by Mr. Gorovitz as follows: "A window breaks and a witness asks a friend, 'What's the explanation?' The friend replies, 'There was a sonic boom'" ([1970], p. 254). Now we have already seen, in 4 above, that neither the question nor the provision of an explanation is required, by my analysis, to produce the requisite explanandum-inscription. But Gorovitz argues from this case further that the above reply fails to "explain what stands in need of explanation. But the reply contains

Explanations, Desires, and Inscriptions · 39

no inscription of the sort Scheffler seeks as object. Hence, that which stands in need of explanation cannot be an inscription in the reply" ([1970], p. 255). We have already remarked that my [1963] seeks to analyse such statements as (2). It is also directed to their denials. Thus, it is capable of interpreting also:

(9) g does not explain why k is blue.

The fact that the reply itself does not contain the requisite inscription does not prevent the explanation-description (9) from containing it. To be sure, the analysis of my [1963] does not interpret such quantifications as:

(10) g explains something

and

(11) Something is explained by g and h.

But such a limitation has already been acknowledged earlier, in the case of desire, and what was said in that connection applies here as well.

Mr Gorovitz, however, feels that we ought to be able to say that two explanatory replies are "aimed at explaining the same thing", that they are "directed at the same object", that there *is* something that both seek to explain, that where one reply fails and a later one succeeds *and also incorporates* an appropriate inscription, the latter inscription, since it was not earlier available, "cannot be the object of the explanation" ([1970], p. 255). In so far as this argument represents simply an insistence on the importance of (10) and (11), I have already commented on it.

But there seems another point here, related to the notion of an *object*. For the expression above noted, i.e. "aimed at", "directed at", etc., suggest that "object" is here taken as a *goal:* it is that which the explainer or explanatory statement strives to explain. Certainly inscriptions are not objects of explanation in this sense; rather they are objects presupposed by the analysans which supplants (2). However, goals are among those very intensional entities which it is the purpose of my analysis to avoid. The entities provided by my analysis are not, as earlier argued, intended to serve as equivalents of the banished intensional objects. Rather, the analysis as a whole is to be judged by its enabling us to circumvent these so-called objects, while achieving the main effects of the obscure originals. It is thus no surprise at all, and no inadequacy of the analysis, that the entities it presupposes are not goal-objects. For what, after all are goal-objects?

NOTES

1. Indeed, this statement (as well as the adverb "logically" in the parenthesised phrase "logically intensional") is explicit enough to indicate that the individuation of facts in my discussion is by logical equivalence rather than by analytical equivalence, despite Gorovitz's misleading suggestion, based on his reference to an independent passage in Quine (Gorovitz [1970], pp. 248-9).

2. [1963] p. 74. For certain critical comments relating to points in this and the succeeding section of the present article I am indebted to Robert Schwartz.

3. It is, in general, not clear that (5) is an important sort of case, for how could we be in a position to assert it with confidence unless we had more specific information as to a putative explanans relative to some particular explanandum? However that may be, the *possibility* of some interpretation of (5) is not ruled out by my proposal.

References

Gorovitz [1970] Inscriptionalism and the Objects of Explanation. *Brit. J. Phil. Sci.* 21, 247-56.

Quine [1960] *Word and Object.* New York: John Wiley & Sons.

Quine [1966] *The Ways of Paradox.* New York: Random House.

Scheffler [1963] *The Anatomy of Inquiry* New York: Alfred A. Knopf.

6. SYMBOLIC ASPECTS OF RITUAL I: RITUAL, MYTH, AND FEELING — CASSIRER AND LANGER

MY TOPIC in this paper and the one to follow is the interpretation of ritual. The topic is so rich and complex and has been cultivated so assiduously by anthropologists, classicists, historians of religion and scholars of other disciplines that I enter with trepidation upon this domain. In extenuation, I assure you that I offer no comprehensive theory, but rather a limited exploratory discussion of its symbolic functioning alone.

The study of ritual is beset with a special contemporary hazard. So dominant is the intellectual prestige of science that theorists tend to depreciate other forms of endeavor as irrational or, at any rate, to segregate them as noncognitive or primarily emotive. In the conviction that to separate knowing from feeling is theoretically as well as practically pernicious, I emphasize symbolic aspects of ritual to promote a more balanced appreciation of its cognitive roles.

An emphasis on the symbolic functioning of ritual requires, however, a preliminary word of caution. For such an emphasis abstracts from the social, historical, and religious contexts in which particular rites are embedded. It ignores, for example, details of the cult, the ideological system, and beliefs as to the efficacy of the rite itself. Such abstraction should not, however, be taken as a denial of the importance of the context nor as a suggestion that any rite can be understood through its semantic features alone. Nor does it imply any judgment as to the truth or warrant of beliefs surrounding the rites to be discussed.

In this paper, I first give the background of my emphasis on symbolic aspects of ritual and then characterize its particular direction through a consideration of the views of Ernst Cassirer and Susanne Langer. In the paper that follows, I explore five modes of ritual symbolization or reference,

---

This paper appears here for the first time.

comprising three modes proposed by Nelson Goodman in and for his study of the arts and two further modes I believe peculiarly relevant to the interpretation of rites.

1. THE DEVALUATION OF RITUAL

In stressing symbolic aspects of ritual, I mean to draw attention to its cognitive roles. Thus, I challenge the view that it is science alone which is cognitive in function, while art, ethics, and religion are altogether matters of feeling. This division, I suggest, underlies much of the modern devaluation of ritual, banishing it, prima facie at least, not only from the realm of cognition, but from the proper sphere of religion as well. The division also underlies contrary philosophical attempts to provide an emotive interpretation for ritual – thus realigning it with religion and leaving the cognitive rule of science untouched. My own view is opposed to the division itself, common to both camps, which splits cognition from feeling, assigning the former wholly to science and the latter exclusively to nonscientific realms such as art, ethics, and religion. In an earlier paper, I attacked it through an analysis of emotional aspects of science.[1] In this and the following paper, I attack it from another side, by elaborating the symbolic, hence cognitive functioning of ritual.

My enterprise is thus opposed to the attitude of William James, whose *Varieties of Religious Experience* begins by dividing the religious field into two parts, the institutional and the personal, proceeding thereafter "to ignore the institutional branch entirely." By contrast, the sort of religion in which James *is* interested gives rise, as he says, to "personal not ritual acts, the individual transacts the business by himself alone, and the ecclesiastical organization, with its priests and sacraments and other go-betweens, sinks to an altogether secondary place. The relation goes direct from heart to heart, from soul to soul, between man and his maker."[2] Ritual is here assimilated by James to institutional machinery, slowing or even obstructing the free flow of religious feeling; it is mentioned only to be dismissed. Clearly affiliated not with science but with religion, it is to be located within the realm of feeling. But in this realm of feeling, it is at best a necessary evil, at worst a positive hindrance.

Where it has not been thus dismissed, ritual has – since it is clearly not mere sentiment – been rather assigned to the opposed realm of cognition – but typically with equally devaluing effects. For here the resulting view – to which my enterprise is also opposed – has predominantly associated ritual with mythology, itself interpreted as defective cognition, bad science, pathological belief. "Among the philosophers," writes Cassirer, "it was especially

Symbolic Aspects of Ritual I • 43

Herbert Spencer who tried to prove the thesis that the mythico-religious veneration of natural phenomena... has its ultimate origin in nothing more than a misinterpretation of the *names* which men have applied to these objects.... For Max Müller, the mythical world is essentially a world of illusion – but an illusion that finds its explanation whenever the original, necessary self-deception of the mind, from which the error arises, is discovered. This self-deception is rooted in language, which is forever making game of the human mind, ever ensnaring it in that iridescent play of meanings that is its own heritage. And this notion that myth does not rest upon a positive *power* of formulation and creation, but rather upon a mental *defect* – that we find in it a 'pathological' influence of speech – this notion has its proponents even in modern ethnological literature."[3] For Lévy-Bruhl, finally, mythical thought is "pre-logical thought" in search of "mystic causes," thought for which "even the law of contradiction, and all the other laws of rational thought, become invalid."[4]

2. CASSIRER ON MYTH

In an effort to redress the balance of such opinion and to give a more sympathetic interpretation of mythical thought, Cassirer proposes to view such thought as in fact continuous with current scientific conceptions, though different in quality. It is not simply bad science but a positive stage in the development culminating in science. Its difference from science, however, consists precisely in its association with ritual – that is, symbolic action and not merely image or representation – such association yielding a form of experience characterized by a distinctive dramatic orientation and unity of feeling.

Cassirer thus denies that there is "an absolute heterogeneity between our own logic and that of the primitive mind", affirming that mythical thought does constitute "a positive power of formulation and creation".[5] It is not to be judged by reference to our current scientific ideas but recognized as a form of conceptualization in its own right, to be studied in the context of actions and appreciated for its characteristic emotional relation to the community and continuity of life. Thus, he writes, "If we wish to account for the world of mythical perception and mythical imagination we must not begin with a criticism of both of them from the point of view of our theoretical ideals of knowledge and truth. We must take the qualities of mythical experience on their 'immediate qualitativeness'. For what we need here is not an explanation of mere thoughts or beliefs but an interpretation of mythical life. Myth is not a system of dogmatic creeds. It consists much more in actions than in mere images or representations.... That ritual is prior to dogma, both in a historical and in a psychological sense, seems now to be a generally adopted maxim. Even if we should succeed in analyzing myth into ultimate conceptual elements, we could, by such an analytical process, never grasp its vital principle, which is a dynamic not a static one; it is describable only in terms of action."[6]

What turns out to be distinctive of mythical mentality, so understood, according to Cassirer, is not its logical character but rather its emotional quality. The charge of prelogicality springs from our contemporary provincialism which seeks but fails "to intellectualize myth – to explain it as an allegorical expression of a theoretical or moral truth."[7] The failure of such attempts at intellectualization shows not that mythical thinking is irrational but at best that its premises may appear so to us. Such attempts, says Cassirer, ignore "the fundamental facts of mythical experience. The real substratum of myth is not a substratum of thought but of feeling. Myth and primitive religion are by no means entirely incoherent, they are not bereft of sense or reason. But their coherence depends much more upon unity of feeling than upon logical rules.... Life is not divided into classes and subclasses. It is felt as an unbroken continuous whole which does not admit of any clean-cut and trenchant distinctions.... Myth is an offspring of emotion and its emotional background imbues all its productions with its own specific color.... To mythical and religious feeling nature becomes one great society, the *society of life.*"[8]

Striving to combat the view of myth and primitive religion as irrational because unscientific, Cassirer defends them as having a distinctive emotive character. His defense tends, however, to reinforce the division between science and emotion, and to cede the critical contention that scientific thought has a monopoly of cognitive virtue. He affirms, indeed, that our "analytical process... is opposed to the fundamental structure of mythical perception and thought," but denies that myth may therefore be regarded as "a mere mass of unorganized and confused ideas." For it is not merely privative, but the root stage of a developmental process from which our "empirical thought" has grown.[9] Its virtue, Cassirer seems thus to say, consists not in its own cognitive character but rather in its linkage with an eventually maturing science.

It is science that first perceives objective features and develops the concept of nature. "What myth primarily perceives are not objective but *physiognomic* characters. Nature, in its empirical or scientific sense... does not exist for myths. The world of myth is a dramatic world – a world of actions, of forces, of conflicting powers.... Mythical perception is always impregnated with these emotional qualities. Whatever is seen or felt is surrounded by a special atmosphere – an atmosphere of joy or grief, of anguish, of excitement, of exultation or depression.... While science has to abstract from these qualities in order to fulfill its task, it cannot completely suppress them. They are not extirpated root and branch; they are only restricted to their own field. It is this restriction of the subjective qualities that marks the general way of science."[10]

Science is the third of three stages distinguishable in the development of human thought. The "first rudimentary stage of our physiognomic experience" is "abandoned and overcome" by the "world of our sense perceptions, of the so-called 'secondary qualities.'" This world is in turn succeeded by "that form of generalization that is attained in our scientific concepts – our concepts of the physical world." Though Cassirer insists that none of these stages is "a mere illusion,"[11] he clearly holds science alone to constitute the field of objective thought. It cannot destroy myth, but it must clear its own ground of it in order to fulfill its task. Conversely, the sphere of myth, dramatic in orientation, is therefore drenched in emotion, allowing therefore no foothold for objective thought. The subjective qualities that form the world of myth must be avoided if our scientific conception of the world is to be attained.

Cassirer's defense of mythical thought attributes to it a positive power of conceptualization and a role in the historical process. But this defense is limited by its contrast of logic and feeling, of emotion and science. Such a defense purports to rebut the view of myth and ritual as mere nonsense and so to reverse their devaluation. But it fails to rebut – and indeed strengthens – the view that cognition, properly speaking, is scientific or it is nothing. And it strengthens equally the false doctrine that scientific cognition is, and must be, devoid of emotion.

3. LANGER ON RITUAL

Commenting on Durkheim and Lévy-Bruhl, Cassirer remarks that one principle of the French sociological school has indeed been "given full and conclusive proof" – that all the "fundamental motives" of myth "are projections of man's social life."[12] But the other principle – that primitive mentality is prelogical – he rejects as contrary to the anthropological evidence. "What is characteristic of primitive mentality," he writes, "is not its logic but its general sentiment of life." Totemism and animal worship exemplify what Cassirer regards as a "general presupposition of mythical thought," that is, "the consanguinity of all forms of life," and he views the "constant and obstinate negation of...death" as constituting the burden of mythical thought.[13] The empirical differences among forms of living things are not overlooked by primitive man; they are rather "declared to be irrelevant in a religious sense", overridden by a characteristic "feeling of the indestructible unity of life."[14]

In Cassirer's view, ritual and myth are both to be interpreted as having their roots in such characteristic feeling. In one important respect, Susanne Langer differs from this account. She separates myth from ritual, treating only ritual as having its origin in *feeling,* while myth is explained as arising rather from *fantasy.* "Although we generally associate mythology with religion," she writes, "it really cannot be traced, like ritual, to an origin in anything like a 'religious feeling.' . . . Ritual begins in motor attitudes, which, however personal, are at once externalized and so made public. Myth begins in fantasy, which may remain tacit for a long time; for the primary form of fantasy is the entirely subjective and private phenomenon of *dream.*"[15]

Despite her separation of myth from ritual, however, Langer takes both to be centered on the concept of life, as does Cassirer. She treats them both, indeed, as being understandable in terms of her concept of a "life-symbol." Objects capable of serving as dream-symbols become vested with emotion in waking life as well; such objects "found and treasured" carry "the imaginative process . . . from dream to reality; fantasy is externalized in the veneration of 'sacra'." That they are phallic symbols and death symbols is to be expected from their dream-related character, but needs no psychoanalytic evidence in support; "any student of anthropology or archaeology can assure us of it. Life and life-giving, death and the dead, are the great themes of primitive religion. . . . Certain animals are natural symbols to mankind: the snake hidden in earth, the bull strong in his passion, the mysterious long-lived crocodile who metes out unexpected death. When, with the advance of civilization, their images are set up in temples, or borne in processions, such images are designed to emphasize their symbolic force rather than their natural shapes. The snake may be horned or crowned or bearded, the bull may have wings or a human head."[16]

Religious feeling, in Langer's account, is "bound . . . to set occasions, when the god-symbol is brought forth and officially contemplated." On such occasions, it evokes excitement, initially "an unconscious issue of feelings into shouting and prancing or rolling on the earth like a baby's tantrum," later "a habitual reaction . . . used to *demonstrate,* rather than to relieve, the feelings of individuals." In the latter phase, the expressive act is no longer self-expressive but rather "expressive in the logical sense. It is not a sign of the emotion it conveys, but a symbol of it; instead of completing the natural history of a feeling, it denotes the feeling, and may merely bring it to mind, even for the actor. When an action acquires such a meaning, it becomes a *gesture.*"[17]

Gestures are symbolic; they are not emotional acts. No longer "subject to spontaneous variation," they are "bound to an often meticulously exact

repetition, which gradually makes their forms as familiar as words or tunes."
It is such formalized behavior in the presence of sacred objects which
constitutes *ritual,* "a complement to the life-symbols." These symbols "present
the basic facts of human existence" while "the rites enacted at their contemplation formulate and record man's response to those supreme realities.
Ritual 'expresses feelings' in the logical rather than the physiological sense ... it
is primarily an *articulation* of feelings. The ultimate product of such articulation is not a simple emotion," concludes Langer, "but a complex, permanent *attitude* ... an emotional pattern, which governs all individual lives.... A
rite regularly performed is the constant reiteration of sentiments toward
'first and last things'; it is not a free expression of emotions, but a disciplined
rehearsal of 'right attitudes.' "[18]

4. VARIETIES OF EMOTIVISM

Now, the interpretations of both Cassirer and Langer may fairly be
characterized as *emotive,* for each gives an account of ritual in terms of
feeling. But, here as elsewhere in philosophy, such characterization must
reckon with the fact that emotivism is an elusive notion and appears in
various guises. Where Cassirer speaks straightforwardly of a "unity of feeling"
underlying myth and primitive religion, Langer speaks rather of feelings
"denoted" and "demonstrated." Where he seeks the distinctiveness of mythical thought in a "substratum of feeling," she emphasizes rather the transition
from "self-expression" to "logical expression," from "an unconscious issue
of feelings" to "meticulously exact repetition" of gestures denoting feelings
and "rehearsing 'right attitudes.' "

It is, in fact, not feeling itself that figures in ritual, according to Langer
but rather the formalization of its outward manifestations. Rituals may have
arisen in the first instance through feelings, but by the time the rituals have
crystallized, the feelings have already evaporated. The symbolism of ritual
gesture is to be understood not as a vehicle of feeling but rather as a vehicle
of reference, the gesture bringing the feeling to mind, denoting, or articulating it. The result of such referential process is the growth of "right attitudes,"
or patterned emotional dispositions, in the individual members of the
group, but the process itself hardly flows from an initial unity or substratum
of feeling. In Langer's view, ritual is not the consequence of a prior unity of
feeling nor is it opposed to the life of feeling; its function is indeed, through
articulation, to help create a common pattern of dispositions to feeling.

Yet, because she contrasts the "spontaneous variation" of "emotional
acts" with the fixity of gestural symbols growing out of such acts and
denoting them, it may seem but a short step from Langer's view to a quite

different one, which I would characterize as an anti-emotive emotivism. This is the view that would explain ritual by its negation of emotion: Such a view supposes that ritual formalism is in fact opposed to emotive freedom, that the fixity of rites is not merely a matter of their symbolic character which, in articulating certain feelings, helps indirectly to shape their patterning in individual lives. Rather, the function of ritual is thought to be to contain emotion, to deaden spontaneity, to stabilize society and conserve the status quo. The rigidity of ritual, in sum, is an aspect not of its symbolic character but rather of its suppressive nature.

A view of this general sort is in fact offered by Henri Bergson, who opposes "Static Religion" and "Dynamic Religion," the former "a product of social pressure," the latter "based on freedom." As the natural world, for Bergson, is characterized by an opposition of the "mechanism of matter" and the "creative power of the *élan vital,*" so the social world "is divided between two opposite forces," the one tending to preserve the present state, the other striving for new forms of life, the one associated with static, the other with dynamic religion.[19]

Now, anti-emotive emotivism is, as I have said, quite different from Langer's symbolic interpretation, despite their shared contrast of ritual with spontaneity. One might have judged such emotivism to be, a fortiori, at odds with Cassirer's view of myth and primitive religion as resting upon a distinctive substratum of feeling. It is thus surprising to find that Cassirer takes a rather sympathetic attitude toward Bergson's anti-emotive interpretation.

He criticizes Bergson only for denying that there could be a "continuous process" leading from "primitive social life" to the "new ideal of a free personal life," and he himself defends such continuity, asserting that "myth is from its very beginning potential religion."[20] But he agrees with Bergson that there is "in all human activities... a fundamental polarity... between stabilization and evolution," agreeing further that myth and primitive religion exemplify stabilization to the highest degree. Thus he writes, "From the point of view of primitive thought the slightest alteration in the established scheme of things is disastrous. The words of a magic formula, of a spell or incantation, the single phases of a religious act, of a sacrifice or prayer, all this must be repeated in one and the same invariable order. Any change would annihilate the force and efficiency of the magical word or religious rite. Primitive religion can therefore leave no room for any freedom of individual thought. It prescribes its fixed, rigid, inviolable rules not only for every human action but also for every human feeling."[21] Here Cassirer's apparently direct emotivism has been transformed into Bergson's anti-emotive doctrine. No longer does he speak of myth as "an offspring of emotion" and of nature itself as imbued by religious feeling with the

Symbolic Aspects of Ritual I

character of "one great society, the society of life." No longer does he emphasize that the coherence of myth and primitive religion "depends more upon unity of feeling than upon logical rules." Here it is just the rigidity of social rules impinging upon and constraining "every human feeling" that Cassirer underscores.

Is his attitude consistent? If it is, there must be a coherent way of characterizing a society both in terms of its distinctive feeling for "the unity of life" and in terms of its rigid constriction of all feeling save, perhaps, the dread of disobeying the rules. The vagueness of the terms in which Cassirer's discussion is cast makes it very difficult to muster any certainty in the matter. But his hospitality to Bergson's sharp contrast of social pressure and human feeling makes it hard to see how he might reconcile them in the interests of his larger account of myth and primitive religion.

In any case, the ultimate consistency of Cassirer's two forms of emotivism is not necessary for us to decide. They are both, after all, *emotivisms:* each characterizes ritual in terms of feeling, the one as springing from or conveying feeling, the other as constraining or suppressing feeling. And what he emphasizes in either case is that the emotive distinctiveness of primitive religion does not preclude its linkage by a continuous process with contemporary thought.

In answer to Lévy-Bruhl he denies "an absolute heterogeneity between our own logic and that of the primitive mind" and emphasizes the growth of science out of mythical thought. In answer to Bergson, he denies that "there exists no continuous process which can lead from ... static to dynamic religions," asserting that "myth, even in its crudest and most rudimentary forms, contains some motives that in a sense anticipate the higher and later religious ideals."[22] His response to the French school is to replace Cartesian dualisms with Hegelian continuities.

But continuity only connects the extremes; it does not abolish them. The sharp division of primitive thought from modern science is replaced by a developmental account that nevertheless leaves primitive thought in a primarily emotional and subjective state. The sharp division of static from dynamic religion, similarly, is replaced by a continuity which leaves the life of primitive man "under a constant pressure ... enclosed in the narrow circle of positive and negative demands, of consecrations and prohibitions, of observances, and taboos."[23] Cassirer's effort to paint a more sympathetic picture of myth and ritual thus leaves intact, indeed reinforces, our complacent belief in progress: the supposition that modern science, exclusively, is cognitive, as in modern "dynamic religion ... religious life has reached its maturity and freedom."[24] Continuity, in short, leads from a state characterized primarily in terms of emotion to a "higher and later" state of science and freedom.

5. RITUAL AS MULTIPLY SYMBOLIC

I should like to extend the pioneering symbolic approach of Cassirer and Langer without, however, relying upon the division of cognition and emotion that I have been criticizing. In the next paper, I shall, to this end, analyze five modes of ritual symbolization or reference. For the remainder of this paper, I want to comment briefly on the underlying idea that ritual may symbolize at once in various modes.

Let me begin by stating that ritual consists of action, though not of mere action. Langer is right in emphasizing both the formalization of ritual behavior and its gestural character – that is, its symbolic function. She is also right in denying the primarily cathartic role of such behavior – its "self-expressive" as distinct from its "logically expressive" aspect.

But her interpretation of the symbolic function of ritual is, I believe, too restricted, both in its understanding of the symbolic *process* itself and in its grasp of the *objects* symbolized. For, in elaborating the process, she speaks of it as *denotation* or *demonstration*, and, as to objects, she limits the things denoted to feelings. The symbolic character of ritual does not, however, respect either limitation. Ritual elements may symbolize anything – and not just feelings. The process of symbolization, moreover, need not be restricted to denotation but may encompass other forms of reference as well. Indeed, ritual is typically symbolic in several modes simultaneously.

The multiply symbolic character of ritual may help to account for one of its often remarked features – its capacity to survive changes in doctrinal interpretation, its *priority over dogma*, as Cassirer expresses it. He attributes such priority to the "vital principle" of myth which is, as he says, "describable only in terms of action. Primitive man expresses his feelings and emotions not in mere abstract symbols but in a concrete and immediate way; and we must study the whole of this expression in order to become aware of the structure of myth and primitive religion."[25]

If this explanation in terms of vital principle is obscure, the multiple symbolism of ritual suggests perhaps an easier account of its staying power. The strength of ritual may lie just in its being anchored by multiple referential bonds to objects. When one or more are cut, the others meanwhile hold fast. When one requires relocation under a new interpretive idea, the untying and retying process does not destroy the whole linkage. Thus it is that rituals change more slowly than creeds, often surviving even drastic alterations of doctrine and entering into new interpretive contexts without loss of vigor. It is in the multiplicity of its symbolic connections rather than its concreteness that the strength of ritual lies. In the next paper, I shall explore such multiplicity.

NOTES

1. Israel Scheffler, "In Praise of the Cognitive Emotions", *Teachers College Record* 79 (1977): 171-86.

2. William James, *The Varieties of Religious Experience* (New York: Random House, 1902, 1929), 30.

3. Ernst Cassirer, *Language and Myth* (New York: Dover, 1946), 3, 5-6.

4. Ernst Cassirer, *An Essay on Man* (New Haven: Yale University Press, 1944), 79-80, and references to Lévy-Bruhl in n. 11, p. 80.

5. Ibid., 80.

6. Ibid., 79.

7. Ibid., 81.

8. Ibid., 81-83.

9. Ibid., 76.

10. Ibid., 76-77.

11. Ibid., 78.

12. Ibid., 79-80.

13. Ibid., 82, 84.

14. Ibid., 83.

15. Susanne K. Langer, *Philosophy in a New Key* (New York: Penguin Books, 1942, 1948), 138-39.

16. Ibid., 121-22.

17. Ibid., 122-24.

18. Ibid., 124.

19. Discussion of Bergson's *Les Deux Sources de la Morale et de la Religion* in Cassirer, *Essay on Man,* 87ff.

20. Cassirer, ibid., 89, 87.

21. Ibid., 224-25.

22. Ibid., 89, 87.

23. Ibid., 225.

24. Ibid.

25. Ibid., 79.

7. SYMBOLIC ASPECTS OF RITUAL II: RITUAL AND REFERENCE — FIVE MODES

MY PURPOSE in discussing symbolic aspects of ritual is to draw attention to its cognitive roles, thus challenging the prevalent division of cognition from emotion and the assignment of cognitive function exclusively to science. Such division, I have suggested, devalues ritual either by excluding it entirely from science and religion, or else by assigning it to religion under an emotive interpretation alone.

My hope is to extend the symbolic approach of Cassirer and Langer by eliminating reliance on the division altogether and attending solely to the cognitive aspects of rites. For, as I have argued, both Cassirer's and Langer's positive portraits of ritual remain emotive in their different ways, Cassirer appealing to a distinctive unity of feeling and Langer to a denotation or articulation of feelings. In contrast to both, I shall emphasize rather the varieties of symbolic function displayed by rites, requiring neither emotive unity nor peculiarly emotive objects, and helping to form the cognition of their adherents. I consider five modes of ritual symbolization or reference, beginning with three varieties proposed by Nelson Goodman in and for his study of the arts,[1] namely, *denotation, exemplification,* and *expression,* and supplementing these with two further modes, *mention-selection* and *reenactment,* which I shall propose as peculiarly relevant to the interpretation of ritual. I shall, in general, make comparisons between ritual and the arts in an effort to bring out what is distinctive in ritual reference.[2]

1. DENOTATION, EXEMPLIFICATION, EXPRESSION

The use made by Goodman of these and related terms is important to note, since it may be unfamiliar. He broadens the notion of denotation so that it comprehends not only verbal description but also pictorial and other sorts of representation, and he employs the notion of reference to encom-

---

This selection is based upon my paper, "Ritual and Reference," *Synthèse* 46 (March, 1981): 421-37.

pass not only denotation but also exemplification and expression. Exemplification is the relation, *being a sample of*: it relates a thing to those of its properties to which it also refers. A tailor's swatch, as Goodman explains, "does not exemplify all its properties; it is a sample of color, weave, texture, and pattern, but not of size, shape, or absolute weight or value."[3] Expression, finally, he treats as implying metaphorical exemplification, that is, reference by an object to a property it metaphorically possesses. Thus a given picture may, at one and the same time, denote a man, literally exemplify certain hues or patterns, and express – that is, metaphorically exemplify – melancholy.

Not only words and pictures but also gestures, as Goodman remarks, may denote, exemplify, and express.[4] Now, we note that ritual gestures, in particular, may denote or represent historical events, or events thought to be historical, they may portray expected occurrences or hoped-for outcomes, they may denote or purport to denote persons, gods or things. They may perform this role through bodily movement, after the manner of mime; they may also employ the voice in song or speech. The range of ritual gestures indeed comprehends verbal gestures; thus any denotative role that can be fulfilled by verbal means is also within the scope of ritual reference. Objects employed in ritual may also stand for, or denote, in a wide variety of ways.

It is important to recognize that, although not every ritual gesture denotes, every such gesture has firm specifications or prescriptions that it must satisfy. These may be transmitted orally or written down or understood in context, but that there is a right and a wrong way of execution is normally evident. What Cassirer says of sacrificial services may be extended generally, "The ... service is fixed by very definite objective rules, a set sequence of words and acts which must be carefully observed if the sacrifice is not to fail in its purpose."[5] Now a proper performance of a rite often functions as a sample of it, that is, it literally exemplifies it. In this way, it may lend itself to auxiliary use as a demonstration in the process of teaching the rite to learners. Rituals may thus be passed on, through participation.

2. RITUAL AND ART: THE QUESTION OF SCORING

Can there be scores for rituals as there are for musical compositions, or is ritual closer to the case of painting or printmaking? Can there be a recipe for producing a valid rite or does validity depend on some particular historical origin? Goodman analyzes musical scoring as a notational system, suggesting that the basic function of a score is "the authoritative identification of a work from performance to performance." As he explains, "Not only must a score uniquely determine the class of performances belonging to the work, but the score ... must be uniquely determined, given a performance and the notational system."⁶ The question, then, is whether rites can be scored.

Goodman allows that, in a trivial sense, anything can be scored. But not everything can be scored in a way that both frees the identification of works from reference to the history of their production, *and* honors the past practice of their identification. Painting, for example, is, in Goodman's terminology, unlike music in being *autographic,* that is, vulnerable to forgery. Yet it can be scored by a system "assigning a numeral to each painting according to time and place of production."⁷ Such a system patently fails, however, to free the identification of paintings from reference to history of production, while any system which did succeed in this regard would violate the past practice of identifying a work with the individual picture alone. Goodman therefore concludes that a nontrivial notational system cannot be devised for painting.⁸

But the case of rituals seems quite different; antecedent practice identifies a rite not with the individual performance but with an appropriate group of performances. To devise a nontrivial scoring system thus seems possible, though certainly not a routine task. Indeed, the motivation for notationality in the case of ritual would appear similar to that suggested by Goodman for certain of the arts. "Where the works are transitory, as in singing or reciting, or require many persons for their production, as in architecture and symphonic music, a notation may be devised in order to transcend the limitations of time and the individual.... The dance, like the drama and symphonic and choral music, qualifies on both scores while painting qualifies on neither."⁹ Ritual would seem to qualify on at least one, frequently on both, scores as well.

Moreover, critical portions of many rites are composed of verbal formulas and these are surely scoreable, as are also musical portions, and even bodily movements – which are theoretically amenable to description by one or another system of dance notation. There seems indeed no bar to increasing notationality for any of these aspects of ritual. Is ritual, then, to be judged as similar to music in this regard – as being, in Goodman's terminology, not autographic but, rather, uniformly *allographic?* The question is not so easily settled. It requires a consideration of differences between art and ritual, to which we now turn.

3. NOTATION AND NUMBER

Two differences between art and ritual, relative to notation, must be addressed. One concerns the number of items to be identified; the other concerns conditions on the performers. Let us consider first the question of number. Recall that the drive for notationality in the arts, according to Goodman, is the need for identification of a work from performance to performance. An additional fact about the arts, however, is the continuing stream of new works to be acknowledged, indeed the inexhaustible number of works to be accommodated by any identifying notation. There is, for example, no limit to the number of musical works to be provided for in a notation that accords with antecedent musical practice. And the standard scoring system for music in fact accommodates an infinite number of works.

Rites are in this respect quite different, at least within any given religious or cultural system. For in any such system, the rites to be identified constitute a finite and, typically, a manageably small number. The problem of scoring seems therefore not nearly so severe as in the case of the arts. Not having to devise a "universal" system – one with infinite potential – one may imagine a restricted language yielding for each rite a correlated description, the whole set of such descriptions satisfying notational requirements.[10] With a list of all the regimented rite-descriptions before us, we apply the system by running through the whole list before making any decision. This sort of system seems to me to approximate more closely than the universal system of standard musical notation the process of authoritative identification of rituals by their adherents. The matter is of course quite different if we think not of adherents but of anthropologists, whose concern is identifying rites cross-culturally and, indeed, in a potentially universal manner. This concern does seek a scoring system with infinite potential, like that of standard musical notation.

4. CONDITIONS ON THE PERFORMER

Consider now the second difference between art and ritual, that concerning conditions on the performer. I said earlier that devising a nontrivial notation for ritual seemed possible because, unlike the case of painting, where prior practice identifies the work with the individual picture, a rite is not identified by such practice with the single performance but with some appropriate group of performances. But this contrast is not, after all, decisive. For printmaking is, like painting, *autographic* even though there are many prints, and not just a single one, normally associated with a given work. The crucial point is that identification of these prints depends upon the history of their production – consists in fact in their linkage to a common source in the original plate.[11] Might it then not be the case, similarly, that identification of the various performances associated with a given rite hinges upon historical origins, thus rendering ritual after all autographic rather than allographic?

To consider this possibility, let us first remark that allographic art, for Goodman, rests on a distinction between constitutive and contingent features of a work, independent of history of production, the constitutive features being singled out by a notation. It is the lack of such a notation that makes autographic art vulnerable to forgery, while the availability of such notation in the case of an allographic art renders forgery of *a work* vacuous. For forgery is deception regarding the circumstances of production, and such deception is powerless to alter identification of a performance as belonging to a given work, decidable as it is solely by reference to the score in question. (Yet forgery of *a performance* is still possible for an allographic work, because it is still possible to deceive as to whether a given performance has certain historical properties – such as whether or not it is the première performance – as distinct from whether or not it is a sample of the work.)

I want now to suggest that a significant contrast between allographic arts and rites consists in this: that rites alone typically impose constitutive constraints on their performers, as well as on their performances. Performers of a rite need to be the prescribed ones, the duly constituted, elected, anointed, or appointed ones, those satisfying the authoritative specifications. While an orchestral performance conforming to the score of a given symphony is an instance of that symphonic work no matter who the players may be, an otherwise proper ritual performance that falsely purports to satisfy the constraints on its performers is in fact a forgery not merely of the given performance but of the rite itself.

We can now return to the question with which we started, namely, whether identification of a rite might hinge on historical origins. Indeed, a positive answer to this question is now forthcoming. For the legitimate performers of a rite might in fact be characterized in terms of a chain of transfers of authority leading back to a specified source. In such an event, identification of the rite itself rests on history of production and the rite in consequence turns out autographic. Gareth Matthews, making this point, adduces the Christian Mass as an example, understood as requiring the celebrant to be a Christian priest ordained by a bishop standing in the Apostolic Succession – that is, a bishop ordained by a bishop, ordained by a bishop,... ordained by one of the Apostles.[12] A feature of the history of production is here constitutive of the rite. Thus deception as to possession of this feature by an otherwise adequate performance forges the rite itself, which must accordingly be judged autographic rather than allographic.

How far can this example be generalized? It might be suggested that, since every rite imposing conditions on its performers requires us to ask, "By whom was the rite performed?" every such rite necessarily appeals to history of production and is thus autographic. I think, however, that this general conclusion does not follow. For the question "By whom was the rite performed?" can be given more than one interpretation. Let me explain.

Goodman writes, "Where there is a theoretically decisive test for determining that an object has all the constitutive properties of the work in question without determining how or by whom the object was produced, there is no requisite history of production and hence no forgery of any given work."[13] But the phrase "by whom the object was produced" covers two sorts of cases – one where there is no distinction, independent of history of production, between constitutive and contingent properties of the producers themselves, and the other, where there is. It is the first case only that Goodman has in mind, and he illustrates it as follows: "The only way of ascertaining that the *Lucretia* before us is genuine is thus to establish the historical fact that it is the actual object made by Rembrandt."[14] The property *being Rembrandt* is the crucial productive property here, and it is not further analyzed into constitutive features shareable by persons other than Rembrandt. Thus, deception as to this property for a given painting constitutes forgery of the work, and the critical question "By whom was the painting produced?" asks, "By Rembrandt or by anyone else?"

But the case of rituals, I suggest, sometimes differs in the following way: The conditions imposed on performers may in certain instances be understood as features not limited to specified persons, but rather as supplementary antecedent performances also scoreable independently of history of production. Such conditions may, for example, require that performers carry out a preliminary cleansing by approved methods, or undergo a prior or concurrent period of silence or of fasting, or execute some other auxiliary procedure. The rite as a whole is thus identifiable by a score specifying not only constitutive features of the focal performance but also constitutive features of the performers – in particular, relevant characteristics of their preparatory or auxiliary performances. The question, "By whom was the rite performed?" does *not* here mean, "By John Doe or by anyone else?" but rather, "By persons satisfying supplementary general specifications or not?" And if the score has been satisfied as a whole, no deception as to further circumstances of production or the identity of the performers will constitute a forgery of the rite as distinct from the *particular performance*.

Of course, forgery of a *particular performance* is still possible, even though the rite is thus allographic. The focal ritual act may, for example, falsely purport on a given occasion to have been preceded by the required preparatory procedures. Analogously, a performance of the fourth movement of a Brahms symphony, itself allographic, may falsely purport to have been preceded by performance of the earlier three. But such forgery of the particular performance is to be distinguished from forgery of the symphony or of the rite itself. Identity of the rite, as of its focal segment alone, is determined (just as is the symphony and its fourth movement alone) wholly by conformity with the relevant score, independently of the history of production. Rites of the sort just considered, inclusive of constitutive conditions on their performers, thus turn out allographic after all.

It appears, in sum, that constraints on performers, which I have suggested are characteristic of rites, lend themselves to allographic as well as autographic interpretations. Now, it is obvious that allographic arts are to be contrasted with autographically interpreted rites. Orchestra members, clearly, do not require historical authorization in order for the work they have played to be identified. But there is, I suggest, a contrast to be made even between allographic arts and rites allographically interpreted or, more precisely, rites whose performers are required to execute supplementary procedures. For the very contrast between the focal and supplementary segments of a given performance, while common to ritual, is foreign to the arts. Certainly the artist is expected to have made various preparations for the performance, but such preparations form no *part* of the performance; they do not enter into its identification. The overt, collaborative result is here separated from the private or individual processes carried out by performers, no matter how essential such processes may be regarded. In the case of rites, by contrast, such private or individual performances may enter into the ritual itself, forming part of its very identity.

Beyond performance altogether, does not intent also enter into ritual identifications and, since intent is notoriously difficult to ascertain, does it not therefore preclude a notational system? We must observe that, in any case, rites do not uniformly require intent. Even where a given rite is in fact considered to require intent, the force of the requirement must be determined: Does its violation actually void the rite or does it merely diminish its value, rendering it, in Austin's terminology, hollow?[15] Finally, let us suppose that a given rite indeed requires intent for its identification and not merely its merit. There is here, I suggest, still no theoretical obstacle to notationality. For notationality in itself does not presuppose ease of application: it is therefore not precluded by difficulty. If it is at least theoretically possible to tell whether a relevant intent has been achieved, the notation may incorporate features of intent among other constitutive features.

5. RITUAL AND EXPRESSION

We have seen that Goodman treats expression as implying metaphorical exemplification. A picture may exemplify not only a certain style or pattern but also a certain feeling or movement, possessing the style or pattern literally, the feeling or movement metaphorically, but in both cases constituting a sample of, hence referring to, the property possessed. A rite expressing a certain feature may, analogously, be taken as metaphorically possessing it and also referring to it. Particular rituals may thus be interpreted as expressing a wide range of features, for example, joy or sorrow, triumph or grief, elation, trust, yearning, contrition, exaltation, steadfastness, supplication, gratitude.

Expression, on this view, is not a matter of what the given symbol denotes or characterizes but of what denotes or characterizes it. The expressive reference made by the symbol is that of exemplification, not denotation. But the multiply symbolic character of ritual should here be recalled. Whatever a given rite may in fact portray, it may simultaneously exemplify, literally or metaphorically, quite different things. Explicitly representing episodes of a sacred story, it may at the same time express, rather than represent, dependence or victory, atonement, or thirst for redemption.

The symbol expressing a feature must, according to the present view, possess it. Must the user or producer or viewer of the symbol also possess it? Not so. "The properties a symbol expresses are its own property," writes Goodman. "That the actor was despondent, the artist high, the spectator gloomy or nostalgic or euphoric, the subject inanimate, does not determine whether the face or picture is sad or not. The cheering face of the hypocrite expresses solicitude, and the stolid painter's picture of boulders may express agitation."[16] Similarly, the feelings, thoughts, or other mental states of performers or spectators of a rite are to be distinguished from the features expressed by the rite itself – at least under the present interpretation of expression.

Yet here we confront a striking contrast between arts and rites, ritual presenting a radically different aspect. For rituals are, in religious as distinct from magical contexts, typically intended to penetrate to the heart. Performers of rites are not actors. The question "Does he truly believe what he is saying?" is relevant to the ritual performer alone, while artful simulation of belief is a feat valued only in the actor. Although both actors and performers of religious ritual may indeed perform flawlessly while their thoughts and feelings are remote from the features expressed, a major point of ritual, though not of drama, is to affect the thoughts and feelings of participants, in part through repeated exposure to such features. Unlike a dramatic performance, a religious ritual usually has a characteristic pattern of recurrence; it is to be repeated with the seasons, or with other units of time, or with the important junctures of a life. Such regular recurrence is intended to pattern the sensibilities of participants, in good part by repeated contact with features exemplified and expressed.

True, not every expressed feature is, even theoretically, to be paralleled in the participant, in ritual as in art. For example, a rite expressing majesty may rather be hoped to induce faith or trust. And even where parallel features are indeed hoped for, successful execution of a rite on any given occasion does not hinge on satisfaction of this hope; that a participant's state of mind is incongruous with the expressed theme of the rite may lower its quality but does not, in general, argue that the rite has not taken place. Yet, quality may indeed be affected, and this is a significant point to be noted. More generally, there is, in the case of ritual, a certain expected linkage between expressed properties and participants' mentality and sensibility; the cognition of expressed features, reinforced by repeated performance, is a major medium of such linkage.

The ideal ritual participant is one whose own character is suitably affected by the role he performs and not simply one who skillfully conveys the character defined by his role. While in painting or drama, the cheering face of the hypocrite may, as Goodman says, express solicitude, hypocrisy being irrelevant, it is absurd to suppose hypocrisy irrelevant to performance of a religious rite expressive, say, of contrition or repentance. While in both cases hypocrisy is independent of what is expressed by the performance, it is only in the ritual case relevant to understanding the whole pattern of associated performances, intended, as it is, among other things, to reduce hypocrisy in the participants themselves.

6. MIMETIC IDENTIFICATION

In the preceding paper, I criticized Susanne Langer's view of rituals as denoting feelings, on the ground that rituals may refer in other ways than by denotation, and that denoted or symbolized things are not limited to feelings alone. She is herself apparently sensitive to the latter point, for despite her limitation of ritual denotata to feelings, she seeks an interpretation of so-called mimetic ritual, which seems rather to denote a wide array of incidents and activities outside the bounds of feelings. Thus, she strives to connect emotions with other things, declaring that "emotional attitudes are always closely linked with the exigencies of current life, colored by immediate cares and desires, by specific memories and hopes."[17] Memory and hope provide two channels linking feeling with incident. The memory of celebrated events in sacred story gives rise to current celebration in which the story is retold; the retelling "soon becomes a formula," its accompanying gesticulations "woven into ritual patterns." Hope and desire, on the other hand, give rise to supplication, an "even more obvious origin of mimetic rites ... an act is to be suggested and recommended to ... the Holy One; the supplicants, in their eagerness to express their desire, naturally break into pantomime. Representations of the act mingle with gestures of entreaty."[18]

Langer emphasizes the increasing schematization of actions in mimetic ritual, and minimizes the role of imitation in mimetic portrayal. "A child's representation of sewing, fighting, or other process will be really imitative at first, but dwindle to almost nothing if the game is played often. It becomes an act of *reference* rather than of representation."[19] If, as I believe, she uses the term "reference" for denotation, then she has here clearly overstepped her restriction of denoted objects to feelings. It is not just that the emotions of ritual are connected with the incidents activated in memory or projected by hope; it is rather that these incidents are themselves also denoted by ritual, or perhaps just by ritual of the mimetic variety. In any case, the primacy of feelings in ritual denotation seems here to be surrendered.

Once such primacy is no longer an obstacle, we can allow ritual freely to denote, through mime or otherwise, all sorts of actions or events directly, while exemplifying and expressing other things as well. Mimetic ritual no longer needs to pose a special problem or be construed as a special category. Nevertheless, the particular problem of interpreting mimetic *identification* remains to be dealt with. The problem is to understand how ritual mimicry may sometimes pass over into identification; and a consideration of it will lead us to a new symbolic mode beyond those already discussed. I introduce the problem in the context of an example from the ancient Near East.

Thorkild Jacobsen describes a cult festival of the end of the third millennium in the city of Isin, then the ruling city of Southern Mesopotamia. Annually, the marriage of the goddess Inanna to the god Dumuzi was celebrated, in a rite in which a priestess and the human king not only took on these respective roles but were *identified with* Inanna and Dumuzi. "Why," asks Jacobsen, "should ... the human ruler and ... a priestess transcend their human status, take on the identities of the deities Dumuzi and Inanna, and go through their marriage?" In answer to this question, he appeals to what he describes as a "tenet of mythopoeic logic that similarity and identity merge; 'to be like' is as good as 'to be.' Therefore, by being like, by enacting the role of, a force in nature, a god, man could in the cult enter into and clothe himself with the identity of these powers, with the identity of the gods, and through his own actions, when thus identified, cause the powers involved to act as he would have them act. By identifying himself with Dumuzi, the king is Dumuzi; and similarly the priestess is Inanna – our texts clearly state this." The phenomenon of identification is one that, according to Jacobsen, recurs in major rites of other sorts as well.[20]

Yet I find the interpretation proposed by Jacobsen not persuasive, in that it rests mimicry upon similarity. But while miming indeed exemplifies certain movements involved in the activity being represented, it does not follow that what the mime is doing is *similar* to what he is representing. Nor does it follow that a three by five inch snapshot of the Grand Canyon, exemplifying many of its hues, is therefore *similar* to the Grand Canyon. What could be meant by saying, as Jacobsen does, that a man was *like* a force in nature? Representation or denotation is to be distinguished from similarity, and the distinction is especially important for mimicry, where certain features of the thing denoted may also be exemplified by the miming action.

Surrendering appeal to similarity, then, how are we alternatively to understand mimetic identification? Let us start from the fact that the mime or his miming is to be considered a denotative symbol. The transition to be explained is one that begins with this fact and ends with taking the mime or the miming itself to be what is denoted.

Some theorists have in fact left the matter thus, in effect promoting the transition itself into an explanatory principle, free of all appeal to similarity. The new principle becomes "the coalescence of a symbol and the thing it stands for," in the formulation of H. and H. A. Frankfort, who offer as an example "the treating of a person's name as an essential part of him – as if it were, in a way, identical with him."[21] I find this alternative also unsatisfying. The similarity theory at least recognized the need for an intermediary notion to ease the transition from the mime to the mimed. The present theory, offering no intermediary notion at all, generalizes the problematic transition into one affecting not only mimetic but all denoting symbols. Since the generalized transition remains, moreover, unmotivated, the theory is driven to attribute a radical confusion to the ancient mind. "For us", say H. and H. A. Frankfort, "there is an essential difference between an act and a ritual or symbolical performance."[22] We, but presumably not the ancients, can tell the difference between a symbol and what it stands for – between a horse and the spoken word "horse," between a picture of a lion and the real thing, between rain and the mere promise of rain.

The problem is indeed to understand the psychological transition between a symbol of something and the thing symbolized, but what is needed for such understanding is some additional idea capable of mediating the transition. This idea should, preferably, also be free of appeal to similarity and should postulate no radical difference between the ancient and the modern mind.

7. RITUAL AND MENTION-SELECTION

Elsewhere I have suggested an idea that may serve here.[23] The idea is that a term is typically used not only to *denote* but also to *mention-select* – to pick out appropriate mentions, related denoting or representing units. Mention-selection is at work in the captioning of pictures and statues. The term "horse" thus serves not only to denote horses but also to caption horse-pictures and other horse-mentions. I have suggested that the functions of denotation and mention-selection interact intimately in the learning of language; a child learns about centaurs not by saying "centaur" and pointing to them but by saying "centaur" and pointing to centaur-representations. Mastery of the conventional use of the term "tree" involves using it properly not only to point out trees but also to select tree-pictures and tree-regions within pictures. That the same term refers *denotatively* to a certain object and *mention-selectively* to itself, among other representations of the object, gives a foothold to the transition we have been seeking. For confusion of these two legitimate functions of the same word, whether by children or adults, ancient or modern, is more understandable than the bare confusion of symbol, considered solely as denoting, with its denotatum. And the confusion of functions in question is facilitated by the central importance of certain images and other representations in a given cultural environment.

My suggestion yields, at any rate, the following interpretation of mimetic ritual: the mimetic gesture portraying the act of a god, is in such capacity denotative. But it also mention-selects representations of the same act, itself included. Then, by confusion of such mention-selection with denotation, the gesture in question is itself taken to be the act of a god, and not just the portrayal of such act. In a related process, the verbal description "act of the god" mention-selects the mimetic *portrayal* which is then, by the same confusion, taken to be the act *portrayed.* Analogously, objects employed in ritual may be regarded not merely as symbolic but, in Langer's words, "as life-givers and death dealers ... not only revered, but also besought, trusted, feared, placated with service and sacrifice."[24]

The matter touches on the interpretation of so-called idol worship. The fierce Biblical polemic against such worship makes it very difficult to fathom the mentality of those who would attribute powers of life and death to mere sticks and stones; indeed the polemic is *intended* to ridicule such mentality. Cannot "idol worshippers" see that their graven images are inert and powerless artifacts? "Their idols," says the Psalmist, "are silver and gold, the work of men's hands. They have mouths, but they speak not; Eyes have they but they see not etc."[25] Modern scholars have offered a more sympathetic view of idol worship, so-called, holding that it was not the images themselves but the gods or forces they symbolized that were the true objects of worship.

Such an interpretation indeed gives a more understandable view of idol worship, but in recognizing only the denotative mode of symbolism, it gives no basis for grasping the phenomenon of identification we have discussed or the causal efficacy ascribed to sacred symbols noted in the above quotation from Langer. Even the Bible suggests such causal efficacy, if not for images of the Deity, then for other sacred objects. As Langer notes, "The sacred ark going up before the children of Israel gives them their victory. Held by the Philistines, it visits disease on its captors. Its efficacy is seen in every triumph of the community, every attainment and conquest."[26] My suggestion of mention-selection as an additional symbolic function beyond denotation is intended also to make such phenomena more understandable.

8. RITUAL AND REENACTMENT

Many religious rituals center on events in sacred story. I call these *commemorative rituals*. The connection between rite and myth, between celebration and story, is indeed so close that it is for the most part exceedingly difficult to disentangle origins. Whether, as some suppose, rite initially derived from myth or whether, as others think, myth originated in rite is a matter we need not try to decide. What is clear is that there are at present intimate connections between rite and story and that, in major cases of religious ritual, the stories are not *mere* stories but believed to relate true and momentous historical occurrences.

Now the relation between a given ritual performance and the event it commemorates is of denotative sort; that is, the performance portrays or represents the event in question.[27] The relation between one performance of a given rite and another performance of the same rite is, rather, that of ritual equivalence or (potential) co-exemplification, that is, each such performance may exemplify the rite in question. I shall here refer to performances of the same rite as *replicas* of one another.

These relations are clearly different. That two performances are ritual equivalents does not imply that there is some one historical event that they denote in common. They may denote something else than a historical event; they may both have null denotation; or they may not purport to denote at all, lacking even null denotation. And that a ritual act denotes a particular historical act clearly does not imply that they are ritual equivalents, even though the former may exemplify certain features of the latter. The historical event is in general not itself a rite; moreover, it is typically denoted or portrayed rather than replicated in ritual, just as the mimed activity is in general portrayed rather than replicated or exemplified by the mime.

Now I propose to reserve the notion of *reenactment* for the relation between a ritual performance and its preceding replicas, rather than using it, as is sometimes customary, for the relation between a performance and its commemorated event. "Reenactment," in my usage, is a reproduction or repetition of the act, a coexemplification of the same rite. Commemorated events, on the other hand, are typically denoted, represented, or portrayed rather than reproduced in ritual, even though the ritual aim may be to promote spiritual union with the historical agents in question. Even mimetic gestures do not, in general, *reproduce* the mimed activity; they exemplify some of its features, but not the activity itself, although they may vividly call it to mind.[28] On the other hand, a performance that replicates earlier performances of a given rite reproduces them in constituting a (potential) sample of – exemplifying – the self-same rite.

In the Jewish Seder feast celebrating the exodus from Egypt, the Haggadah text is recited, one passage of which reads, "In every generation one ought to regard oneself as though one had personally come out of Egypt."[29] The whole Seder ritual is indeed intended to foster identification with the liberated Israelites of the exodus and to kindle in participants a vivid sense of the joy of redemption from slavery. But the various symbolic means through which the ritual strives to accomplish this goal do not add up to a literal reenactment of the portrayed historical exodus. Rather, the story of the exodus is described, elaborated, and emphasized, the exodus portrayed as a key event in history. The empathic identification sought is intended not only to lay down a particular past event as a major temporal marker but to make that event come alive now, to bring some of its main features into the temporal foreground. It is in order to promote the contemporary appreciation of freedom that the Haggadah declares, "Not only our forefathers did the Holy One, Blessed be He, redeem, but also ourselves did He redeem with them."[30] The actions comprising the ritual nevertheless do not reenact but rather portray the historical redemption celebrated.

My proposed use of the term "reenactment" may indeed, however, properly describe the relation of a ritual performance to its past replicas. And it has the advantage of calling attention to a further mode of ritual reference, which the familiar use neglects. For each new performance of a rite not only reproduces earlier replicas but, as I now suggest, refers indirectly to them, alludes to them, that is, while independently denoting whatever it may denote, and symbolizing in the other modes so far distinguished. In the regular recurrence of a rite, a sense builds up, in each new performance, of the prior performances that have taken place through the lifetime of the participants, but normally, beyond as well, to the time of the ritual's origin nearest the commemorated historical event. The ritual thus calls to mind not only the commemorated event but the sequence of vehicles of its commemoration.

Reenactive reference, operating thus allusively, constitutes a further symbolic mode, beyond denotation, exemplification, expression, and mention-selection. The relation of one performance to a replica is a relation holding between performances denoted by, and exemplifying, the same ritual specifications. These performances are, so to speak, on the same symbolic level. If we picture denotation as running downward from symbol to object, then exemplification and expression will run upward from denoted object to (certain) symbols. Mention-selection, in this picture, will run laterally from symbol to parallel symbols. And the replica relation involved in reenactment will also run laterally, from object to parallel objects, from performances to others of the same kind. Such replication may be thought of as transmitted through a chain of symbolic links already distinguished. A given present performance is linked to the ritual specification that it exemplifies. This specification is in turn linked to past performances exemplifying it. The allusion by the present performance to earlier ones thus may be construed as transmitted through a two-link chain of exemplification.[31]

Although such chains are widely available, they become referentially operative only in certain cases. Thus, the notion of reenactment, as here interpreted, plays no role, or virtually no role in the arts, at least by comparison with religious ritual. A given performance of a musical work typically makes no reference to past performances of the same work any more than it denotes a significant historical event. By contrast, a ritual performance alludes to its own past kin, just as it may point back to a commemorated event. The sense of reenacting, reexperiencing, an important procedure is strong here. The relevant chain is referentially activated, and it is perhaps a likely symptom of the religious consciousness that it *is* thus activated.[32]

Such activation gives some body to the notion of tradition, so strong in religious contexts. A tradition is not merely a repeated sequence of acts, no matter how well defined. What is needed is some sense of the fact, with each repetition, that it *is* a repetition, some sense of its predecessors. And that I would interpret in terms of reenactive reference.

The marking out of commemorated events defining a temporal matrix, and the concomitant reenactive reference to a ritual tradition serves also to form a conception of community. For the performers of past ritual replicas constitute a body of actors to which present performers relate themselves through the reenactment in question and, hence, indirectly to one another contemporaneously. The community thus defined bears, like all communities, not only common bonds to the past but also common orientations in the present and outlooks for the future. Thus, an organization of time, as well as of the space occupied by a historical community, is facilitated.

This perhaps, is the root of the emphasis on stabilization in primitive religion, in the work of Bergson, Cassirer, and others. In this vein, Cassirer writes, agreeing with Bergson, "primitive religion can... leave no room for any freedom of individual thought. It prescribes its fixed, rigid, inviolable rules not only for every human action but also for every human feeling."³³ But, as I have pictured it, the general phenomenon of ritual is no mere squelching of emotion, no cage of the feelings. Rather, we have to do with a cognitive ordering of categories of time, space, action, and community.

Since I have dwelt here largely on religious and ancient example, I close with a contemporary and secular illustration. "Every enterprise," writes R. S. Peters, "must develop its own appropriate rituals. The importance of such rituals is that they convey atmosphere; they link the past with the present and mark the value of what is being passed on without anything being explicitly stated."³⁴ And, speaking of the rituals of Parliament, Peters says, "Such rituals help to unite the past with the future and to convey the sense of participation in a shared form of life. They do something to mitigate the feeling any rational being must have about the triviality and transience of... life upon earth. They do much, too, to develop that feeling of fraternity which is the life-blood of any effective institution."³⁵

NOTES

1. Nelson Goodman, *Languages of Art*, 2d ed. (Indianapolis: Hackett Publishing, 1976).

2. See also my "Reply to Gareth Matthews," which follows.

3. Goodman, op. cit., 53.

4. Ibid., 61.

5. Ernst Cassirer, *The Philosophy of Symbolic Forms; vol. 2: Mythical Thought* (New Haven: Yale University Press, 1955), 221.

6. Goodman's discussion of notationality is in *Languages of Art*, ch. 4. The passages quoted here are on pp. 128-30.

7. Ibid., 194.

8. Ibid., 198.

9. Ibid., 121-22.

10. Syntactic requirements could clearly be satisfied. But even crucial semantic requirements could be met. That is, (1) the class of performances answering to a given rite-description would share no common elements with the class associated with any other, and (2) determination, for any performance not satisfying both of two descriptions, that it does not satisfy the one or the other, would be theoretically possible.

11. Ibid., 114-15, 118-19.

12. Gareth Matthews, "Comments on Israel Scheffler," *Synthèse* 46 (1981): 439-44.

13. Goodman, op. cit., 122.

14. Ibid., 116.

15. J. L. Austin, *How To Do Things with Words* (Cambridge: Harvard University Press, 1962), 16.

16. Goodman, op. cit., 85-86.

17. Susanne K. Langer, *Philosophy in a New Key* (New York: Penguin Books, 1942, 1948), 124.

18. Ibid., 125.

19. Ibid., 127, italics in original.

20. Thorkild Jacobsen, "Mesopotamia", in H. and H. A. Frankfort, John A. Wilson, and Thorkild Jacobsen, *Before Philosophy* (Baltimore: Penguin Books, 1946), 214-15.

21. H. and H. A. Frankfort, op. cit., p. 21. See also Cassirer, *Language and Myth* (New York: Dover [Copyright 1946, Harper]), ch. 4, esp. p. 49.

22. Ibid., 22.

23. I. Scheffler, *Beyond the Letter* (London: Routledge and Kegan Paul, 1979), 31ff., 45ff.

24. Langer, op. cit., 124.

25. Psalm 115. On the Biblical polemic against idolatry see Y. Kaufmann, *The Religion of Israel* (Chicago: University of Chicago Press, 1960), 13ff., 19-20, 146, 236-37, 387.

26. Langer, op. cit., 125. See also, for example, I Samuel 5.

27. For brevity's sake, I do not here treat cases of null denotation, that is, cases where there was no event of the sort purportedly represented in the ritual act in question. Such cases would need to be treated nonrelationally as involving certain historical-event symbols of denotative kind but with null denotation.

28. Goodman, Op. cit., 63-64.

29. Passover Haggadah, numerous editions. (For a general account, see Theodor H. Gaster, *Festivals of the Jewish Year,* New York: William Sloane Associates, Publishers, 1952, 1953.)

30. Ibid. My main point in this paragraph disagrees with the position of Gaster, op. cit., who apparently takes the view that the goal of personal identification is associated with reenactment of the historical event commemorated. He writes that "when the Jew recites [the Haggadah], he is performing an act not of remembrance but of personal identification in the here and now" (p. 42), and he writes also, "Those present at the Seder ceremony are expected to adopt a casual, reclining posture, symbolizing that of freemen at ancient banquets. In some parts of the world, however, everyone appears in hat and coat, with satchel on back and staff at hand, thus *re-enacting* the Departure from Egypt" (p. 40, my italics).

31. The concept of chains of reference has been noted in Goodman, op.

cit., 92, and elaborated in Goodman, "Routes of Reference," Second Congress of International Association for Semiotic Studies, Vienna, 1979.

32. Some people have suggested that I here underestimate reenactive reference in the arts. Murdoch Matthew has told me, "Opera buffs take value from perceiving *this* Tosca as one of a line running back through Callas to Jeritza." I agree that reenactive reference is not limited to religious ritual but hold that it is considerably more prominent there. And I would distinguish ritual reenactment, as gathering all related performers into a community, from cases (perhaps like that of the opera) in which special performers or performances are singled out as landmarks for comparison. (I thank Matthew for other helpful comments on ritual and art.)

33. See discussion of Bergson in Cassirer, *An Essay on Man* (New Haven: Yale University Press, 1944), 87, 89, 224-25. The cited passage is on p. 224.

34. R. S. Peters, *Ethics and Education* (London: George Allen and Unwin, 1966, 1970), 260.

35. Ibid., 318-319.

8. REPLY TO GARETH MATTHEWS ON RITUAL AND REFERENCE

I. CONTAMINATION VS. PURITY

PROFESSOR Matthews argues that "one's account of the semantics of ritual is inevitably contaminated by one's own religious or political beliefs or disbeliefs." But what does he take as an "account of the semantics of ritual"?

He describes the misguided Puritan as attempting to "stand outside our ritual observances and the beliefs they express and tell us what those observances really refer to." On the other hand, he allows that the semanticist of ritual might, instead, "provide us with an analytic framework in terms of which we could think out for ourselves what *we* take to be the semantics of our own ritual observances." The latter task apparently does accord with Matthews' Purity Principle but, for a reason I cannot determine, does not lead him to modify his unqualified defence of the Contamination Thesis.

Now I in fact intend the phrase "the semantics of ritual" to refer to the latter, analytic, task, aiming to give a general account of modes of ritual symbolism, not to tell anyone what his rituals really refer to. I explicitly denied "any suggestion that ritual can be understood exclusively in terms of semantics." Thus, I cheerfully acknowledge that any effort to interpret the reference of particular rituals will be affected by a variety of independent beliefs. Yet the general characterization of the mode of reference need not be similarly affected. It may, indeed, as Matthews himself suggests, offer a common conceptual tool for formulating the points of substantive difference.

Consider denoting. If I disbelieve in unicorns, I will bar them as denoted elements in my theory. If you believe in them, you will not equally bar them from yours. Our disagreement is zoölogical not semantic, and may be formulable, indeed, by reference to the common principle that to be denoted is to be. The latter principle does not itself decide zoölogical issues, but it is not therefore inappropriately assigned to the semantics of theory.

Professor Matthews perhaps restricts the phrase "semantics of ritual" to deciding the reference of *this or that* ritual. He argues, for example, that

---

This paper appeared in *Synthèse* 46 (1981): 445-48.

whether the Eucharist is to be taken as a case of literal or metaphorical exemplification is a theological question, concluding that "the semantics of *this* Christian rite" is a religious issue. But while the relevant interpretation of the Eucharist may be described as belonging to the semantics of *this* rite and therefore subject to the Contamination Thesis, there is no reason to think of semantics exclusively in this way. The general analysis of exemplification, upon which this very example depends, surely belongs to the semantics of ritual, as of art, and is in fact used by Professor Matthews to formulate both sides of the theological issue.

2. AUTOGRAPHIC OR ALLOGRAPHIC

Professor Matthews argues that if the constraints on celebration of the Mass require reference to the Apostolic Succession, the validity of the rite hinges on the history of its production, making the Mass autographic rather than allographic. I accept his argument, insisting only that rites, even when bound by constraints on their performers, vary – some interpretable as autographic, others rather as allographic. The details are set out in Section 5 of my paper.

3. COMMEMORATIVE RITUAL

Professor Matthews takes me to be denying that a ritual act may denote a historical event that is itself a rite. In this he is mistaken. My intent was to separate two relations: (a) the relation of *historical representation* between a ritual act and the historical event it commemorates; and (b) the relation of *replication* between one ritual act and an equivalent ritual act. In separating these relations, I argued that neither implies the other, that is, replication does not imply historical representation, nor does historical representation imply replication.

Now in the course of making the latter point, I said that the historical event in question is in general not itself a rite. This statement certainly allows that some such events may indeed be rites. So long as this is not always the case, the implication in question fails and the intended separation of the relations is made. Actually my claim was stronger than it need have been. Minimally, I needed only to claim that the historical event is not uniformly a rite; in fact I proposed the stronger claim that it is in general not a rite. The stronger claim is, I believe, true in any case. And either claim allows that *some such events may be rites*.

In the course of the same argument that historical representation does not imply replication, I said that the historical event in question is typically denoted or portrayed rather than replicated in ritual. Again, I need only have claimed that the event is not typically replicated but I put forward the stronger and, I believe, true claim that it is typically not replicated. In any case, I surely did not exclude instances of replication. Professor Matthews' interpretation of the Last Supper as a case of commemorative replication is thus in fact *not* ruled out by my treatment in the passages cited.

However, I am independently inclined to treat the Mass not as a case of replication of the Last Supper but rather as representation. For while the Mass exemplifies certain features of the original event, e.g. the breaking of bread, serving of wine, and saying the words "This is my body" etc., it does not follow that such *exemplification* comprises ritual *replication*.

That it does not comprise replication is suggested by the following consideration. The initial utterance "Do this in remembrance of me" *prescribed* a constitutive feature of the rite thus inaugurated which the very process of inauguration could itself not *satisfy*, being a process of demonstration and instruction rather than remembrance. If this consideration is correct, only *later* exemplifications of the demonstrated actions and words could also satisfy the constitutive prescription of remembrance, thus comprising not *replicas* but rather *representations* of the original act.

4. REENACTMENT

Professor Matthews asks for a clarification of reenactment, referring in particular to Freud's *Totem and Taboo*. Does "persistent unconscious memory" of a "pre-historic rite" constitute reenactment of such rite? My answer is no, even should it turn out possible to make good psychological sense of the idea of unconscious memory.

What I hold centrally important is the reference to ritual specifications or prescriptions, written or orally transmitted, and understood in context by adherents of the rites in question. An act falling outside the range of such specifications or prescriptions is excluded from the scope of "reenactment" in my use of this term, no matter how similar to the ritual performances in question by independent criteria, and no matter how tightly associated to them by psychological or causal links. What counts is just the co-exemplification of ritual specifications by a given performance and earlier ones. The allusion of reenactment thus relates the performance of a rite to its earlier ritual replicas.

Must adherents of the rite be conscious of the allusion "at the actual time of the performance", asks Matthews. Could they be wholly unconscious of it? No conscious awareness is required but surely something more than an altogether unconscious process opaque to each adherent. Rather, what Matthews calls "a general sort of realization" is what I had in mind, and this seems to me sufficiently different from the case of artistic performances to be a fact worth noting.

9. FOUR QUESTIONS OF FICTION

Four questions relating to fiction are treated: (1) The first question is "How can null singular terms be meaningful?" and it is argued that Russell's theory of descriptions and Quine's theory of proper names together show that meaninglessness does not follow upon failure to name; fictional discourse does not, therefore, lack sense simply because it is about nothing. (2) The second question is "How can null terms differ in meaning?", and it is argued that while bare significance needs no reference, variation in significance may plausibly be given a referential interpretation by taking parallel compounds into account. (3) The third question, "How can null replica-inscriptions differ in meaning?" concerns variation in meaning for inscriptions rather than word-types; here appeal to constituents and use of the relation of mention-selection are offered as allowing a suitable interpretation. (4) The fourth question is "What truth is there in fiction?" and it is argued that fiction, though literally referring to nothing, may metaphorically refer to anything, thus removing from fiction the general threat of falsehood. Further, an interpretation of varying metaphorical behavior by literally null terms is suggested.

1. HOW CAN NULL SINGULAR TERMS BE MEANINGFUL?

Russell's (1920: ch. 16) theory of descriptions and Quine's (1953) theory of proper names free us from bondage to fictional reference by null singular terms – at least, on the score of meaning. For, together, they show that meaninglessness does not follow upon failure to name. Russell paraphrases

(1)   The minotaur lives in a labyrinth

as:

(2)   Something is a minotaur and lives in a labyrinth and nothing else is a minotaur,

thus eliminating the null singular descriptive phrase of the original and providing no naming unit as a counterpart, though retaining (through

---

This paper appeared in *Poetics* 11 (1982): 279-84.

deployment of the quantifiers "something", "nothing") all the content of the original. (2) is clearly false, hence meaningful and, since its equivalent (1) must be equally false and likewise meaningful, the connection between meaningfulness and reference is here effectively severed: (1) has meaning though its singular description fails to refer; the denial of (1) is, moreover, true under the very same condition.

Quine extends the Russellian treatment to proper names by construing each such name as a predicate entering into a singular description. Thus

(3)   Zeus dwells on a mountaintop

becomes:

(4)   The thing that is-Zeus dwells on a mountaintop.

which in turn is transformed into:

(5)   Something is-Zeus and dwells on a mountaintop and nothing else is-Zeus.

Again, since no singular referential unit remains in (5) although it is clearly false and therefore meaningful, the same must hold for its original equivalent. (3): It has meaning despite the referential failure of its proper name, and the same holds for its true denial.

The severance of meaning from reference frees us from having to affirm, under threat of loss of meaning, that our names or singular descriptions are invariably satisfied by actual things. Fictional discourses do not collapse into meaninglessness upon referential failure of their descriptions. Whether we talk about something – or about nothing – we may make perfectly good sense.

2. HOW CAN NULL TERMS DIFFER IN MEANING?

An assumption of both Russell's and Quine's devices is that general terms, certainly, may fail to refer without failure of meaning, (2) and (5) are meaningful even though nothing satisfies either the general term "is a minotaur" or the general term "is-Zeus". But how account for the patent *difference* in meaning of these terms? Bare significance needs no reference, but *variation* in significance seems to require appeal to reference if we are to avoid invoking either Platonic entities (e.g. forms, concepts, attributes, intensions) or psychological entities (e.g. thoughts, ideas, conceptions, images) underlying general terms.

Here the direction of our interest has changed. Our earlier concern to rid ourselves of unwanted references of fictional terms has given way to the finding of references explaining the meaning differences among these very terms. The strategy of Goodman (1972: 221-230) is to look to the references of suitable compounds. Though nothing is-Zeus and there are no minotaurs, there are Zeus-pictures and minotaur-pictures, Zeus-descriptions and minotaur-descriptions. Moreover, the compound "is a Zeus-description" differs referentially from its parallel compound "is a minotaur-description". There is, in fact, some Zeus-description that is not a minotaur-description and some minotaur-description that is not a Zeus-description. Similarly, there are Zeus-pictures that are not minotaur-pictures, and minotaur-pictures that are not Zeus-pictures. Taking the likeness of meaning of two terms to consist, then, not only in the sameness of their own references (their primary extensions, in Goodman's terminology) but also in the sameness of reference of their parallel compounds (their secondary extensions), we can hope to explain how null terms (such as "is-Zeus" and "is a minotaur") may differ in meaning though uniformly referring to nothing [1].

3. HOW CAN NULL REPLICA-INSCRIPTIONS DIFFER IN MEANING?

Extensionalism is not yet inscriptionalism. The parallel compounds that differentiate meaning through their own varying extensions serve thus only for syntactically different terms, i.e. for terms construed as word-types spelled differently, as are "is-Zeus" and "is a minotaur". For the compounds of the one are easily separable (by spelling) from the compounds of the other. Such appeal to compounding fails, however, when we seek meaning-differentiation for null concrete inscriptions with identical spelling. For, being replicas of one another, such identically spelled inscriptions share all their compounds, and the notion of distinguishable "parallel compounds" collapses [2]. Thus, in

(6)   A green centaur is a more naive consumer than an experienced centaur

and

(7)   A green centaur is harder to spot in the forest than a red or yellow centaur,

there are two "green centaur" inscriptions, different in meaning, but incapable of having separate groups of parallel compounds assigned to them on the basis of syntax, since themselves syntactically indifferent.

An answer that suffices here is one that appeals not to compounds but rather to constituents. For the two "green" inscriptions comprising, respectively, the first word-constituents of our "green centaur" inscriptions in (6) and (7) are themselves different in extension, the first referring to inexperienced things, the second to things of a certain color. Thus we can again understand meaning differences among null terms without reversion to Platonism or psychologism – even taking such terms nominalistically, i.e. as inscriptions rather than repeatable word-types. We need, for this purpose, to take into account the references not only of the inscriptions themselves and of their compounds, but also of their constituents.

This plan fails, however, where no word-constituents are available. "The child Linus of Argos must be distinguished from Linus, the son of Ismenius, whom Heracles killed with a lyre" (Graves 1957: 212, section 147). This distinction has to be made in the face of the fact that neither Linus-inscription refers, that they have, moreover, no separable groups of compounds nor word-constituents. To this end, I have suggested use of a relation of *mention-selection* (see Scheffler 1979: part I, sections 9 and 10), whereby a term is employed to caption mentions, or representations, rather than to denote objects. Thus a given Linus-inscription may mention-select, i.e. serve as a suitable caption for, a certain range of pictures or descriptions different from those mention-selected by a replica Linus-inscription. Analogously, a student producing a Jones-inscription purportedly to discuss a fictional character in a certain novel may be considered thereby to have mention-selected certain Jones-descriptions in that novel (or elsewhere) and thus to have produced something differentiable in its meaning from a replica applied in discussing another novel [3]. It is worth noting, incidentally, that although mention-selection was proposed originally for the case of inscriptions lacking word-constituents, it has considerably wider range, capable of yielding also the differentiations earlier effected by recourse to parallel compounds and to constituent extensions.

4. WHAT TRUTH IS THERE IN FICTION?

Fictional, or null, terms, though literally referring to nothing, may metaphorically refer to actual things of any sort. As Goodman has said (1978: 103), "Whether a person is a Don Quixote ... or a Don Juan is as genuine a question as whether a person is paranoid or schizophrenic and rather easier to decide". Literally fictional works may thus express metaphorical truths, or contain literally fictional terms metaphorically applicable to things. Thus, not only is the threat of meaninglessness eliminated from the literally fictional, but also the general threat of falsehood – a point beyond the reach of the Russell-Quine strategy applied to literal expressions. (Recall that (2), and therefore (1), are meaningful but, alas, false.)

To say this much is, however, not sufficient for we need still to account for the differential metaphorical behavior of terms that are uniformly null taken literally. If we analyze

(8)   Hamlet is neurotic,

taken literally, in accordance with the Russell-Quine treatment discussed above, we get:

(9)   Something is-Hamlet and is neurotic and nothing else is-Hamlet,

which is literally false. Moreover, if we try to interpret (8) metaphorically, as applying to Hamlet-like persons, we face this difficulty: How can we liken anyone to Hamlet if there is no Hamlet? The clues to a metaphorical interpretation cannot lie in a null literal reference shared with every other null term.

Goodman suggests recourse to compounds here again (1978: 104, fn. 10): "In sum, 'Don Quixote' and 'Don Juan' are denoted by different terms (e.g. 'Don-Quixote-term' and 'Don-Juan-term') that also denote other different terms (e.g. 'zany jouster' and 'inveterate seducer') that in turn denote different people".

Put in terms rather of mention-selection, "a person metaphorically described as Don Quixote is not literally likened to Don Quixote nor does he share the satisfaction of important predicates with the literal Don Quixote; rather he satisfies certain important predicates constituting [mention-selected] Don-Quixote descriptions" (Scheffler 1979: 142, fn. 97). Where, as in (8), a null term is the grammatical subject of an attribution with a non-null predicate, some important description mention-selected by the null term literally refers to actual things to which the predicate is ascribed. Thus, "vacillating person" is mention-selected by "Hamlet" and literally refers to actual individuals to whom "is neurotic" is applied.

In a somewhat related way, when, during the course of a performance, a member of the audience says, "There's Hamlet, coming on stage now!", he is not to be understood as merely uttering a literal falsehood; he is saying something accurate. I take his "Hamlet" utterance to be mention-selecting a Hamlet-representation, that is, the actor playing Hamlet. While there is in such cases no fusion of literal reference with mention-selection, I have suggested that such fusion may play a role in so-called mimetic identification in certain religious contexts (Scheffler 1981: esp. 429-431). Perhaps it has a more general role also in play.

NOTES

1. My special concern in this paper is with fiction, hence with null terms. But all the meaning-differentiating devices referred to throughout the paper are of course applicable also to non-null coextensive pairs, e.g. the singular terms "Morning star" and "Evening star," and the general terms "rational animal" and "featherless biped."

2. This section is based on part I of Scheffler (1979). Details of the argument here may be found especially in sections 7 and 8 of part I. The compounding of an inscription is not its literal embeddedness in a larger one but rather the embeddedness of any of its replicas therein. Since the replica relation is transitive, it follows that replicas share the same compounds.

3. Catherine Z. Elgin has called to my attention that non-null terms in fiction also function mention-selectively in an important way, the name of an actual historical figure appearing in a novel serving to select relevant portrayals in the novel – to which, indeed (and whether true in fact or not), primary interest may attach, rather than to the denotation. I am grateful to Dr. Elgin for her helpful comments on the initial version of this paper.

References

Goodman, N. 1972. Problems and projects. Indianapolis: Hackett.

Goodman, N. 1978. Ways of worldmaking. Indianapolis: Hackett.

Graves, Robert. 1957. The Greek myths, vol. 2. New York: Braziller.

Quine, W.V. 1953. 'On what there is.' In W.V. Quine, From a logical point of view. Cambridge: Harvard University Press.

Russell, Bertrand. 1920. Introduction to mathematical philosophy. 2d ed. London: Allen & Unwin.

Scheffler, I. 1979. Beyond the letter. London: Routledge & Kegan Paul.

Scheffler, I. 1981. Ritual and reference. *Synthèse* 46: 421-37.

PART TWO
Science & Reality

THE PAPERS in this part all concern topics in the philosophy of science. It is clear from what has already been said, however, that I do not hold science to have a monopoly of cognitive function. Reference is larger than science; ritual, art, and fiction purport also, in their several ways, to indicate, symbolize, represent, and inform – and so, in an ordinary sense of the phrase, they too "relate to reality." What distinguishes science is not its mere symbolic character but the special form of such character as well as the systematic and critical nature of its method.

As to symbolism, science is primarily linguistic and mathematical rather than pictorial or gestural, assertive and not merely indicative, denotative in aim rather than expressive, tending toward the robustly literal rather than the allusive, metaphorical, or oblique. Methodologically, its lifeblood is criticism, its assertions viewed not as dogmas but as claims to truth to be tested by fair criteria of logic and observation. Publicity of argument, openness to evidence, willingness to view its doctrines as fallible hypotheses and to revise them for cause – these are the hallmarks of scientific thought. A main function of philosophy of science is to analyze such thought in detail, to give a philosophically intelligible account of its various forms.

The first nine selections to follow offer analyses of such forms as explanation, prediction, confirmation, induction, and teleological description. The rest address ontological questions relating to science: how theory affirms things; what, in describing such affirmation, we ourselves affirm; whether coherence is the only alternative to certainty; how credibility underlies estimation of truth; how objectivity grows from the soil of historical inconstancy; how theoretical relativity relates to talk of the world or worlds. I am grateful to my collaborators, Nelson Goodman, Noam Chomsky, and Robert Schwartz, for their permission to include our respective joint papers here.

In *Science and Subjectivity*,[1] first published almost twenty years ago, I offered a reinterpretation and defense of scientific objectivity in the face of the severe attacks then rising against it from various quarters. These attacks challenged the independence of observation from theory and, hence, its capability to provide a test of theory; they also challenged the independence of meaning from theoretical framework and, therefore, the possibility of intelligible discourse between theoretical opponents. Finally, they challenged the role of logical deliberation in the process of historical change in science, substituting for talk of such deliberation the "conversions," "gestalt switches," and "paradigm shifts" that Thomas Kuhn introduced into the vocabulary of the history of science. I took these attacks as a serious threat to the ideal of objectivity in science, without which, as I believe, there can be no science or, indeed, any rational enterprise whatever. Responding to this threat, I analyzed both what I described as the "standard view" of science and the charges brought against it, offering a reinterpretation of scientific objectivity capable of withstanding the critical storm.

By far the widest notice received by *Science and Subjectivity* was directed to its critique of Kuhn's *The Structure of Scientific Revolutions*.[2] That attention, welcome as it has been, has overshadowed the book's ensuing discussion of the opposition between *coherence* and *certainty* as criteria of adequacy for scientific systems. The view I work out in earlier portions of my book is, however, incomplete without a consideration of what Schlick calls the "contact between knowledge and reality," discussed in connection with his debate with Neurath, in Chapter 5. I have thus decided to include that chapter on the "epistemology of objectivity" in this collection.

I have also included "Vision and Revolution: A Postscript on Kuhn." This paper summarizes my initial arguments against Kuhn's views and offers a rejoinder to his replies. Further, it expands upon the discussions of *Science and Subjectivity* to consider critically the metaphors of *vision* and *revolution*, with which Kuhn replaces traditional talk of deliberation in describing scientific change.

I offer now some further comments on my controversy with Nelson Goodman regarding world-making.[3] Let me begin with a preliminary remark: I don't much like the elastic term "world" and do not want to be taken as defending some doctrine about the world – arguing that there really is one world, or that the world is the touchstone of truth, or independent of mind, or the like. I should not wish to express any of my philosophical convictions by using this term in a primitive, literal, and essential way. My references employing the term are wholly addressed, in critical vein, to Goodman's uses, or else are to be cashed out by terms denoting more limited and more comprehensible entities. For this reason, I introduced reference to stars, about which sensible and scientifically sound things can be said, for example, that in any case stars were not made by men.

Another preliminary point is this: I do not dispute the sort of relativism, or pluralism, propounded in Goodman's *The Structure of Appearance*,[4] for which, given any pre-philosophical subject matter, there are likely to be conflicting though adequate systematizations for it, the points of conflict falling in the region of "don't cares." The existence of such systematizations underlies Goodman's espousal of extensional isomorphism rather than identity as a criterion of adequacy for what he calls "constructional systems." Thus, a systematic definition of points as certain classes of lines does not establish that points are identical with such classes but only that, relative to our purpose to preserve certain pre-philosophical "cares," they do not need to be construed as nonidentical with them. We can, indeed, compatibly say something similar concerning a conflicting systematic definition of points as certain classes of volumes. There is, in this sort of account in *The Structure of Appearance*, no talk of worlds at all and certainly no talk of world-making, although the same form of relativism shines through.

What I criticize in my paper is not such relativism, but the later, accreted talk of worlds and their making, construed "objectually" and not simply "versionally." I find no difficulty in taking worlds to be made, *if* by "worlds" one means versions. But I cannot see how one can suppose worlds to be made, if by "worlds" one means things "answering to true versions" – including, as Goodman says, "matter, anti-matter, mind, energy, or what not . . . fashioned along with the versions themselves."[5] Now Goodman does not define "world" in his book, and he uses it ambiguously, drawing what I can only consider cold comfort from the alleged fact that physicists' talk is also ambiguous. But when he insists that worlds are literally made, in *both* of the interpretations he gives to this claim, I conclude that he can avoid outright falsity only by such an unnatural construal of "made" as to cause high philosophical mischief. My paper offers a variety of considerations in support of my argument, to which Goodman offers five main replies, as follows:

First, he admits to the ambiguity in his use of the term "world," arguing that, though conflicting, the versional and objectual interpretations are equally right and often interchangeable.[6] But I do not object to mere ambiguity, which can as a rule be cleared up with sufficient care and the refinement of terminology.

Second, he says, "We cannot find any world feature independent of all versions. Whatever can be said truly of a world is dependent on the saying – not that whatever we say is true but that whatever we say truly . . . is nevertheless informed by and relative to the language or other symbol system we use. No firm line can be drawn between world-features that are discourse-dependent and those that are not."

The trouble with this reply is that it appeals to the notion of a feature. But what *is* a feature? I presume that, for a nominalist such as Goodman, features will not be properties or classes but terms or predicates, construed as, or constituted by, tokens of one or another sort. Then of course features will obviously be dependent on the saying – that is, brought forth by the process of token production. Indeed, whatever we say, whether truly *or* falsely, will in this sense be dependent on the saying, informed by and relative to our language or symbolism. However, whether a feature or predicate of our making is *null or not* is not in the same way dependent on the saying; whether a statement is true or not is, as Goodman agrees, independent of our saying. Thus if by a *world*-feature, Goodman means a feature that is not null in fact, then that any given feature *is* a world-feature is indeed independent of our version. Its status *as* a world-feature is *not* discourse dependent.

Third, Goodman suggests that it is fallacious to assume "that whatever we make we can make any way we like." I agree in rejecting this assumption. I certainly do not deny the difficulty of making a true or right version. What I deny is that by making a true version we make that to which it refers.

In his book, Goodman speaks of "actual worlds made by and answering to true or right versions." Now, whether a world answers to a version of our making is, in general, not up to us. Thus, if an "actual world" answers to a version of our making, we can hardly be supposed to have made it do so. Moreover, if a version of our making turns out to be true, it hardly follows that we have made its object. Neither Pasteur nor his version of the germ theory made the bacteria he postulated, nor was Neptune created either by Adams and Leverrier or by their prescient computations.

Fourth, Goodman asks me "which features of the stars we did not make" and challenges me "to state how these differ from features clearly dependent on discourse." Surely we made the words by which we describe stars; that these words are discourse-dependent is trivially true. But the fact that the word "star" is non-null is not therefore of our making; its discourse-dependence does not imply our making it happen that there *are* stars, or in short, our making the stars: It doesn't imply that the *stars* are *themselves* discourse dependent. Goodman writes, in *Languages of Art*, " 'Sad' may apply to a picture even though no one ever happens to use the term in describing the picture; and calling a picture sad by no means makes it so." Analogously, "star" may apply to something even though no one ever happens to use the term in describing it; and calling something a star by no means makes it one.

Finally, Goodman tries to dispel the absurdity of supposing that we made the stars by arguing that we made "a space and time that contains those stars.... We make a star as we make a constellation, by putting its parts together and marking off its boundaries." I find this singularly unconvincing. We have surely made the scientific schemes by which we formulate temporal and spatial descriptions, but to say that we have therefore made space and time can be no less absurd than to say we made the stars. Nor did we make the Big Dipper or Orion merely by defining their respective boundaries.

Goodman concludes by saying, "We do not make stars as we make bricks; not all making is a matter of molding mud. The worldmaking mainly in question here is making not with hands but with minds, or rather with languages or other symbol systems. Yet when I say that worlds are made, I mean it literally.... Surely we make versions, and right versions make worlds." The suggestion here is that my critique of worldmaking construes it as a physical rather than a symbolic process.

But my argument is altogether independent of this contrast. My claim is that in any normal understanding of the words, we did not make the stars, whether by hand, mind, or symbol. Certainly, we make things with minds; we thus make words, symbols, versions. The issue is whether in thus making star-descriptions, we also make stars. To propose, as Goodman does, that we may be said to make something whenever we devise a true description for it is certainly possible, even if wildly unnatural; we can certainly make language mean anything we want it to mean. But such a proposal seems to me unusually mischievous in inviting confusions, paradoxes, and misunderstandings – and encouraging an overblown voluntarism. And it blurs the ordinary distinction between making an omelet and writing a recipe for one. Rather than Goodman's "We make versions, and right versions make worlds," I would rather adopt the slogan "we make versions, and things (made by others, by us, or by no one) make them right."

NOTES

1. *Science and Subjectivity* (Indianapolis, Bobbs Merrill, 1967; now Hackett Publishing, 1982).

2. Thomas S. Kuhn (Chicago: University of Chicago Press, 1962).

3. See selection 14 Below, "The Wonderful Worlds of Goodman," and Goodman's response, given in *Synthèse* 45 (1980): 211-15, and again in his book *Of Mind and Other Matters* (Cambridge: Harvard University Press, 1984).

4. Goodman, *The Structure of Appearance*, 3d ed. (Dordrecht:Reidel, 1977).

5. Goodman, *Ways of Worldmaking* (Indianapolis: Hackett Publishing, 1978).

6. This and the following passages quoted from Goodman are taken from *Of Mind and Other Matters*.

1. EXPLANATION, PREDICTION, AND ABSTRACTION

IN RECENT philosophy of science, three basic views concerning explanation and prediction have received wide support, attaining almost canonical status. They are (A) the view that explanation and prediction share a *common structure,* with but the pragmatic difference that an explained event antedates the statement of its explanation while a predicted event can only follow its prediction, (B) the view that explanation and prediction represent the *central purpose* of science and are epistemologically basic, and (C) the view that explanation and prediction are abstract in reference, their objects being not concrete things but *idealistic or intensional entities* like phenomena, facts, or states-of-affairs. I shall argue, in what follows, that these three views are untenable, and I shall affirm instead that explanation and prediction are structurally distinct, that, associated with control, they are subsidiary to the primary concern of science with comprehensive relationships among events, and that they require no abstract, idealistic entities as objects.

1. THE STRUCTURAL IDENTITY CLAIM

The notion of explanation has a variety of uses both in ordinary speech and in scientific contexts. In both spheres we speak alternatively of explaining concepts or terms, laws or generalisations, and concrete occurrences or events. It is interesting to observe, at the outset, how this very variety of uses contrasts with those of the notion 'prediction'. For while we speak of predicting occurrences or events, we surely do not speak of predicting concepts or terms, nor, in any obvious sense, laws or generalisations. The claim of structural identity is made, however, not with reference to all patterns of scientific explanation, but specifically regarding explanation of events, usually described somewhat as follows.[1]

Let a and b be distinct events, described by the sentences A and B respectively, and let L be a law or conjunction of laws. Suppose, also, that B

This paper appeared in the *British Journal for the Philosophy of Science* 7 (No. 28, 1957): 293-309.

is a logical consequence of A and L, but not of A alone. If A and L are true, while b has already occurred, we may say that b has been accounted for or explained by the conjunction of A and L, or that this conjunction satisfies the requirements for an explanans of the explanandum B.

This sketch incorporates the four conditions listed by Hempel and Oppenheim in their discussion of the logic of explanation,[2] which we take here as a model:

> ($R1$) The explanandum must be a logical consequence of the explanans.
> ($R2$) The explanans must contain general laws required for the derivation of the explanandum.[3]
> ($R3$) The explanans must have empirical content.
> ($R4$) The sentences constituting the explanans must be true.

Justifying ($R4$) as contrasted with an alternative requirement of high confirmation for the explanans, Hempel and Oppenheim cite the case of a purported explanans highly confirmed at time t_1 but later highly disconfirmed at time t_2, in which event we should not wish to say that what was an explanans at t_1 ceased to be one at t_2 but should rather prefer to assert that, while its truth had been probable relative to available evidence at t_1, its falsity was probable at t_2, and correlatively, its inadequacy as an explanation at any time.

It is this pattern which is alleged to be identical with that of scientific prediction. As Hempel and Oppenheim put it, 'the same formal analysis, including the four necessary conditions, applies to scientific prediction as well as to explanation'. The pragmatic difference as they, in agreement with several other authors, formulate it consists in the fact that for explanation, B is given, b having occurred, and the conjunction of A and L is provided afterwards, while for prediction, this conjunction is given and B is derived prior to the occurrence of b. Explanation, as they put it, 'is directed towards past occurrences', prediction 'towards future ones'.

Now this account implies that every explanation, if stated prior to the event described by its explanandum, would be predictive, while every prediction, stated after the event in question, would be explanatory. The first of these consequences is indeed explicitly drawn in the study of Hempel and Oppenheim, 'It may be said, therefore, that an explanation is not fully adequate unless its explanans, if taken account of in time, could have served as a basis for predicting the phenomenon under consideration.' The second consequence, though equally necessary for the structural identity in question, is not further elaborated.[4] In view of the following considerations, it seems to me that it is untenable.

(a) First, note that 'is a prediction' is not properly applicable to abstract sentences or propositions, since the same sentence 'It rains on

Explanation, Prediction and Abstraction · 89

May 8, 1952' is or is not a prediction depending on the temporal circumstances of its utterance. Or, more accurately, since what predicts must have an appropriate time relative to what is predicted, and since abstract sentences or propositions are non-temporal altogether, they cannot be properly denoted by 'is a prediction'. We may distinguish the abstract sentence from the uses made of it at various times and denote uses as predictions, if we like. Alternatively, we shall here construe 'is a prediction' as predicable of concrete utterances or inscriptions (i.e. tokens) with temporal boundaries, but the point to be made can be readily put in terms of the other usual analyses.

Consider now any utterance or inscription of declarative, non-compound form. In accordance with the dominant ordinary notion of prediction, any such utterance or inscription is a prediction if it explicitly asserts something about some time later than any of its own. But it is clearly false that restating each such prediction following the time of its predicted occurrence, explains this occurrence even when both prediction and restatement are true. Thus, no inscription like 'Eisenhower is elected President on November 4, 1952' *explains* Eisenhower's election, though every such inscription before November 4, 1952 is a prediction in the ordinary sense, and true at that. The point then is that, in the usual sense of 'prediction', not every restatement of a prediction after the event is explanatory, even though every statement of an explanation before the event is predictive.

Nor will it do to invoke epistemology at this point by asking how a prediction of Eisenhower's election could have been made without the use of general laws and statements of relevant antecedent conditions. That the methodological genesis of a prediction does not meet rational or scientific requirements may involve irrational behaviour by its producer, but is no bar to its predictiveness in the ordinary sense, nor even to its truth. Clairvoyants, prophets, and news commentators all predict in the sense under consideration, just as do scientists. For pragmatists and positivists in particular, who justify scientific method by success in prediction, a restriction of the latter to *scientific* prediction would reduce their justification to triviality.

(*b*) Suppose, however, that structural identity is interpreted as holding between explanation and *rational* prediction *as practised in the sciences,* involving reference to general statements and specific condition statements.[5] Even so, it will appear that not all predictive restatements after the event are explanatory, since predictive success involves the possibility of predictive failure, i.e. false predictions. But no explanation is false, since it consists of an explanans, which by ($R4$) must be true, and an explanandum which, being a logical consequence of the latter, cannot be false either. This divergence is related to the use of scientific predictions in testing the body of assumptions at a given time; for such testing to occur, it must make sense to judge a derived prediction false, thereby forcing a revision in its ground-premises. Indeed, to the extent that predictive test is involved in confirming the truth of general laws, themselves required for explanation by ($R2$), to that extent the possibility of falsifying predictions is presupposed by the confirmation of explanations.

If this divergence between scientific explanation and prediction is granted, one might still attempt to reinterpret the structural identity claim as holding between rational prediction (by deduction from general statements and condition statements) and *proffered* explanation, which, of course, may be false. Without artificial restriction of the latter notion, however, such reinterpretation fails. For in its ordinary sense, 'proffered explanation' refers not only to certain explanations which are false, i.e. violate ($R4$), but also to some which fail to exhibit the required logical character as specified by ($R1$), fail to contain general laws as required by ($R2$), or lack the empirical content demanded by ($R3$). Artificial expansion of this notion would, moreover, also be necessary since we would not, ordinarily, consider any account to be even a proffered explanation unless it purported to explain some *fact*, i.e. unless at least its explanandum were true; derived *predictions*, on the other hand, may clearly be false. To specify then explicitly that we are to require fulfilment of ($R1$), ($R2$), and ($R3$), but neither ($R4$), nor the truth of the explanandum, (while certainly legitimate and often convenient for other purposes), renders trivial the claim of structural identity between prediction and explanation. For, if true, this claim appears no longer a surprising description of two antecedently-known patterns which happen to correspond, but rather a consequence of our deliberate theoretical tampering. What is finally correlated to the term 'prediction' is a technical artifact independently related neither to the ordinary sense of 'explanation' nor to that of 'proffered explanation'.[6]

We conclude, then, that the structural identity claim should be rejected; far from differing only in pragmatic relationships, explanation and prediction have different logical characteristics: explanations are true, predictions need not be; making predictions is part of one way of confirming the existence of explanations; predictions may be made with or without rational grounds, and some rational grounds adequate for prediction fail to *explain* the predicted occurrences. If, then, the structural identity claim is to be approved for implying that every explanation must have been capable of prediction, it is no less censurable for glossing over these important differences in structure and rôle.

2. THE CENTRALITY CLAIM

If explanation and prediction are structurally distinct, the usual claim of centrality in scientific procedure is at least ambiguous: are they both central, or is one more important than the other? The following considerations will specify some distinctive temporal asymmetries peculiar to each, arguing for their irrelevance to general scientific inference, and hence the inadvisability of incorporating them into typical or central models of such inference.

(*a*) Consider first prediction. A necessary condition for the predictive character of an utterance or inscription is its asserting something about some time later than its own. This is the force of future-tense indicators often taken as a sign of predictive character in ordinary usage, though, of course, not essential to such character. If, now, we examine the four requirements of Hempel and Oppenheim, we find no temporal conditions among them.[7] To be sure, this pattern of requirements is intended to reflect the inferential process of making (scientific) predictions, but the inferences admitted by the pattern include other types as well, which cannot be classed as explanations either. Thus, even if this pattern is not exhaustive, but represents one scheme of scientific inference, it is much wider and more general than inferences of predictive or explanatory nature.

Thus, as the distinction between explanation and prediction is drawn by Hempel, Oppenheim, and others, it would be said, with reference to our earlier example, that if B is given, i.e. if we know that the phenomenon described by B has occurred, and a suitable set of statements A and L is provided afterwards, we have an explanation, while if the latter statements are given, and B is derived prior to the occurrence of the phenomenon it describes, we speak of a prediction. Note, however, one way in which this description fails to exhaust the inferences allowable by the pattern: If A and L are given rather than B, thus precluding explanation, their logical consequence B may be derived not prior to, but simultaneous with or after the occurrence of b. For example, b may have occurred prior to B's derivation but later than a, or it may have occurred prior to a.[8]

For an illustration of the first case (i), consider an astronomer who, from statements describing the *beginning of an ancient eclipse a*, plus the appropriate laws, deduces a statement describing *its end b*. For an example of the second case, (ii) imagine the same astronomer who, from appropriate laws plus statements describing *some relevant configuration of heavenly bodies at some time during his own personal experience a*, deduces statements describing *some eclipse in former times b*. In neither case do we have a prediction, yet both inferences fulfil the pattern in question.[9] What is common to both and to the predictive inference in question is not any temporal relation between statements and described events, but rather the givenness of A and L and the later derivation of B. The latter sequence, however, bears no simple relation to the sequence of described occurrences. It seems reasonable, then, to avoid the partial notion of prediction altogether in this connection and suggest the full potentialities of the pattern when A and L are given, by assigning a temporally neutral term to the derivation of B, say 'positing'.[10] From assumed laws and information about some spatio-temporal regions, we posit phenomena at other such regions, in any spatial or temporal relations to our assumed phenomena or our own utterance. Some positing is also predicting, but prediction has no more primacy for the pattern in question than positing events to the left of us in space has. Whether, aside from the pattern, there is independent reason to consider prediction scientifically or epistemologically primary is a question which we shall discuss at a later point.

(*b*) Now consider explanation, i.e. with B given, and a suitable set of statements A and L provided afterwards, fulfilling Hempel and Oppenheim's four requirements. Once again we find that, since no temporal criteria are to be found among these requirements, they define a wider class of inferences than simply explanatory ones. A non-explanatory instance which fits the pattern is afforded by any case where *b* precedes *a*. A concrete illustration is at hand in any situation analogous to our previous example (ii). Thus, given a description of *some eclipse in former times b,* an astronomer who provides appropriate laws L and statements describing *some relevant configuration of celestial bodies during his own lifetime a* from which B is deducible, is fitting the pattern, but is surely not *explaining* or *accounting for b*. For explanation, we require, in addition to our four desiderata, that *a* must not temporally follow *b*.[11] What is common to our above instance and explanation is not the temporal order of *a* and *b*, but rather the givenness of B and the later provision of A and L. To suggest the full potentialities of the pattern in such a case, we ought to drop the partial notion 'explanation' here and again assign a temporally neutral term to the provision of A and L, e.g. 'substantiating'.[12]

(*c*) If both explanation and prediction are characterized by temporal asymmetries which are irrelevant to the generality of scientific inference, is there any independent epistemological ground for considering explanation or prediction as central to scientific procedure?

(i) It may be noted that the interpretation of explanation in question is generally taken as a reflection of causal notions, and that the peculiar temporal asymmetry of explanation is identical with the temporal asymmetry of cause and effect. To consider the latter notions central to science is to justify treating explanation as scientifically primary.

A number of authors have however remarked the fact that causal notions come to be used less and less by an advancing science, while they remain of relatively constant importance in practical affairs. The point often made in this connection is to stress the relation of causal notions to interest in control by voluntary action. As Braithwaite has recently expressed it,[13]

> If an earlier event's occurring is a nomically sufficient condition for a later event to occur, we can (in suitable cases) ensure that the later event should occur by taking steps to see that the earlier event does occur. For this purpose it is irrelevant whether or not the later event's occurring is a nomically sufficient condition for the earlier event to occur.... But, if a later event's occurring is a nomically sufficient condition for the earlier event to occur, we cannot indirectly produce the earlier event by producing the later event, since by the time that we should be producing the later event the earlier event would irrevocably either have occurred or not have occurred. This difference between the case of regular sequence and that of regular precedence is, I think, the reason why we are prepared to call a nomically sufficient condition for an event a cause of that event if it precedes the event but are not prepared to call it a cause if it succeeds the event.

If this general account is true of causal notions, it would seem to apply equally to so-called causal explanation, which is characterised by the same temporal asymmetry. If scientific inference, however, unlike voluntary control of the future, may be based on temporally backward as well as forward nomic regularities,[14] it is misleading and partial to view science from the vantage point of such control, and of its cognate notions, 'cause' and 'explanation'. It would seem a better reflection of the full generality of scientific reasoning if we view it as concerned with comprehensive nomological relations among events and abstract from causal explanation entirely. Science may then be compared, in Toulmin's apt analogy,[15] to a route-neutral map, quite general as regards direction, but capable of guiding variant itineraries for those with practical purposes.[16]

(ii) Much of what has been said of the relation of explanation to voluntary control holds in an obvious way for prediction. In addition, however, the scientific primacy of prediction is often supported by reference to the acceptance or confirmation of statements.

It may, for instance, be granted that we posit events both past and future to our posit-utterances but it is pointed out, for any present posit-utterance, it is peculiarly contingent on the future since there is a possibility that it is reasonably rejected then, owing to future rejection of some confirmatory sentence asserting a future occurrence. But, in the first place, such future rejection may be due to future rejection of some confirmatory sentence asserting an occurrence prior to the present posit-utterance. And, in the second place, it is equally true for any present posit-utterance that there is a possibility of its reasonable rejection in the past. That something is a posit-utterance at t implies neither its acceptance at all times following t nor its acceptance at all times preceding t.

It may then occasionally be suggested that the predictiveness of any given posit-utterance is not a matter of its own future acceptance but involves rather the fact that it is false if any confirmatory sentence asserting a future occurrence is false. Obviously, however, it is also false if any confirmatory sentence asserting a prior occurrence is false. It may, of course, be held that a posit-utterance, all of whose confirmatory sentences asserting prior occurrences are true, is false only if some confirmatory sentence asserting a future occurrence is false. But the obvious converse is equally the case.[17]

It may, however, finally be countered that we cannot now voluntarily choose to carry through a past test, while we can now decide to institute a test of a specified posit-utterance in the future. This claim, true enough, is as trivial as the general truth that voluntary control at a given time is of later phenomena, never of earlier; we cannot now choose to institute any past event, *a fortiori* we cannot now choose to institute a past test. There is no special relevance to science in this truism. Furthermore, we have seen that a given posit-utterance is false if a confirmatory sentence asserting a prior occurrence is false, even if all confirmatory sentences asserting future events are true. Hence, even if all its tests which are voluntarily choosable (hence in the future) at t are positive, i.e. yield true confirmatory sentences asserting later events, the posit-utterance at t may still be false.

Pragmatists and positivists have championed the further doctrine that the meaning or content of a physical-object statement *is* its future verifiability. Hence, e.g. even an apparently retrodictive posit-utterance of historian H in 1950, 'Caesar crosses the Rubicon' is really *about* future possible confirmations or disconfirmations in experience. This doctrine of meaning is however, ambiguous, since 'future' is unclear.[18] In our example above, are the future confirmations future to 1950 or future to Caesar's crossing the Rubicon? Only if they are future to 1950 is it plausible to construe the content of H's assertion as its testing future to the assertion, but this interpretation leads to quite undesirable results, e.g. a replica of H's utterance in 1954 has a quite different content. If, on the other hand, 'future' here means 'future to Ceasar's crossing the Rubicon', then it refers also to confirmations prior to H's utterance, and the ground for considering its predictive content primary disappears. It goes almost without saying that once we are prepared to admit confirmations prior to H's utterance, there is no longer any advantage in excluding those which precede the historical event itself.

But the general difficulties of this dictum on meaning far outweigh in importance the ambiguity mentioned.[19] Abandoning it in favour of some other criterion of meaning, we remove a reason for considering prediction as primary which has been dominant in recent philosophy.

3. THE CLAIM OF IDEALISM

It is commonly said that scientific inquiry abstracts from its raw subject-matter, and more seems often intended by such statement than the harmless truths that science is embodied in language, that it deals with only some of the concrete entities available for study at any given time, and that it selectively analyses complexes into parts. Often the point seems rather to be that the true *objects of,* e.g., scientific explanation[20] are not concrete events or things at all, but supposed abstract, intensional features, or properties of the latter, idealistic entities like phenomena, facts, or states-of-affairs.

Now this view might plausibly be construed as claiming that every adequate analysis of locutions like '... scientifically explains – ', or '... scientifically accounts for – ' presupposes an abstract, intensional ontology. Independently of whether the sciences themselves afford evidence of the existence of idealistic entities, that is, the *theory of* scientific explanation is asserted to require them.

It will be recalled that in our previous discussions of the supposed inference-pattern of scientific explanation, we followed the conventional description of it, i.e. we spoke of accounting for or explaining the *event b* by providing appropriate sentences A and L which have its description B as a logical consequence. If the idealism claim is correct, however, all talk of this pattern as concerned with explaining or predicting the concrete events mentioned in the explanandum is, taken literally, false (unless it is merely abbreviatory of appropriate idealistic talk). In what follows, we shall (a) discuss grounds for the idealism claim as formulated above (b) propose an inscriptional alternative, and (c) specify how this alternative enables us to speak of science as abstractive, though rejecting abstract entities.

(a) Consider that events (taken in the ordinary sense as concrete occurrences with spatio-temporal boundaries) are describable in alternative, logically independent ways and hence that it is false to suggest that B is a *unique* description of b. With a specified stock of individual predicates 'Q', 'R', 'S', etc., we could truly describe b by either affirming or denying for *each* of these predicates that it is true of b. If so, how can we speak of having explained the *event b* when we have fulfilled our pattern's requirements in providing appropriate deductive grounds for just one of these true descriptions, say 'Rb'? Have we both explained and not explained b in providing such grounds for 'Rb' but not for the equally true description 'Qb'? Clearly, a particular fulfilment of the pattern explains not simply the event b, but b as qualified in a certain specified way, or the fact that b is so qualified.

Suppose that, regarding such an event as Cicero's birth (in 106 B.C. on the particular day D), we provide specific information to the effect that the relevant conception occurred 280 days prior to D, that it was followed by a normal pregnancy, was human, etc., as well as the generalisation that, for human beings, births follow their respective conceptions, in normal cases, by exactly 280 days. Have we now explained Cicero's birth? But this birth could be characterised in innumerable ways other than in terms of when it occurred, e.g. it followed a specific period of labour, it had individual obstetrical characteristics, etc. Clearly our particular explanans does not explain Cicero's birth as such, but rather Cicero's-birth-as-occurring-on-D-in-the-year-106 B.C., or as often expressed, the fact that Cicero was born on D in 106 B.C. We ought then to discard simple contexts like 'Explanans E explains...', (where the dots are replaceable by names of concrete occurrences) in favour of indirect locutions calling for appropriate qualification of the occurrences in question.

For our above case (calling its specific explanans 'E'',) we might say:

(i) E' explains why Cicero's birth occurs on D in 106 B.C.

Now in view of the fact that:

(ii) Cicero's birth = Tully's birth

is true, whatever we say of Cicero's birth ought to be truly sayable of Tully's birth. However, when we replace 'Cicero's birth' by 'Tully's birth' in (i), we turn its truth into falsehood, since though E' logically implies 'Cicero's birth occurs on D in 106 B.C.', it does not logically imply 'Tully's birth occurs on D in 106 B.C.' Following Quine,[21] we may say that the occurrence of 'Cicero's birth' in (i) is not purely referential, i.e. the truth of (i) depends not merely on the particular *event*, but on the *way* in which the event is denoted.

Since, moreover, the identical (as well as any appropriately analogous) inter-change fails to alter the truth of the subsentence of (i), i.e.:

(iii) Cicero's birth occurs on D in 106 B.C.

when it stands alone, but does alter truth when it is embedded in (i), we may, again following Quine,[22] call the contexts exemplified by (i) referentially opaque, to signify that, like quotation contexts, they can change referential occurrences into non-purely-referential occurrences.

The important thing to note about referentially opaque contexts is that we cannot quantify *into* them from the outside, just as we obviously cannot do so for quotation contexts. Thus to apply existential generalisation to the occurrence of 'Cicero's birth' in (i), we should get:

(iv) $(\exists x)$ (E' explains why x occurs on D in 106 B.C.)

or:

(v) Something is such that E' explains why it occurs on D in 106 B.C.

But what event is it whose time of occurrence is explained by E'? Cicero's birth, or equivalently, Tully's birth? But we have already seen that to put 'Tully's birth' into (i) makes it false. The failure of (iv) and (v) means thus that we cannot construe sentences like (i) as being *about* explained events; we cannot, that is, analyse them in quantificational form so that the bound variables take the events mentioned in the explananda as values. Correspondingly, we cannot then describe our pattern as concerned with explaining these concrete events at all.

We may thus be inclined to construe the objects of scientific explanation as so-called facts or states-of-affairs, which cannot be denoted by non-synonymous units in the nature of the case, and hence do not give rise to the difficulty of referential opacity as elaborated above. We may, for example, construe the whole subsentence as it occurs in (i) as designating an abstract fact or state-of-affairs (Cicero's birth occurring on D in 106 B.C.) not equally designated by 'Tully's birth occurs on D in 106 B.C.' in the same context, and existentially generalise as follows:

(vi) $(\exists x)$ (E' explains why x)

or:

(vii) Something is such that it is explained by E' (i.e. the state-of-affairs Cicero's birth occurring on D in 106 B.C.).[23]

(*b*) The referential opacity of contexts represented by (i) precludes analysing them as referring to concrete entities mentioned in their respective explananda. In this sense, then, we cannot take these concrete events, processes, or other things to be the objects of scientific explanation. If we persist in this sort of interpretation, we prepare the ground for those familiar and puzzling philosophies of science according to which science cannot ever explain its own true object of study.

Nevertheless, an idealistic interpretation such as is given in the paragraph before the last is not the only alternative open to us. We do indeed have to avoid quantifying into referentially opaque contexts in the manner of (iv) and (v), but we can, anyhow, suggest an analysis of (i) which remains concrete in ontology throughout, though referring to linguistic rather than extra-linguistic entities. This analysis rests on the fact that what the explanation-pattern under discussion actually exhibits is a logical relationship among certain statements, i.e. explanantia and explananda. Assume, then, that the whole why-clause in (i), i.e. 'why Cicero's birth occurs on D in 106 B.C.' functions as a single indivisible predicate in this context, applicable *not* to any so-called state-of-affairs, but rather to every concrete inscription which is a rephrasal or translation of the subsentence following the 'why'.[24] Now analyse (i) as:

(viii) $(\exists x)$ (x is a why-Cicero's-birth-occurs-on-D-in-106 B.C., and x is logically implied by E' which fulfils $(R2)$, $(R3)$, and $(R4)$).

The variable here ranges over concrete inscriptions exclusively, and 'E'' can be taken to name some such inscription as well (with the relation of logical implication appropriately understood). Construing the why-clause as applying to *rephrasals* rather than simply to replicas of 'Cicero's birth occurs on D in 106 B.C.' (like direct quotes), enables (viii) to cover cases where the explanans is formulated in some language other than that of (i).

As we focus our attention on some particular branch of science, moreover, taken as formulated in some specific constructed language, and where our purpose is to build a theory of explanation for this language rather than to explicate ordinary statements like (i), it becomes possible to avoid interlinguistic reference (as in (viii)), together with such rather imprecise notions as those of translation or rephrasal.[25] We may then always specify particular inscriptions in the language in question, and interpret explanation in terms of the relation of logical implication between some such inscriptions. In effect, we may look upon both (viii) and the latter approach as the assimilation of event-explanation to the explanation of laws or generalisations, mentioned earlier in section (1). For the explanation of laws is clearly no more than a question of relating law-*statements* to appropriate other statements.

(*c*) This assimilation of event-explanation to that of laws or generalisations serves to underline the fact that *abstractiveness* is distinct from and need not involve *admission of abstract entities,* i.e. for a mode of inquiry to be said to abstract from raw experience or from the world, it need not be thought to take abstract entities as its objects. Clearly, the explanation of laws or generalisations is no less abstractive relative to experience than event-explanation, though conceivable in straightforward fashion as taking concrete law-sentences for its objects and concerned with showing appropriate relationships among certain of these objects. Its abstractiveness relative to entities referred to by its explananda consists just in the multiplicity of explananda with the same reference (i.e. correlated by reference with just the same entities) and the consequent need of *selection* in any case of explanation. If every law, that is, makes the same reference to everything via the universal quantifier, each law-explanation abstracts from everything in selecting as explanandum only one out of many sentences, all equally referring to everything. Such selection in no way depends on these initially choosable sentences being abstract.

Analogously, though event-explanation be taken (in accordance with (viii) and the paragraph following) as not requiring abstract entities for its objects, it is yet thoroughly abstractive relative to the events mentioned in its explananda. For each such event is correlated with many sentences, i.e. all those expressing true descriptions of it. In the selection of one of these as explanandum on a particular occasion lies the abstractiveness of event-explanation. As before, the crux of the matter is the correlation of a multiplicity of choosable units with the same entities rather than the abstractness of these units.

NOTES

I thank Professors C. G. Hempel, N. Goodman, W. V. Quine, and N. Chomsky for critical comments on this paper.

1. The description here discussed (as well as the structural identity claim) is given in various forms by a number of authors. Among recent empiricist writings, the following should be especially mentioned: K. R. Popper, *Logik der Forschung*, Wien, 1935, pp. 26 ff., and *The Open Society and Its Enemies*, London (first published 1945) 1947, Vol. 2, p. 249 and pp. 342-43, C. G. Hempel, 'The Function of General Laws in History,' *Journal of Philosophy* 39 (1942): 35-48, and C. G. Hempel and P. Oppenheim, 'Studies in the Logic of Explanation,' *Philosophy of Science* 15 (1948): 135-75.

2. Hempel & Oppenheim, op. cit., 137-38.

3. Hempel and Oppenheim do not also require the explanans to include one statement which is not a law, for 'to mention one reason', they wish to consider an explanation of generalisations a *bona fide* explanation. In the light of what was observed above, regarding the restriction of the notion of 'prediction' to events, it would seem unlikely that they wish to extend their statement of the explanatory pattern to the *prediction* of generalisations, though they do not make this point explicit.

4. This sentence does, however, appear (p. 138) though it does not figure importantly in the later treatment of the authors: 'only to the extent that we are able to explain empirical facts can we attain the major objective of scientific research, namely not merely to record the phenomena of our experience, but to learn from them, by basing upon them theoretical generalisations which enable us to anticipate new occurrences and to control, at least to some extent, the changes in our environment.'

5. One might, incidentally, raise the question whether scientific prediction always takes the form mentioned, in view of predictions only inductively well-grounded, but not implied by any relevant conjunction of universal generalisations and condition statements. For example, suppose we predict that the 5,000th ball drawn at random from an urn will be red, since all have

heretofore been red with, say, the exception of the first drawn. If such prediction is acknowledged as rational, surely its full restatement after the event is non-explanatory even if true. Thus, suppose our prediction is fulfilled and we are told that the explanation why the 5,000th ball was red is because the previous 4998 were. This seems unacceptable as an explanation though it restates the rational grounds for having made the prediction previously. Indeed, were there not a single exception, the fact that, e.g. 4,999 balls were red does not explain why the 5,000th is red though it does rationally ground the prediction that it will be. This example serves to illustrate a distinction, of some general importance, between asking 'Why P?' in the sense, 'What is an explanans for P?' and in the often divergent sense, 'What are rational grounds for asserting P?' The distinction is sufficient to invalidate a widespread identification of the *explanation of A* with *showing that A was to be expected*, as suggested, e.g. by Toulmin, *The Place of Reason in Ethics,* Cambridge, 1950, pp. 122 ff., and many others.

6. Furthermore, it is doubtful if even this explicit delimitation of just what is supposed to correspond to prediction is sufficient. For on one widely-held view (shared by Hempel) abstract, partially interpreted theories are an integral part of advanced sciences and, as such, are essential to rational prediction in those sciences. It is unclear, however, in what sense we may speak of the truth or falsity of *partially interpreted* formal systems. If we cannot, then such theories, though predictive, are not even proffered explanantia in our explicit sense. For though this sense does not *require* truth, it does require that the explanans be, in point of fact, either true or false.

7. Often, indeed, terms like 'consequence', 'derivation', 'antecedent condition', 'presuppose', etc., are ambiguously employed with occasional temporal reference. It does not appear, however, that such interpretation is here intended.

8. b may, it goes without saying, have occurred at any time relative to the givenness of A and L as well.

9. We might call one or both of these inferences 'postdictive', following Reichenbach in *Philosophic Foundations of Quantum Mechanics,* California, 1944, but, contrary to Hempel and Oppenheim, op. cit., p. 138, 'postdiction' is not applicable to explanation.

10. Our use of this term should be clearly distinguished from other uses in the literature, especially that of Reichenbach in his many discussions of probability and confirmation, e.g., in *Experience and Prediction,* Chicago, 1938.

11. This does not imply that all these requirements are sufficient even if necessary. For example, suppose symptom S precedes and is a lawfully sufficient condition for contraction of cancer, enabling prediction. Yet we do not, it might plausibly be said, *explain* contraction of cancer by the presence of S. It may be noted that, while we require that a not follow b, we have not also required that b precede the explanation-utterance accounting for it. Since Hempel and Oppenheim interpret the givenness of B as implying that b precedes its explanation-utterance, it may be worthwhile to justify here our departure from this view: The givenness of B, it seems to us, means in practice, merely that we are fairly confident in its truth, but such confidence is surely not limited to statements about the past. If common use is any guide here, then if asked 'Why will the sun rise tomorrow?' I may reasonably be said both

to be predicting and explaining the sun's rising when I offer the appropriate astronomical information. Of course, I cannot here be certain of the *truth* of *B*, which truth is necessary if I am truly to explain. But the same uncertainty holds for a *B* which refers to the past, though I am confident in its truth. It seems then that there is here no sharp temporal difference crucial to explanation, though it is required that *b* must not precede *a*, whenever *b* occurs.

12. The distinction between explanation and our case cannot be easily discerned by attending to the ordinary use of 'Why...?' as a clue. For while explanations often answer the question 'Why *B*?' in the sense, 'In accordance with what laws and following what conditions does *b* occur?', our case answers the question 'Why *B*?' often asked in the sense, 'What rational grounds are there for asserting '*B*?'. Perhaps the confusion of these two senses is at least partially responsible for the notion that 'functional' or 'teleological' explanations are explanatory in the ordinary sense, whereas they actually share the logical form of our nonexplanatory instance. For Nagel, in 'Teleological Explanation and Teleological Systems' in *Vision and Action,* ed. S. Ratner, 1953, for example, the statement, 'The function of chlorophyll in plants is to enable plants to perform photo-synthesis' is equivalent to 'A necessary condition for the occurrence of photosynthesis in plants is the presence of chlorophyll'. In his words, 'A teleological explanation states the *consequences* for a given biological system of one of the latter's constituent parts or processes; the equivalent non-teleological explanation states some of the *conditions* under which the system persists in its characteristic organization and activities'.

Now it is clear that in Nagel's use here, 'necessary condition' means 'necessary non-subsequent condition' and 'consequence' is temporal in reference. Otherwise, knowing that all breathing organisms die, we might say that since death is a necessary condition for breathing in organisms, the function of death is to enable organisms to breathe. Hence, for his view presumably, to explain *b* functionally or teleologically is to refer to a suitable *later a* such that *A* and *L* imply *B*. Actually, his equivalent non-teleological statement is simply *L* and does not alone provide deductive grounds for asserting the presence of chlorophyll, and hence for explaining it. But assuming that we add the appropriate *A*, i.e. 'Photosynthesis occurs at *t* in plant *P*', this does not *explain* why chlorophyll occurs at some earlier time in *P*, though it *substantiates* such occurrence. We ought therefore not to speak, as Nagel does here, of 'equivalent non-teleological *explanations*'. Given a case of photosynthesis at *t* however, we may ask 'Why must chlorophyll have been present?', though we do not intend to ask, 'Following what antecedent conditions and in accordance with what laws did chlorophyll occur?'. Confusion of the two questions may partly account for the misleading idea of functional *explanation* as an answer to some 'Why...?'

13. R. B. Braithwaite, *Scientific Explanation,* Cambridge, 1953, p. 313.

14. History, geology, archaeology, astronomy are only some striking instances of scientific use of backward regularities. Backward inference, whether postdictive positing or the substantiation (by use of information about the present) of past events, is a partial goal of all sciences, it seems to me, including physics.

15. S. Toulmin, *The Philosophy of Science,* London, 1953, p. 121.

16. In addition, the connection of explanation with control may indicate why, as pointed out above, temporal requirements alone may not be sufficient to explicate it.

17. I am throughout this passage following the usual type of argument in talking of confirmatory sentences, etc., but this does not commit me to such an epistemology.

18. For a discussion of this point, in relation to C. I. Lewis's analysis of historical statements, see the writer's 'Verifiability in History: A Reply to Miss Masi,' *Journal of Philosophy* 47 (1950): 164 ff.

19. For a critical review of these difficulties leading to an alternative proposal in terms of translatability into some empirical language, see C. G. Hempel, 'Problems and Changes in the Empiricist Criterion of Meaning,' *Revue internationale de philosophie* 11 (1950): 41-63.

20. For clarity, I here discuss explanation specifically, but the transfer of this treatment to prediction is readily made.

21. W. V. Quine, *From a Logical Point of View*, Cambridge, Mass., 1953, p. 140.

22. Ibid., p. 142, and W. V. Quine, 'Three Grades of Modal Involvement', *Proceedings of the XIth International Congress of Philosophy* 14 (1953): 67.

23. In 'The Function of General Laws in History,' Hempel seems indeed to be taking some such view. He there construes events or phenomena not as concrete minimal spatio-temporal chunks, but rather as 'kinds or properties of events', while descriptive sentences like A and B figuring in explanantia and explananda express the *fact that* such properties occur at specific places and times. One consequence of this particular analysis seems to be a revision of the ordinary use of 'event' to denote properties of what are usually called 'events'. Insofar as the distinction between the two may be blurred, this consequence may be thought undesirable. Also, it would seem that 'event' in this revised use is not appropriate for denoting causes or effects, a goal which some proponents of the explanatory pattern discussed seem to have had at the back of their minds.

24. This interpretation was proposed originally for indirect discourse generally in my 'An Inscriptional Approach to Indirect Quotation,' which appears as article 1 in Part 1 above.

25. For a discussion of some of the difficulties in using synonymy and translation to explicate indirect discourse, see my 'On Synonymy and Indirect Discourse,' which appears as article 2 in Part 1 above.

2. THOUGHTS ON TELEOLOGY

TWO STRATEGIES IN INTERPRETING TELEOLOGY

How shall we relate teleological notions, referring events to certain of their consequences, with causal notions, referring events to certain of their antecedents? Two hard-headed strategies are discernible in the recent literature – hard-headed because in neither one are teleological statements taken to refer to special sorts of entities such as entelechies nor to embody special sorts of explanation by means of final causes. One such strategy is to interpret teleological statements as descriptions of plastic or self-regulating behaviour, in principle explainable in ordinary causal terms. The other strategy is to construe ostensible mention of goals future to an action as referring rather to ideas of such goals prior to the action, and hence capable of figuring in its normal causal explanation.

Each of these strategies has occasionally been used in such a way as to provoke the charge of misapplication. The first has thus, for the most part, been criticised for failing to do justice to purposive action in the higher animals and man, the second for unplausibly ascribing goal-ideas to the lower animals and to non-purposive human behaviour. These criticisms, however, leave unscathed a moderate use of either strategy within well-defined limits: so long as you do not try to reduce *purpose* to self-regulative behaviour, nor attribute *ideas, wishes, or beliefs* to behaviour that is not consciously purposive, you may proceed without hindrance.

It is this moderating assumption that I shall examine in the two main sections of the present paper: (i) First, with respect to two recent interpretations of teleology in terms of self-regulative behaviour, I shall argue that they are inadequate not merely for purposive, but also for certain sorts of non-purposive behaviour, for which I will suggest the possibility of an alternative treatment. (ii) Secondly, I wish to indicate some serious logical-semantic difficulties attending purported references to beliefs about and desires for a goal even in accounting for fully purposive human action, and I want to explore a way of handling such difficulties.

This paper appeared in the *British Journal for the Philosophy of Science* 9 (No. 36, 1959): 265–83.

1. THE SELF-REGULATION STRATEGY

A. *Negative Feedback.* Perhaps the most basic recent paper applying the notion of self-regulation to the interpretation of teleology is that of Rosenblueth, Wiener, and Bigelow.[1] Convinced that 'purposefulness (is) a concept necessary for the understanding of certain modes of behaviour',[2] and that its importance has been slighted as a result of the rejection of final causes, these authors propose to explain purpose, in largely engineering terms, as behaviour 'that may be interpreted as directed to the attainment of a goal, i.e. to a final condition in which the behaving object reaches a definite correlation in time or in space with respect to another object or event'.[3] In order to accomplish this explanation, the authors introduce the terms 'input', 'output', and 'feedback', as follows. 'Input' applies to events external to an object that modify the object in any manner. 'Output', on the other hand, refers to changes produced in the surroundings by the object. 'Feed-back' may be used in two different ways. It may be applied to objects (such as electrical amplifiers), some of whose output energy is returned as input. In such cases feed-back is positive in that the output re-entering the object 'has the same sign as the original input signal',[4] thus adding to this signal rather than correcting it. On the other hand, feed-back is negative when the object's behaviour is 'controlled by the margin of error at which the object stands at a given time with reference to a relatively specific goal'.[5] In such cases, 'the signals from the goal are used to restrict outputs which would otherwise go beyond the goal'.[6]

With the notion of negative feed-back at hand, the authors propose that teleological behaviour be construed as behaviour controlled by negative feed-back.[7] 'All purposive behavior', they say, 'may be considered to require negative feed-back. If a goal is to be attained, some signals from the goal are necessary at some time to direct the behavior'.[8] We may, of course, construct a machine that will impinge on a luminous object although the machine is insensitive to light, as well as to other stimuli emanating from the object. It would however, be a mistake to consider such impingement behaviour purposive inasmuch as 'there are no signals from the goal which modify the activity of the object *in the course of the behavior*'.[9]

By contrast, some machines are 'intrinsically purposeful', for example, 'a torpedo with a target-seeking mechanism'.[10] The behaviour of such objects involves 'a continuous feed-back from the goal that modifies and guides the behaving object'.[11] The path followed by the torpedo, for example, is controlled by the signals it receives from the moving target.

The foregoing conception of teleology is one whose scientific and practical importance is generally acknowledged and is surely not in question here. That it nevertheless fails adequately to characterise clear instances of purposive behaviour is an independent point that emerges from the following considerations.

The authors interpret purposive behaviour as directed toward 'a final condition' in which the object achieves correlation with respect to some other entity. Since they describe the goal of such behaviour as something that emits signals guiding the behaving object, the goal cannot be identified with the final condition of correlation mentioned above. For one thing, this final condition, if it occurs at all, is later than the behaviour in question and cannot therefore be supposed to guide it if the notion of final causes acting upon earlier events is rejected. For another, the authors describe cases of 'undamped feedback', in which a purposive machine, by increasingly larger oscillations, overshoots the mark and fails to attain the final condition toward which its behaviour is directed, though such behaviour has presumably been modified by signals from the goal. The goal, as described by these authors, cannot, therefore, be identified with the final condition of correlation and must rather be construed as that with which correlation is supposed to occur, i.e. what we may call 'the goal-object'. This goal-object, even where correlation fails to occur, emits signals that modify the purposive activity of the behaving object.

The difficulty of the missing goal-object

Now it is just the existence of such a goal-object that cannot be assumed in all cases of purposive behaviour. As pointed out in a critical paper by R. Taylor,[12] a man's purpose in groping about in the dark may be to find matches that are not there, his purpose in going to the refrigerator may be to obtain a non-existent apple, he may seek the philosopher's stone, the holy grail, the fountain of youth, or a live pulsing unicorn. In every such case his behaviour is clearly purposive and yet in none is this behaviour guided by signals emitted from a goal-object, correlation with which represents the final condition toward which the behaviour is directed. Generally, if a man's purpose is to obtain object O, we cannot infer the existence of something x, identical with O, such that he seeks x, nor can we infer, from the fact that a man's purpose is to obtain something of kind K, that there exists some object of kind K for which he strives.[13]

This criticism, it will be noted, establishes the inadequacy of the Rosenblueth-Wiener-Bigelow interpretation only with respect to purposive cases. Indeed, Taylor supposes that we have here an 'irreducible difference' between human beings and machines,[14] suggesting thereby that what may be called 'the difficulty of the missing goal-object' does not arise for non-purposive cases on the Rosenblueth-Wiener-Bigelow account. As against this suggestion, I now argue that the same difficulty arises also with respect to non-purposive teleological behaviour in non-human as well as human organisms, and conceivably in machines too.

Consider a standing passenger thrusting his foot outwards suddenly in order to keep his balance in a moving train, a rat depressing the lever in his experimental box in order to secure a food pellet, a small infant crying in order to attract mother's attention. To describe each of these cases as has just been done is to provide a teleological account relating the behaviour in question to some selected end. It is not, however, to attribute purpose to the organism concerned, in the sense of our previous instances, for example. If the knight's purpose in travelling far and wide is to find the holy grail, he has chosen to search for it in the hope of finding it. If a man's purpose in opening the refrigerator door is to get an apple within, he opens it with the intention of getting an apple within. It would in most cases be peculiar, however, to describe the train-rider's sudden kick as a product of choice or as an expression of intention. Psychologists, similarly, often feel uncomfortable in describing the rat as choosing to depress the lever, or as depressing it with the intention of securing a pellet. Finally, we should not generally say of a small infant, even when it cries in order to attract mother, that it has chosen to cry with the intention of attracting mother – though it may of course do the latter at a later stage, when its crying will be interpreted quite differently by its elders. I suggest these cases merely as examples of a large group for which we willingly offer teleological accounts but withhold ascriptions of intention, choice, and purpose.

It does not take much reflection to see that the difficulty of the missing goal-object arises even for the moderate attempt to apply the Rosenblueth-Wiener-Bigelow interpretation to this set of non-purposive cases. For we cannot plausibly suppose our train-rider to be receiving guiding signals from some region with which his foot is to be correlated. Neither, when the psychologist stops replacing the rat's pellets, can we describe the rat as receiving directive signals from some such pellet. Nor, finally, when mother, expecting baby to sleep, steps out to the corner store, is she available for the issuing of signals guiding the infant's behaviour toward final correlation with herself. The Rosenblueth-Wiener-Bigelow analysis cannot therefore be judged generally adequate even for non-purposive teleological behaviour, nor does it seem to suffice even when we restrict ourselves to non-human cases.

Teleology and learning

A suggestion: For at least *some* non-purposive cases, I suggest that an interpretation in terms of learning may throw some light on teleological description, where analyses in terms of self-regulation fail. In the case of the infant, the suggestion is that our 'in order to'-description of its present crying reflects our belief that this crying has been learned as a result of the consequences of like behaviour in the past – more particularly, as a result of having received mother's attention. Having initially cried as a result of internal conditions C, and having thereby succeeded in attaining motherly solace, representing a type of rewarding effect E, the infant now cries in the absence of C, and as a result of several past learning sequences of C followed by E. The infant's crying has thus been divorced from its original conditions through the operation of certain of its past effects.[15] These past effects, though following their respective crying intervals, nonetheless precede the present crying interval which they help to explain. The apparent future-reference of a teleological description of this present interval is thus not to be confused with prediction, nor even with mention of particular objects in the current environment, toward which the behaviour is directed. Rather, the teleological statement tells us something of the genesis of the present crying, and in particular, of the prominent rôle played by certain past consequences in this genesis. Such an account is perfectly compatible with normal causal explanation, and it indicates why goal-objects may very well be missing in some cases for which teleological description is appropriate. It shows also that we have in this respect no irreducible difference between human beings and machines, since machines capable of learning through the effects of their own operations are equally subject to such teleological description.

B. *Plasticity of Behaviour.* The interpretation of teleology proposed by R. B. Braithwaite,[16] though it resembles the Rosenblueth-Wiener-Bigelow approach, is considerably different in detail and restricts itself explicitly to what Braithwaite calls 'goal-directed behaviour' as distinct from 'goal-intended behaviour',[17] thus avoiding criticisms such as that of Taylor, based on purposive cases. Braithwaite's use of the self-regulation strategy is deliberately moderate; he wants to interpret 'all teleological explanations which are not reducible to explanations in terms of a conscious intention to attain the goal'.[18]

That the notion of *causal chains* is fundamental in the analysis of teleology is, for Braithwaite, shown by the fact that the organism's behaviour does not directly produce the goal; it constitutes, rather, part of a causal sequence of events progressing toward the goal. What, however, distinguishes teleological causal chains from all others with which science is concerned? Here Braithwaite follows E. S. Russell in proposing as the main clue 'the active persistence of directive activity towards its goal, the use of alternative means towards the same end, the achievement of results in the face of difficulties'.[19] Such plasticity of behaviour is 'not in general a property of one teleological causal chain alone: it is a property of the organism with respect to a certain goal, namely that the organism can attain the same goal under different circumstances by alternative forms of activity making use frequently of different causal chains'.[20] To specify this plasticity further, Braithwaite provides a schematic description.

We have, let us suppose, some object b which, in the normal case, is to be taken as a physical or organic system. Every event at any time in b is assumed determined by the whole preceding state of b taken together with the set of actual field or environmental conditions causally relevant to this event. Consider now a causal chain c, comprised of events in the system b during a particular interval and following immediately upon the initial state e of the system. Suppose also that the set of actual field conditions causally relevant to the events comprising the chain is f. Then c is uniquely determined by the initial state e taken together with the set of field conditions f.

Causal chains may now be related to a goal gamma as follows. Every chain ending in a gamma-event without containing any other such event is a gamma-goal-attaining chain. Relative to a given initial state e of b, we may consider the class of all (possible) sets of field conditions f uniquely determining chains that are gamma-goal-attaining. This class is the variancy Phi, with respect to b, e, and gamma, comprising the 'range of circumstances under which the system attains the goal'.[21]

Plasticity may be attributed to the system when the variancy has more than one member, so that the goal may be attained under alternative environmental circumstances, though not necessarily by means of alternative causal chains. Teleological explanations assert that the behaviour in question is plastic, and are intellectually valuable to the extent that this plasticity is not asserted on the basis of known causal laws of the system's mechanism, but rather on the basis of past observation of the conditions under which similar behaviour has taken place. Further, teleological explanations predict the occurrence of some gamma-goal-attaining chain on the basis of knowledge of the system's plasticity; they predict, in other words, that the set of field conditions that will in fact occur is a member of the appropriate variancy Phi, on the ground that Phi includes the class Psi of all sets of field conditions likely to occur.

It is at once apparent that, since Braithwaite does not construe goal-direction in terms of a relation to some goal-object but rather in terms of the variability of appropriate field circumstances, the difficulty of the missing goal-object does not arise in his scheme. Fido's pawing at the door need not be controlled by signals from outside; it need only lead to his being let out under alternative circumstances, say when his master is at home or when only his master's young son is at home. The rat need not, as he depresses the lever, be receiving any stimuli from the pellet that drops into his cup only after the lever is depressed; it is enough that his depressing of the lever results in his obtaining the pellet under a variety of experimental conditions.

The difficulty of goal-failure

Braithwaite's interpretation is, however, subject to other difficulties. For he appears to hold that particular teleological explanations do not merely ascribe plasticity to the behaviour being explained but also predict goal-attainment in the circumstances that will in fact ensue. It is perhaps the least of the difficulties of this view that its use of the term 'explanation' is strange, bearing little relation to normal causal explanation. Of crucial importance, however, is the fact that teleological descriptions, whether or not they qualify for the title 'explanation', do not generally predict the attainment of the goal indicated. It is interesting that, despite the requirement of Rosenblueth, Wiener, and Bigelow for a *goal-object,* they allow for non-attainment or *goal-failure* in their examples of undamped feed-back resulting in overshooting the target. But it is easy to multiply other examples of goal-failure. If Fido, trapped in a cave-in, is in fact never reached, is it therefore false that he pawed at the door in order to be let out? If, as in the case previously described, the psychologist stops replacing the consumed pellets with new ones, is it false to say the rat continues to depress the lever in order to obtain a pellet? To suppose, with Braithwaite, that all teleological explanations carry with them an 'inference that the set of relevant conditions that will in fact occur in the future will fall within the variancy'[22] is to answer the preceding questions affirmatively and quite unplausibly.

The difficulty of goal-failure may, of course, be avoided if the prediction of goal-attainment is eliminated from the content of teleological descriptions. Retaining the rest of Braithwaite's scheme, we should then take such descriptions simply as attributing plasticity to the behaviour in question with respect to the goal indicated. We should be saying of the system b exhibiting the state e that there is more than one possible set of field conditions f, such that, were f conjoined with e, some suitable goal-attaining chain would ensue. We should not, however, be predicting that one of these sets will in fact be presently realised, and our statement would therefore be protected from falsification through goal-failure.

The difficulty of multiple goals

This proposal is, however, immediately confronted with a new difficulty that we may call 'the difficulty of multiple goals'; the proposal becomes too inclusive to differentiate between acceptable and unacceptable teleological descriptions in an indefinitely large number of cases. These cases may be indicated schematically in Braithwaite's terms. Imagine that for the system b, and relative to the present state e of this system, the class of those sets of field conditions f uniquely determining gamma-goal-attaining chains has more than one member. The system in state e is thus plastic with respect to gamma. Suppose, now, that the class of those sets of field conditions f uniquely determining delta-goal-attaining chains relative to b and e also has more than one member. The same system in state e is thus also plastic with respect to delta. If our teleological description of the system's present behaviour embodies neither a prediction of the attainment of gamma nor a prediction of the attainment of delta but restricts itself merely to an assertion of the plasticity of this behaviour, we should be able, with equal warrant, to frame our teleological description either relative to gamma or relative to delta. We should, that is, be able to say either that b exhibits e in order to attain gamma or that b exhibits e in order to attain delta. This is, however, exactly what we cannot generally say.

The cat crouching before the vacant mouse-hole is crouching there in order to catch a mouse. Since no mouse is present, there will in fact be no goal-attainment. Nevertheless, the cat's behaviour is plastic since there are various hypothetical sets of field conditions, each set including one condition positing a mouse within the cat's range, such that, in conjunction with the cat's present behaviour, each set determines a mouse-attaining causal chain. On the other hand, there are also various other hypothetical sets of field conditions, each set including one positing a bowl of cream within the cat's range, such that, conjoined to the cat's present behaviour each set determines a cream-attaining causal chain. It should therefore be a matter of complete indifference, so far as the present proposal is concerned, whether we describe the cat as crouching before the mouse-hole in order to catch a mouse or as crouching before the mouse-hole in order to get some cream. The fact that we reject the latter teleological description while accepting the former is a fact that the present proposal cannot explain.

A limiting case of this difficulty occurs when the state e is one for which we should reject *every* positive teleological description. Imagine, for example, that e is the state of physical exhaustion and that the relevant class of field-condition sets is the one just considered in connection with mouse-attainment, suitably supplemented by conditions stipulating the cat's recuperation and assumption of a crouching position before the mouse-hole. The present proposal would then warrant us in saying, absurdly, that the cat is physically exhausted in order to catch a mouse.

Other examples of the difficulty of multiple goals arise in cases where every gamma-goal-attaining chain is also a delta-goal-attaining chain or contains such a chain as a part, so that if the variancy with respect to gamma has more than one member, so has the variancy with respect to delta. Consider, for example, an infant crying in order to get mother's cuddling, which is always preceded by the sound of mother's footsteps. Since every set of field conditions determining a cuddling chain in this situation also determines a footstep chain, and since we may assume the class of such sets determining cuddling chains to have more than one member, we may infer that the class of sets determining footstep chains also has more than one member. Thus we should again be warranted by the present proposal in saying either that the infant cries in order to receive its mother's cuddling or that it cries in order to hear mother's footsteps. The proposal again proves itself too inclusive to differentiate between teleological descriptions we accept and those we reject.

The learning suggestion again: a caution

Can the wanted distinctions be made on the basis of the learning interpretation suggested earlier? The answer seems to be 'yes' with regard to the examples we have considered. Thus we do not believe the cat has learned to crouch before the mouse-hole as a result of having been rewarded with a bowl of cream for having done so in the past. Nor do we suppose the infant's crying to have been learned as a result of having heard its mother's footsteps following past crying intervals. Nevertheless, I *surely do not* wish to propose the learning interpretation as an adequate analysis simply on the basis of a few examples and I am convinced, moreover, that several lines of interpretation will be required to account for the variety of statements commonly labelled 'teleological'. Furthermore, to appeal to our beliefs as to how the behaviour in question has been learned is, I think, to provide no firm philosophical answer: it is rather to suggest the importance of addressing our critical attention to the field of learning.

2. THE GOAL-IDEA STRATEGY

The second strategy mentioned at the outset consists, it will be recalled, in replacing reference to *goals* future to an action by reference to prior *ideas of such goals* functioning as causes of the action. What is the motive for this strategy? Suppose we explain John's decision to take a pre-medical course by saying that he decided to do so in order to enter medical school. Of what explanatory value is the apparent reference to John's future entrance? If it indeed comes to pass, it follows his present curricular decision and so cannot be among its causes. On the other hand, if it never comes to pass – for example, if John fails to complete his pre-medical training, we surely cannot count such entrance as a cause or partial cause of his actual decision. If the brick in fact did *not* strike the window, we cannot well say the window shattered because of the impact of the brick. What sense shall we then make of the fact that, although John's entrance to medical school is possibly fictional and, in any event, future to his decision, he clearly did choose his pre-medical course in order to enter medical school at a later date? The natural, and quite plausible, suggestion is to say that John's prior *desire* to enter, and his prior *belief* that entrance is contingent on choice of a pre-medical course, jointly determine his choice. Even if his desire is in the end thwarted and his belief in fact false, they are nevertheless realities of the situation preceding his choice, and they contribute, moreover, to its causal determination.

This strategy, quite plausible in the case just considered, has been rightly criticised in numerous others for projecting belief and desire into phenomena without adequate justification. As Ducasse has put it,

> the disrepute into which teleological explanations have fallen is doubtless due to their having been so frequently thus put forth in cases where the existence of the agent appealed to and of his beliefs and desires was not already known, but invented outright and purely *ad hoc*... But when antecedent evidence for their existence is present (e.g., when the hypothetical agent is a human being), a teleological explanation is methodologically quite respectable, although, like any other, it may in a given case not happen to be the correct one.[23]

Accordingly, Ducasse suggests the following as essential elements in genuine cases of purpose, rendering teleological explanation possible in the form exemplified by our recent illustration:

'1. *Belief* by the performer of the act in a law ... e.g. that if X occurs, Y occurs [or that Y is contingent upon X].
2. *Desire* by the performer that Y shall occur.
3. *Causation by that desire and that belief jointly*, of the performance of X'[24]

The sort of rule allowing explanation of the occurrence of X is further suggested by Ducasse as 'If an agent believes that Y is contingent upon X and desires Y, then that agent is likely to do X.'[25]

Now, even if we restrict the present strategy to clearly purposive human action in the sense specified by Ducasse, for example, we face serious problems in formulating the required explanatory statements. By way of illustration, let us try to explain John's curricular choice according to some such schema as that suggested by Ducasse. We ask, 'Why does John choose a pre-medical course?', and we receive the following statements in reply:

(i) John desires John's entrance to medical school.
(ii) John believes that John's entrance to medical school is contingent on John's choice of a pre-medical course.

(iii) Whenever someone desires something, believing that it is contingent on something else, he performs this other thing.

In symbols:

$(x)(y)(z)((x \text{ desires } y \cdot x \text{ believes that } y \text{ contingent on } z) \supset x \text{ performs } z)$,

(with the range of the variable 'z' restricted to acts or choices).

This sample explanation is obviously crude and oversimplified, raising numerous questions peripheral to the logico-semantic difficulties that are my present concern. Thus, for example, to say 'John performs John's choice of a pre-medical course', in concluding our explanatory argument here, is indeed a violation of English but one that provides a uniform technique for describing action. More important is the fact that (iii) requires much qualification before it can be taken seriously as a truth of conduct. Nobody can perform every act upon which the fulfilment of his desires is believed to be contingent, for some such acts are beyond his power. Nor does anyone actually choose to do everything within his power that he believes necessary for the attainment of every one of his desires, even if the dictates of the latter are compatible. Nevertheless, if John's desire and belief are to figure in the causal account of his choice, as the strategy we are considering requires, some generalisation incorporating these three elements but considerably more complicated than (iii) must be supposed as a premise. Since any such generalisation faces the same difficulties that I wish to indicate with reference to (iii), the latter statement may here be taken as a simple model for a general problem.

Logico-ontological difficulties: objects of desire

We may begin by asking, 'What is the intended range of the variable 'y' in (iii)?' Surely this range is not limited to actual things for, as mentioned earlier, the argument is intended to hold even if John's entrance to medical school never comes to pass. As in the case of purpose discussed earlier, we cannot infer existence of the objects of desire. To expand the range of 'y' to include possible but non-actual entities is not, moreover, sufficient, for the squaring of the circle and other impossibles have undoubtedly been desired. Allowing the variable 'y' to range over impossible so-called things as well is a step not many are prepared to take. It is bad enough to have to say, 'There is a possible but non-actual pot of gold at the end of the rainbow which has been desired by men', but to say, 'There exists a non-actual and, moreover, impossible squaring of the circle which has been desired by many' puts an impossible strain on the word 'exists', to say the very least.

Logico-ontological difficulties: belief

The variable 'y', further more, appears not only in the context 'x desires y', but also in the context 'x believes that y contingent on z', and both 'y' and 'z' refer back to prefixed quantifiers. Such quantification into belief contexts is, as Quine has repeatedly warned, a problematic procedure.[26] The occurrence of singular terms within belief contexts is, as he puts it, not purely referential in that the truth of such contexts taken as wholes depends not merely on the objects named by such terms but also on the manner of naming; belief contexts are, in short, referentially opaque. To use Quine's example, Philip may believe that Cicero denounced Catiline without believing that Tully denounced Catiline, though Cicero and Tully are identical. To replace either the name 'Cicero' or the name 'Tully' with a variable ranging over concrete objects such as people, in quantified belief sentences about Philip, thus yields an unintelligible result, since Philip's belief cannot be interpreted as a relation between Philip, Catiline, and some man to whom the names 'Cicero' and 'Tully' equally apply.

Quine has indeed admitted with reference to modal contexts, which are also referentially opaque, that a suitable limitation of the ontology to objects not nameable by non-synonymous names renders unrestricted quantification legitimate once more.[27] This result is due to the fact that interchangeability of synonymous names preserves truth value even within modal contexts, thus rendering the manner of naming again irrelevant and allowing modal statements to be true or false depending simply on the objects referred to. If, however, as I have elsewhere argued,[28] synonymy is not sufficient to guarantee interchangeability with preservation of truth value within belief contexts, the ontological restriction above mentioned will not legitimise quantification into *belief* contexts despite its adequacy for *modal* contexts. To be safe, we should need to restrict our ontology to objects not nameable by non-identical names within our language. Instead of the concrete object Cicero, for example, we would now recognise as many objects in our new ontology as there are different names or descriptions of Cicero, whether synonymous or not. This seems a desperate expedient indeed, not to say a transparent projection of our language into the world of things.

Notice, finally, that the variable 'z' occurs once within the belief context and once outside it, subject to the same external quantifier, thus serving to tie together the agent's act with his view of his act, as the variable 'y' serves to tie together the agent's desire and his beliefs about the conditions of fulfilment of that desire. Waiving all questions as to the nature and individuation of such entities as acts, it should be noted that these connections between act, desire, and belief are essential not only to the strategy under consideration, but also to every developed conception of rational behaviour. Indeed, if we have no way of formulating these connections in such a way as to avoid the difficulties discussed, we lack a satisfactory account of human action. The following remarks are devoted to the exploration of a possible solution.

Suggestion for an inscriptional interpretation

Suppose we try to construe desire, belief, and performance, at least with respect to the present schema, as relations between agents and inscriptions. The initial advantage to be hoped for is, of course, the relative clarity of inscriptions as concrete physical objects in comparison with the possibles, impossibles, and other strange entities posited by the approaches we have so far considered. The existence of inscriptions is, furthermore, completely independent of the existence of their purported objects. Though no pot of gold dangles from the end of any rainbow, we cannot infer that no inscription affirms that there is such a pot of gold. Where an object of desire fails to exist, its purported description need not, correspondingly, fail to exist. Even where something lacks not only being but possibility, its nonexistence does not generally carry over to its purported description. To assert the existence of something both square and not square is a serious and offensive thing to do; to attribute being to some inscription asserting such existence is something else and quite innocuous. Can we reconstrue the explanatory argument (i)–(iii) in such a way as to capitalise on the advantages of an inscriptional formulation?

Desiring true

Suppose we begin by transforming (i) into the following statement, converting the singular term naming the object of John's desire into a clause:

(I) John desires that John enter medical school.

Let us now construe this statement as:

(i') John desires-true some that-John-enters-medical-school inscription.

Desiring-true is here to be taken as a relation between people or other suitable agents on the one hand and inscriptions on the other, and does not require that the agent produce or even understand the inscription that he desires-true. To interpret this first premise of the explanatory argument as we have done can, moreover, be no more obscure than to construe it as relating John to some such entity as a hypothetical state of affairs named by the that-clause of the premise. For it is always possible to satisfy proponents of the latter construal by explaining the desiring-true of a given inscription as the desiring of that state-of-affairs which is named by it or by the proposition it expresses. The present interpretation, it should also be noted, treats the that-clause of (I) as a single predicate that applies to inscriptions peculiarly related to itself – inscriptions constituting rephrasals of its own sentential content, *including this very content as well.*[29] Thus (i') tells us that John desires-true something which is a that-John-enters-medical-school. Every such thing, further, is an inscription rephrasing the 'John enters medical school'-inscription that forms the content of the that-clause predicate of (i'). We shall, finally, make the assumption that to desire-true any given inscription denoted by a that-clause predicate is to desire-true every such inscription.

Believing true

Turning now to the second premise (ii) of our explanatory argument, let us first transform it into the following statement, eliminating reference to contingency in favour of a conditional of the type suggested by Ducasse:

(II) John believes that if John enters medical school then John chooses a pre-medical course.

Let us now interpret this statement as:

(ii') John believes-true some that-if-John-enters-medical-school-then-John-chooses-a-pre-medical-course inscription.

Believing-true[30] is, analogously with desiring-true, here taken to be a relation between agents and inscriptions, not implying that the agent produce or understand those inscriptions he believes-true. Such an interpretation is at least as clear as the usual construal of belief as a relation between a person and a proposition, for the believing-true of a given inscription is explicable to the intensionalist as the believing of the proposition expressed by it or by the statement it designates. The that-clause of (II) is to be treated also as a single unanalysable predicate applicable to rephrasals of its own sentential content, *inclusive of such content.*[31] Finally, we are to assume that to believe-true some inscription denoted by a given that-clause predicate is to believe-true every inscription denoted by that predicate.

Making true

Turning now to the third premise required to complete our explanatory argument, let us replace the original (iii) with the following:

(iii') Whenever anyone x desires-true any inscription v and believes-true another inscription z, such that z is the conditional of v and a third inscription w, then x makes-true w.

(Here the range of the variable 'w' is restricted to English sentence-inscriptions of the form 'x chooses. . . .')[32]

Making-true, similarly to desiring-true and believing-true, is a relation between agents and inscriptions and does not require the agent to produce or even to understand the inscriptions he makes-true. Making-true a given inscription can always, furthermore, be explained as performing the act described by the inscription or by its associated proposition; in the present case, John's making-true of a 'John chooses a pre-medical course'-inscription is explicable as his performing of the choice described, as his doing of the act. We shall, further, suppose that to make-true a given inscription is to make-true every rephrasal of it. It should especially be noted that, unlike inscriptions desired-true or believed-true, every inscription made-true is in fact true, together with all its rephrasals.

Reformulation of the teleological argument

How shall we now explain John's choice of a pre-medical course with the aid of our reformulated premises? From (i') we learn that John desires-true some inscription x to which the that-clause predicate of (i') is applicable. This inscription x, we know, is a rephrasal of the content of this predicate itself. (ii') conveys that John believes-true some inscription y denoted by the that-clause predicate of (ii'). Looking at this latter predicate of (ii'), we note that its sentential content is a conditional of which the antecedent is a *replica* of the content of the that-clause predicate of (i'). On the assumption that (i') and (ii') belong to the same language and contain no indicator terms, this antecedent must therefore also be a *rephrasal* of the that-clause content of (i'). If so, it must also be denoted by the that-clause predicate of (i'), that is, this antecedent must also be a that-John-enters-medical-school. Since, moreover, (i') says that John desires-true *some* inscription x which is a that-John-enters-medical-school, he must also desire-true the antecedent in question. This follows because of our assumption that to desire-true any inscription denoted by a given that-clause predicate is to desire-true every such inscription.

Now y, which John believes-true according to (ii'), is denoted by its that-clause predicate. But we know that the content of this predicate itself is also so denoted. Thus we may conclude that John believes-true this very content itself, i.e. the conditional inscription of (ii'). This conclusion, it will be recalled, is warranted by our assumption that to believe-true any inscription denoted by a given that-clause predicate is to believe-true every inscription denoted by that predicate.

The conditional inscription of (ii') believed-true by John is, however, the identical one of which we already know that John desires-true the antecedent. This conditional may thus be taken as the conditional z of (iii'), its antecedent may be taken as v in (iii'), and its consequent may then be taken as w of (iii'). The conclusion may now be drawn that John makes-true the consequent of the conditional in question as well as every one of its rephrasals. Since the consequent is made-true, it is in fact true, together with every one of its rephrasals. We may now, in particular, write down a replica of this consequent in the same language as our three premises. On the assumption that it contains no indicators, it will then also be a rephrasal of the consequent, hence true. It will thus represent the needed conclusion of our argument:

(iv') John chooses a pre-medical course.

The foregoing analysis is only exploratory. It remains to be seen to what extent the treatment of our example can be generalised. If it proves sufficiently general, it will, by avoiding reference to desired states-of-affairs, believed propositions, and performed acts, have managed also to escape the semantical difficulties plaguing the second strategy we have considered. It does not, however, provide an 'operational' definition of 'belief', 'desire', and 'performance' idioms; thus believing-true, desiring-true, and making-true are, for all that we have said, no clearer *in application* than these original idioms themselves. Their explication presents an important problem left unsolved by our above treatment, which has addressed itself to specific semantic difficulties.

SUMMARY

1. Two strategies for interpreting teleology have been examined: the self-regulation strategy, and the goal-idea strategy. Criticism has been directed explicitly against the moderate use of each strategy, i.e. the use of the first solely for non-purposive behaviour and the use of the second solely for purposive behaviour.

2. Two basic variants of the self-regulation strategy have been criticised as inadequate to account for non-purposive (let alone purposive) teleological behaviour. (A) The inadequacy of the 'negative feedback' variant has been discussed in connection with the difficulty of the missing goal-object. (B) The inadequacy of the 'plasticity' variant has been argued by reference to the difficulty of goal-failure and the difficulty of multiple goals. (C) These inadequacies are not, of course, taken as proving that the self-regulation strategy is never applicable, but only as showing that it is not sufficient, even for non-purposive teleological behaviour.

3. With respect to non-purposive cases, another interpretation of

teleology, in terms of learning, is suggested. (A) This interpretation construes the future-reference of teleological statements not as predictive nor as descriptive of objects in the current environment, but as pointing to the rôle of consequences in the genesis of the learned behaviour in question. (B) The learning interpretation seems to avoid the difficulties previously discussed, but the matter both requires and deserves further study. (C) The learning interpretation is, in any event, proposed as only *one* among several that are likely to be needed.

4. The second (i.e. goal-idea) strategy has been argued to involve logico-ontological difficulties in its natural formulation, which makes use of the notions of belief, desire, and performance. These difficulties hold even when the strategy is applied to clearly purposive cases.

5. An inscriptional proposal for avoiding these difficulties is explored. (A) This proposal makes use of the new notions 'desiring-true', 'believing-true', and 'making-true', holding between agents and inscriptions. (B) A sample teleological explanation is reformulated in terms of these notions, to show how the argument may be carried through in a way that is both materially adequate and free of the logico-ontological difficulties mentioned earlier. (C) This proposal, too, requires further study, and is, in any event, not intended to give an *explication* or operational analysis of belief, desire, and performance.

NOTES

For certain critical suggestions, I am indebted to Professors N. Chomsky, N. Goodman, and M. Mandelbaum, who should not, however, be assumed to share the views presented here.

1. A. Rosenblueth, N. Wiener, and J. Bigelow, 'Behaviour, Purpose and Teleology,' *Philosophy of Science* 10 (1943): 18ff.
2. Ibid., 23.
3. Ibid., 18.
4. Ibid., 19.
5. Ibid.
6. Ibid.
7. Ibid., 24.
8. Ibid., 19.
9. Ibid., 19-20, italics in original text.
10. Ibid., 19.
11. Ibid., 20.

12. R. Taylor, 'Purposeful and Non-Purposeful Behavior: A Rejoinder,' *Philosophy of Science* 17 (1950): 329.

13. Such inferences, invalid in cases where we have an apparent reference to a goal-object, cannot even be formulated naturally for purposive cases lacking such reference. The mystic striving to become more humble through spiritual training is not moving toward correlation with some object, guided by signals emanating from it.

14. Op. cit., 330.

15. R. S. Peters, 'Symposium: Motives and Causes,' *Arist. Soc. Suppl.* 26 (1952).

16. R. B. Braithwaite, *Scientific Explanation*, 1953, pp. 319ff.

17. Ibid., 325.

18. Ibid.

19. Quoted from E. S. Russell, *The Directiveness of Organic Activities*, p. 144, by Braithwaite, op. cit., 329.

20. Braithwaite, op. cit., 329.

21. Ibid., 330: I do not discuss at length the problem of determining the *beginnings* of goal-attaining chains, on this view. Strictly, according to the definition given, a cat's capture of its first mouse ends a single mouse-attaining chain that began with the birth of the cat.

22. Ibid., p. 334.

23. C. J. Ducasse, 'Explanation, Mechanism, and Teleology,' *Journal of Philosophy* 23 (1926). Reprinted in Feigl and Sellars, *Readings in Philosophical Analysis*, New York, 1949, pp. 543-44.

24. Ibid., p. 543. Addition in brackets mine; this seems quite in line with Ducasse's intent.

25. Ibid.

26. W. V. Quine, 'Notes on Existence and Necessity,' *Journal of Philosophy* 40 (1943): 113ff.; Chap. 8 of *From a Logical Point of View*, Cambridge, Mass., 1953; and 'Quantifiers and Propositional Attitudes,' *Journal of Philosophy* 53 (1956). 177ff.

27. W. V. Quine, *From a Logical Point of View*, 150ff.

28. I. Scheffler, 'On Synonymy and Indirect Discourse,' in Part 1 above.

29. See I. Scheffler, 'An Inscriptional Approach to Indirect Quotation,' in Part 1 above, for a fuller explanation of this general approach. The fact that the sentential content of the that-clause predicate is included in its application guarantees the existence of an appropriate inscription by the very existence of the stated premise.

30. This relation is suggested by W. V. Quine, 'Quantifiers and Propositional Attitudes,' p. 186. Quine uses believing-true as a relation between agents and

sentences, and thus expands it from a dyadic to a triadic relation, relativising it to language.

31. As with desiring-true, the existence of an appropriate inscription is guaranteed by the existence of the whole premise.

32. Strictly, other restrictions on this range would be needed to prevent awkward results. It might be best, for example, to restrict the vocabulary and syntax of such inscriptions in certain ways. But I am concentrating on the general idea here and so do not enter into details.

3. PROSPECTS OF A MODEST EMPIRICISM

THE HEART of modern empiricism has been its doctrine of empirical meaning, with its sharp line between the verifiable and the unverifiable and its rejection of non-analytic,[1] non-experiential statements as nonsense. This doctrine has, however, fallen on evil days. Increasing logical precision in philosophy has weakened rather than strengthened it, while advanced theoretical physics seems more and more a living counter-example, to be accommodated only by stretching the doctrine out of all shape. Once a proud polemical tool, the doctrine has thus come to be treated as a problem or a proposal. Yet the label "empiricist" continues to function as a philosophical symbol, and it is not sufficiently realized that the decline of the doctrine calls for new examination of the label. Indeed, it is not too much to claim that without such new examination, the idea of empiricism may be thought empty; or, shall we say, meaningless.

I want to address myself, then, to the task of such examination, offering first a brief review of the philosophic career of the doctrine and a critique of a recent revision, and going on to formulate a modified empiricist thesis, and to consider its basic problems and some general approaches to them. In Section I, I shall survey early attempts to state an empirical or verifiability criterion of cognitive significance and to indicate difficulties encountered and philosophic consequences. In Section II, I shall criticize the recent revision in terms of translatability. In Section III, a modified empiricism in terms of inclusion will be outlined and its consequences explored. Section IV will be devoted to a consideration of two outstanding problems faced by such empiricism: the problem of dispositional terms and the problem of theoretical or transcendental terms.

This paper appeared in the *Review of Metaphysics* 10 (March-June, 1957).

I. THE EMPIRICIST SEARCH FOR A VERIFIABILITY CRITERION OF COGNITIVE SIGNIFICANCE: EARLY ATTEMPTS AND DIFFICULTIES

1. Material adequacy.

Early formulations of the verifiability principle of meaning were intended to obviate the need for weighing the truth-claims of every purported doctrine or thesis, since some supposed theses were held clearly recognizable in advance as meaningless, i.e., neither true nor false. Thus, a basic criterion of adequacy governing formulations of the principle demanded that every sentence satisfying it be true or false and, conversely, that every true or false sentence satisfy the principle:

Criterion of (Material) Adequacy (I): $(S) ((S \text{ true} \lor S \text{ false}) \equiv S \text{ meaningful})$

C. A. (I) could clearly not have served as the wanted principle itself for it would not have enabled the bypassing of issues of truth or falsity in determining the meaningfulness of disputed theses. What was wanted, rather, was some *independent* criterion which would enable us to pick out the neither true nor false sentences, so that a point-by-point study of their claims to truth might be spared, and which would at the same time enable us to select for serious investigation just those sentences which are, as a matter of fact, true or false. In the Introduction to his 2nd edition of *Language, Truth, and Logic*,[2] Ayer states, for example, "I suggest that it is only if it is literally meaningful, in this sense, that a statement can properly be said to be either true or false."

2. Empiricist adequacy.

Not every principle satisfying C. A. (I) would have met *empiricist* demands, however. The desired reduction of the notion of meaningfulness had to be, for non-analytic sentences, a reduction to experience, i.e., in terms of predicates applying to observables or experienceables or their semantic parallels ("is an observation-predicate," "is an observation-sentence," etc.), and auxiliary semantical notions such as logical implication or deduction, and denotation or application. Of course, variant formulations seemingly ignored this restriction on their basic terms by employing solely notions of verification, confirmation, or prediction. But it will be no distortion if we take them in every case to refer to verification or confirmation by, or prediction of, observable occurrences or experiences. The very use of notions of verification and prediction (variously explained, partly in terms of logical deduction) serves to epitomize the core idea of early formulations, viz. that some deductive relationship with observation-sentences is the necessary and sufficient condition of *empirical* significance. Because "is a possible experience" and other such object-language terms are more obscure than some of their always-available semantic counterparts, it seems preferable to state our second criterion of adequacy semantically. Moreover, instead of referring to deductive relationships, we will use "is a deductive-relation term" to apply to any relative predicate definable (with the use of truth-functional and quantificational logic) by "logically implies." Non-semantical formulations of C. A. (II) will, however, readily suggest themselves to those who find them preferable.

>*Criterion of (Empiricist) Adequacy (II):* $(\exists x)\,(S)\,x$ is a deductive-relation term $\cdot$ (S meaningful $\equiv$ $\cdot$ S analytic $\vee$ $(\exists y)\,(y$ is an observation-sentence $\cdot$ x applies between S and $y.$))

As Ayer phrased it, "Let us call a proposition which records an actual or possible observation an experiential proposition. Then we may say that it is the mark of a genuine factual proposition, not that it should be equivalent to an experiential proposition, or any finite number of experiential propositions but simply that some experiential propositions can be deduced from it in conjunction with certain other premises without being deducible from those other premises alone."[3]

3. Non-universality.

Finally, since the whole point of the principle was to generalize the conditions under which it is otiose to seek to decide the truth-claims of putative theses, meaninglessness could not turn out vacuous under analysis, i.e.,

>*Criterion of Adequacy (Non-universality) (III):* $\sim (S)\,(S$ meaningful)

To quote Ayer again, "Our charge against the metaphysician is not that he attempts to employ the understanding in a field where it cannot profitably venture, but that he produces sentences which fail to conform to the conditions under which alone a sentence can be literally significant. Nor are we ourselves obliged to talk nonsense in order to show that all sentences of a certain type are necessarily devoid of literal significance. We need only formulate the criterion which enables us to test whether a sentence expresses a genuine proposition about a matter of fact, and then point out that the sentences under consideration fail to satisfy it."[4]

Joint fulfillment of these three criteria would, then, have served two major purposes: (a) to characterize generally every true-or-false sentence in terms of easily applicable predicates, enabling us thereby to eliminate, rather than try to decide, the truth-claims of every other sentence, and (b) to exhibit the domain of non-analytic significance as experiential in purport, to show, e.g., that (and how) every non-analytic truth-question relates to observation-questions, and that non-analytic questions which do not so relate lie outside the boundaries of inquiry. The philosophic traditions of Hume and of Kant were thus to be jointly vindicated and given rigorous logistic statement: Kant's insistence on the exclusion of some issues as beyond the limitations of human reason (C. A. III) was, in effect, to be fused with Hume's derivation of knowledge from experience (C. A. II).

4. Difficulties with completeness requirements.

The fate of early attempts to construct a definition of significance meeting these three criteria has been critically reviewed by Professor Hempel.[5] Very briefly, definitions requiring what he calls "complete verifiability in principle" or "complete falsifiability in principle" (by asking of every empirically meaningful S [in Hempel's reconstruction] that it or its denial, respectively, be non-analytic and be logically implied by some finite, consistent class of observation sentences) violate C. A. (I) (material adequacy) in both directions. For (a) each such definition judges as empirically meaningless either all universal or all existential statements, and mixed quantifications, in the face of the fact that we have empirical truths and falsehoods of each type, and (b) each such definition judges as meaningful either some (truth-functional) disjunctions or some conjunctions with nonsensical components, none of which compounds we regard as true or false. Furthermore, if C. A. (I) is to hold, then since the denial of every true or false sentence is itself true or false, denials of meaningful sentences must themselves be meaningful. But in ruling empirically meaningless either all universal or all existential statements, each definition we are now considering withholds this reciprocal property from either the corresponding (non-analytic) existential or universal denials which it, respectively, admits as significant, thus again violating C. A. (I). It may here be noted that a combining definition which proposed either-complete-verifiability-or-complete-falsifiability as a definiens for "empirically meaningful," thus admitting both purely universal and purely existential sentences, would still violate C. A. (I) in the left-to-right direction by ruling mixed quantifications all empirically meaningless, while continuing to violate it in the opposite direction by ruling significant both certain disjunctions and certain conjunctions with nonsensical components.

5. Difficulties with incompleteness requirements.

Definitions abandoning completeness requirements in the above sense have typically required instead, of every empirically significant S, that it be a component of some conjunction logically implying an observation-sentence not also implied by the other conjuncts alone. In essentially this form, verifiability principles have figured strongly not only in positivistic but also in pragmatist writings. However, as above described, they run afoul of C. A. (III), barring significance to no sentence, as Ayer himself admits in acknowledgment of a criticism by Berlin.[6] Thus, any sentence N can be conjoined with $N \supset O$ to imply O, where O is an observation-sentence, and does not follow from $N \supset O$ alone. Such definitions also violate C. A. (I) in granting significance to certain conjunctions with nonsensical components. The latter violation characterizes also every attempt to revise such definitions by restricting the domain of admissible other conjuncts, while for Ayer's particular revision in the second edition of his *Language, Truth, and Logic*, Church[7] proves that it virtually violates C. A. (III), (i.e., for every statement S, it renders either S or its negation significant), provided only that there are three observation-sentences such that none implies any other.

6. Philosophic consequences of above difficulties.

In the face of such developments, it seems impossible simply to reiterate early formulations such as we have been discussing. It is natural that the more central some early formulation has been for a given mode of philosophic practice, the more disastrous will an outright relinquishing of it appear and the more tenacious will be the search for *ad hoc* repairs. Goodman's[8] reminder that philosophic criticism of specific meaningless discourses does not depend on a *theory* of meaninglessness is thus quite to the point, as is the example of other modes of critical philosophy which have not been prevented from giving potent, piecemeal critique by their traditional lack of interest in stating a general theory underlying their methods. After all, one could not construct such a general theory unless the notions of empirical significance and meaninglessness had some clear predefinitional applications anyhow, while the very defects of the proposals we have discussed are relative to such predefinitional applications. But if particular philosophic critiques do not depend on a general theory of significance, neither can they serve in place of such a general theory. The moral of the recent developments sketched above seems to be this: Early formulations of the type discussed must be dropped, both as theoretical analyses of cognitive significance and as critical tools of the applied philosopher. Philosophic critiques can however employ other methods and tools, while the search for a general theory is an independently legitimate undertaking which is not proved quixotic by recent failures. What is, however, shown by the latter is that, construing empiricist requirements on a definition of significance after the manner of C. A. (II), we have so far not been able to avoid violating either C. A. (I) or C. A. (III), and have been able, moreover, to see why certain types of avoidance-proposals are foredoomed to failure, e.g., no ingenious restrictions on the other conjuncts, for theories of the Ayer type, can reasonably be expected to render meaningless all conjunctions with some meaningless components and also avoid virtual universality of significance.

While the possibility of a general definition of cognitive significance is, then, not in question, it may plausibly be doubted if an adequate definition is likely which takes seriously the empiricist requirement C. A. (II). If no alternative requirement can be given which is equally empiricist in spirit and also more promising, then we have a failure of empiricist philosophy, unless, of course, some flaw in the other two requirements can be shown. Professor Hempel[9] takes the first course, abandoning the attempt to satisfy C. A. (II) and proposing instead a translatability requirement.

II. THE TRANSLATABILITY CRITERION OF COGNITIVE SIGNIFICANCE: ITS RATIONALE AND DIFFICULTIES

7. The translatability formulation of empiricist adequacy.

Hempel's alternative idea is simply to specify a vocabulary containing logical terms plus a finite number of observation-predicates, and a (customary) syntax governing sentence-formation (and inference), and then to characterize cognitively significant sentences as those translatable into this restricted language. Control over vocabulary would eliminate from this so-called empiricist language all nonsensical predicates, while control over syntax would eliminate from it all nonsensical sentences formed exclusively out of logical signs and remaining predicates. Presumably this syntax would need to be rather complicated to exclude such sentences: consider "Galaxies divided by 7/8 are sad." Yet, there is theoretically no reason to suppose that it could not be constructed. Thus, given the exclusion from such a language (let us call it "E" for "empiricist") of elementary sentences presystematically deemed meaningless, the problem of compounds with such components disappears, while the customary logic incorporated in E guarantees sentencehood for all types of quantified statements and for denials of all sorts, thus obviating a basic difficulty with completeness proposals above discussed. At the same time, since not all sentences are translatable into E (i.e., go over into well-formed sentences in primitive notation in E, when suitably transformed in accord with appropriate definitions), universality is avoided. The idea is, then, to characterize cognitive significance generally in terms of a specific relation T (translatability) to a well-specified language meeting stated conditions, much as logical truths can be described as those bearing a complex relation R to listed logical constants, i.e., such that free reinterpretation of all their components other than these constants fails ever to produce a falsehood. Hempel's alternative to C. A. (II) may thus be written as:

Criterion of (Empiricist) Adequacy (IIA): (S) (S meaningful $\equiv$ $(\exists L)$ (L is an empiricist language $\cdot$ S translatable into L)).

Though, of course, a definition still needs to be constructed satisfying C. A. (IIA), i.e., explicitly characterizing some empiricist language, reasons are already apparent for thinking C. A. (IIA) more likely than its predecessor to be capable of joint fulfillment with C. A. (I) and C. A. (III). These reasons, summarized briefly in the preceding paragraph, hinge essentially on the fact that C. A. (IIA) requires a fixed relation to a language subject to our deliberate manufacture: controlling this language ourselves, we can by careful design eliminate objectionable predicates and syntax and avoid universality.

8. Difficulties with the translatability formulation.

Nevertheless, there is a subtle difficulty in C. A. (IIA) which is independent both of the fact that we can exercise total control over E, and of the problems involved in designing E so that it will be materially adequate.[10] The latter problems have been widely discussed (and will be here examined in subsequent paragraphs): they relate to dispositional and abstract theoretical terms. Our present difficulty has, however, gone unnoticed, though sufficient in itself to incapacitate C. A. (IIA). This difficulty relates to the notion of translatability.

Presumably, what is involved is *legitimate* translatability; the translation must not be based on merely nominal definitions. Were this not so, every sentence would turn out meaningful, since any purported term can be arbitrarily correlated by nominal definition with some primitive term or complex of terms within any system, provided formal requirements are met. Indeed, any whole sentence could, in this manner, be arbitrarily correlated with some systematic sentence in primitive notation, and since thus translatable, would be meaningful. To object that such arbitrary correlation would really be taking as definiendum not the disputed term or sentence at all, but only a completely different homonym, is to acknowledge precisely the difference between nominal definition of a term and definition which is controlled also by extra-formal requirements relating to the term's familiar meaning or extension; it is in effect to withhold the label "definition" from correlations which flout such requirements. Such merely terminological variance, however, leaves the basic point intact: translatability requires some definitional matching of systematic with presystematic meanings if any sentence is to be ruled meaningless under any formulation satisfying C. A. (IIA).

Now the precise nature of such matching is a matter of some controversy into which we need not enter, beyond noting that certain very strong requirements on it have been shown unsuitable by Goodman.[11] Let us take here the weakest requirements proposed, i.e., those stated by Goodman under the label "extensional isomorphism."[12] Detailed consideration of the latter is unnecessary for our purpose here. All we need to note is that a proposed definition meeting the minimal demands of extensional isomorphism must still enable truth-value-preserving translations of certain sentences which we are particularly interested in. Indeed, Goodman considers the latter condition, without further specification of sentences of particular interest, to be a criterion of adequacy for any satisfactory view of the definitional matching in question.[13] And while other proposals put forth even stronger conditions, this minimal condition is unquestioned.

Thus, for a sentence S to be legitimately translatable into a system Y, S must be transformable into a well-formed sentence S' in Y, when all suitable components of S are replaced by their primitive counterparts of Y in accord with definitions which are legitimate, i.e., which, at the very least, meet the demands of extensional isomorphism. And every such definition must be truth-value-preserving for at least some translated sentential contexts in which the definiendum figures presystematically.

Now, when a presystematic sentence is syntactically jumbled, as, e.g., "Galaxies divided by 7/8 are sad," (or, worse, the same words in reverse order) there may be legitimate definitions for all its component predicates in terms of the primitives of Y, and yet appropriate replacements in accord with such definitions throughout may fail to yield a well-formed sentence in Y. In such a case, our original sentence is not legitimately translatable into Y. Now if our sentence is not thus translatable into any *empiricist* system, it is clearly meaningless under C. A. (IIA). Thus, the ability of C. A. (IIA) to exclude at least some cases of meaninglessness, hinging on syntactic jumble, seems clear. Indeed, there is a sense in which our sample sentence (but not its reversal) *is* intuitively grammatical (or unjumbled) in form, so that the notion of syntactic jumble *as we have used it* relates not to intuitive grammaticalness but just to the possibility of excluding a sentence by reference to a deliberately restricted syntax, coupled with this sentence's independent obnoxiousness. If a sentence is not jumbled because not obnoxious, we would not want to exclude it, and could easily so frame our empiricist syntax as not to exclude it. But are there sentences which, though obnoxious, are not jumbled, i.e., excludable by the method just outlined?

Consider the case of nonsense words which look like predicates, e.g., "is a glame," "is froomy," and the case of words like "is spiritually ectoplasmic," "is an entelechy," "is identical with the Absolute." The former we unhesitatingly declare meaningless and rule every given sentential context containing any to be neither true nor false. The latter, with many sentential contexts seriously defended, we wish nevertheless to declare cognitively meaningless, having no context with truth-value. For the purposes of our present discussion it makes no difference if we disagree over the precise composition of the list of terms of which this is true. It is enough that we agree on its being true of some nonsense words, and it is more than enough, as well as quite likely, that we also believe *some* terms to have no sentential contexts with truth-value though having currency at some time for some member of the language community. Consider now a sentence containing such a term, e.g., "Some persons are spiritually ectoplasmic." Can we validly rule this sentence meaningless under C. A. (IIA)?

To do this, we need to establish its non-translatability into every empiricist system E, via legitimate definitions by means of E's primitives. These primitives we may be sure, do not include "is spiritually ectoplasmic." We have, however, to decide not simple inclusion, but legitimate definability. This involves, as we have seen, some presystematic context of the definiendum with some truth-value which is preserved upon replacement by its proposed definiens. Suppose that "some – s are . . . " in the sentence above is acceptable as a correlate of some basic locution in E, e.g., the existential quantifier, parentheses, and the dot of conjunction, while "person" is legitimately definable by some extralogical primitives of E, as tested by reference to truth-value-preserving translation of some of its contexts. We already at this point note a restriction in applying C. A. (IIA). For in testing the meaninglessness of our present sample sentence, we refer to the truth-value, hence meaningfulness, of some other sentence, namely some context of "person." If circularity and infinite regress are both to be avoided, some sentences must be judged meaningful, then, independently of our proposed test. Given some listing of such sentences, we may subsequently test some others without infinite regress and, moreover, without begging the question, since even if all of a sentence's extra-logical terms are legitimately definable by reference to our independent listing, the sentence may prove meaningless as a whole through lack of well-formedness in E, when all legitimate replacements are made.

Getting back to our sample sentence, we have left only "is spiritually ectoplasmic," and we must decide whether or not it is legitimately definable by E's primitives. Now to say whether any of its contexts is translatable with preservation of truth-value, we must be able to say whether any such context has truth-value, e.g., by reference to our independent listing. But the peculiarity of "is spiritually ectoplasmic" is precisely that none of its presystematic contexts has truth-value at all. And to determine this is ipso facto to determine that the sample sentence before us is meaningless through containing such a term. Thus to rule our sample sentence meaningless on grounds of untranslatability requires us to know in advance that it is meaningless.

Suppose someone were to suggest that it is possible that the sentence before us is meaningful even though no context of "is spiritually ectoplasmic" appears on our independent list. In such an eventuality, our test would provide no method for determining this, since all determinations of legitimate translatability have to be carried out by reference to this list. Thus, either the translatability test is incapable of deciding the significance of our sample sentence or else it is completely superfluous in ruling it meaningless, since such ruling requires as a precondition knowledge of the meaninglessness of this very sentence, as reflected in the composition of our list. While inclusion in this list of a particular predicate P in some context c does not determine the significance of some other of its contexts c', c'', etc., such determination to be made by translatability, exclusion of any purported predicate P' from the list either leaves every P'-context undecided as to truth-value, or determines its meaninglessness without translation. Thus, for the whole class of sentences containing nonsensical would-be predicates, the translatability test is either insufficient or superfluous as a criterion for exclusion. Everything rests on the antecedent composition of our list.

Thus, the use of C. A. (IIA) as a test of significance for presystematic sentences involves two difficulties: (a) it requires, over and above the explicit construction of an empiricist language E, the independent specification of truth-value (and hence significance) for some *presystematic* context of every *bona fide presystematic* predicate, to serve as a test of legitimate definability, and while such specification will not automatically decide the significance of sentences containing as predicates *only* specified items, (b) such specification, if adequate, must independently exclude as meaningless all sentences containing nonsensical, would-be predicates, which it was the very purpose of our criterion to exclude; application of the test to these sentences requires that the job be already completed.

III. INCLUSION AS A SUFFICIENT CONDITION OF COGNITIVE SIGNIFICANCE: TOWARD A MORE MODEST EMPIRICISM

9. Inclusion in some empiricist language.

The failure of C. A. (IIA) is not a failure in its basic conception, i.e., to characterize the domain of significance by reference to a fixed relationship to some given language. It is rather a consequence of the use of translatability as the fixed relation in question. This use leads us to say that S is meaningful only if every suitable component of S has some context with truth-value which is preserved upon translation into E, and hence that S is meaningless, i.e., neither true nor false (by C. A. (I)) if either every context of some component has its truth-value altered by translation into E, or if some component lacks any context with truth-value at all. But of course, in the case of the first alternative, if E is to serve as a criterion of significance and fails to preserve a truth-value (hence a clearly significant unit), we would declare E inadequate. Much more important, however, is the case of the second alternative, since if true, it is tantamount to the meaninglessness of S itself and renders the test pointless in such an instance.[14]

The basic conception of C. A. (IIA) is preserved if we employ *inclusion* in place of *translatability* as the relation in question, i.e., if we characterize significant sentences as those belonging to empiricist languages. Indeed, the difference between the two relations in this connection has apparently been unnoticed and the latter seems to have been at least partly in the minds of proponents of translatability. Thus, e.g., Professor Hempel, in expounding C. A. (IIA), says, "In effect, therefore, the translatability criterion proposes to characterize the cognitively meaningful sentences by the vocabulary out of which they may be constructed, and by the syntactical principles governing their construction."[15] Actually, of course, this characterization of vocabulary and syntax relates to the sentences of E, while sentences translatable into E may differ in either vocabulary or syntax, or both. Consider now a criterion in terms of inclusion or well-formedness in some empiricist language.

Criterion of (Empiricist) Adequacy (IIB): (S) (S meaningful $\equiv$ ($\exists L$) (L is an empiricist language $\cdot$ S is well-formed in L))

This criterion recalls the usual way in which systems are characterized, i.e., as comprising all and only sentences formed in accordance with stated rules out of a fixed vocabulary. In its avoidance of the notion of translatability, C. A. (IIB) is, moreover, an improvement over C. A. (IIA).

10. Consequences of formulation in terms of inclusion.

Owing to abandonment of the notion of translatability, however, a serious restriction in generality results, which needs to be considered at this point. Once having constructed some empiricist language E, the idea behind C. A. (IIA) was to assert translatability into E as a necessary and sufficient condition of cognitive significance for all sentences whatever. E being a clearly empiricist language, such a definition would have implied, hence fulfilled, C. A. (IIA). Not to have constructed E, and to have rested content with C. A. (IIA) itself as a *definition*, would have meant resting the empiricist theory of meaning on the notion "is an empiricist language" and this would have been rather unsatisfactory. Although we might recognize a clear case to which this predicate applies, e.g., E itself, the predicate itself is generally vague and likely to be highly variable in use. As an element in a criterion of adequacy guiding us to the construction of E, its function is rather narrow and it serves well in this way, but such construction is needed to render the theory in precise and unambiguous form, much as the general predicate "is a logical constant" needs to be supplanted by a finite list before theses concerning logic can take on precision and be discussed with rigor. White[16] has made this point quite generally, and indeed, thinks such a process of finitizing philosophical theses is always salutary.

Having given up C. A. (IIA) in favor of C. A. (IIB), however, we must face the fact that a similar course of finitization (i.e., constructing some E) yields no general significance-principle. For while *translatability into E* applies to some sentences outside E, *inclusion in E* applies to no sentence outside E. Hence, even if E is constructed to meet most liberal requirements, it will hardly suffice to include *all* significant sentences of *all* languages. And if this is the case, then inclusion in E will at best be a sufficient, but not a necessary, condition of cognitive significance generally.

If we are willing to rest content with C. A. (IIB) as an ultimate *definition* of significance itself, we, of course, do not face this problem and we salvage our coveted generality. But the price is extravagant, involving acceptance of the vague predicate "is an empiricist language" as our ultimate tool of analysis. Further refinement of this predicate by provision of a general characterization of empiricist languages would not represent substantial progress, since such characterization would presumably rest, in part, on the use of "is an observation-predicate," rather than a finite listing of acceptable predicates. But this general term is as loose as "is an empiricist language," and requires finitization no less than its predecessor. We might think, finally, of listing all observation-predicates of all languages as included in E's vocabulary. But such a pooling of predicates, aside from its practical impossibility, could theoretically be carried out presently only for languages already known. In such restricted form, this pooling would be patently unfair to as yet unknown languages, while a method for generally characterizing observation-predicates for these languages is precisely what we lack; if we had such a method, pooling would be unnecessary.

11. A restricted problem for empiricism.

Thus, adopting C. A. (IIB) as our (empiricist) criterion of adequacy, we arrive (through construction of E) at a sufficient but not a necessary condition of cognitive significance. Again, this result does not show that a general definition of cognitive significance is impossible: if we did not mind taking "is an empiricist language" as primitive, we could use C. A. (IIB), or, if we could construct an adequate psychological theory of observationality for predicates, we might refine C. A. (IIB) acceptably. The point is, rather, that the finitizing of C. A. (IIB) by construction of a particular language E yields at best only a sufficient but not a necessary condition of cognitive significance. Having rejected C. A. (II) and C. A. (IIA), and having decided on finitization for the reasons mentioned, we have at best a sufficient condition of cognitive significance, unless, indeed, an alternative empiricist requirement is forthcoming to replace C. A. (IIB).

In constructing E, we are, then, committed to the significance of its sentences; on the other hand, we do not thereby rule all excluded sentences meaningless. Furthermore, though mere exclusion is not tantamount to meaninglessness, we may be guided in the process of construction by the desire to exclude certain sentences independently deemed meaningless. Given the possibility of constructing an observational E purified of meaninglessness as independently recognized, but not including all significant sentences, can we make it commodious enough to house all our respectable beliefs in scientific domains? This problem, which naturally grows out of the initial search for a general empiricist significance-principle, is nevertheless distinct, once we give up C. A. (IIA). For, even if we now make E commodious enough, we cannot dismiss excluded sentences as meaningless without independent judgment, since it at best provides only a sufficient condition of significance. Nevertheless, though less ambitious in aim, this heir to the earlier problem seems certainly important and has perhaps the advantage of proving more manageable. This new problem, once more, is not to provide a general rule for excluding meaningless sentences, but to specify the construction of some empiricist or observational language E containing no meaningless sentences, and adequate for the formulation of all our sincere beliefs of specified types.

IV. PROBLEMS OF A MODEST EMPIRICISM: DISPOSITIONAL AND TRANSCENDENTAL TERMS

12. The problem of disposition terms.

We may begin by characterizing the basic logical apparatus of E as comprising joint or alternative denial, universal quantification, and overlapping[17] or membership, with suitable rules for sentence formation. Next, we may provide a list of (observational) predicates which are clearly acceptable, i.e., for which we recognize clear applications. Our rules will guarantee the inclusion of all modes of quantification, will guarantee the sentencehood of all denials of sentences in E, while our vocabulary and rules

Prospects of a Modest Empiricism · 135

will exclude sentences clearly recognizable as meaningless presystematically. Most recent discussions have followed Carnap[18] in treating so-called dispositional predicates as a special problem. Hempel's treatment,[19] for instance, holds that the language just described, with all extra-logical predicates clearly observational, is inadequate, since it leaves no room for dispositional terms not definable within it. Thus, for example, "is magnetic," "is at temperature 100° C.," "is irritable," etc., though predicable of observable entities, and important for everyday and scientific discourse, are not definable by those observation-predicates describing usual test-operations determining their applicability. Why this is so may be seen from the following examples, incorporating Carnap's well-known arguments. Suppose we wish to define "x weighs 5 lbs. at time t" by "If x is placed on scale S at t, S's pointer moves to '5' at t." This proposed definiens, formulated as a material conditional, is equivalent to "x is not placed on scale S at t, or S's pointer moves to '5' at t." But clearly the latter (and hence, by equivalence, the former) open sentence is true of every object o, taken as value of "x," such that o is not placed on scale S at t. Hence this definition would assign a weight of 5 lbs. at t to, among other things, Mt. Vesuvius, the Eiffel tower, and every ant not on S at t.

To replace the material conditional of the definiens by a subjunctive locution, i.e., "If x were placed on scale S at t, S's pointer would move to '5' at t" is indeed intuitively more adequate, but such a locution is itself not included in the repertoire of E as thus far described, while its reducibility to this repertoire has not been shown. Thus if dispositional terms cannot be directly defined in E, their subjunctive interpretation is simply irrelevant, at present, to their reduction to E.

Carnap's notion of reduction-sentences has been widely acclaimed as a solution to the problem of introducing disposition terms into E. Illustrating this notion by reference to our present example, we would replace our whole definitional equivalence by the (bilateral) reduction-sentence, "If x is placed on scale S at t, then x weighs 5 lbs. at t if and only if S's pointer moves to '5' at t." The difficulty originally encountered by our first proposed definition is here successfully avoided. For, though the whole reduction-sentence is indeed true of every object o not placed on S at t, e.g., Mt. Vesuvius, we cannot therefrom infer that every such o weighs 5 lbs. at t.

Certain difficulties, however, soon become apparent. Let "T" stand in place of some predicate of E applying to x when and only when x fulfills certain test conditions, let "R" stand in place of another predicate of E applicable to x when and only when x exhibits a specified reaction, and let "D" stand in place of the dispositional term to be introduced. It is clear that in:

$$(R1)\ Tx \supset (Dx \equiv Rx)$$

"Tx" cannot be universally false, and that in:

$$(R2)\ Tx \supset (Rx \supset Dx)$$

"$Tx \cdot Rx$" must be true of some x, for [assuming that in each case (R1) or (R2) is offered as the sole reduction-sentence for "D"] if this were not so, the application of "D" would remain undetermined for every x. But nevertheless, (R1) or (R2) can be fulfilled trivially by some would-be predicate otherwise objectionable.

E.g., "is spiritually ectoplasmic" will presumably not be reducible into E because applying to no x, but "is a paper clip and is in desk d at t or is not a paper clip and is spiritually ectoplasmic" is reducible by (R1) or (R2) if we put "is a paper clip" in place of "T" and "is in desk d at t" in place of "R." From the point of view of someone who wants to use "is spiritually ectoplasmic" freely in purported description of certain human beings, the reducible would-be predicate will serve equally well and hardly represents a concession. Moreover, though eliminable in one trivial context, it is ineliminable in all other contexts, notably those in which such a person was anxious to use "is spiritually ectoplasmic" in the first place, and for which we deemed such use objectionable, i.e., in application to some non-paper-clips.

Of course, introduction via (R1) or (R2) does not positively specify a use for our objectionable term in these wider contexts either. But then, if we care only about the narrower use which it does determine, we are also able to *define* such use, in effect replacing the objectionable "D" by "is a paper clip and is in desk d at t," a point recently made by Goodman.[20] To replace such definition by reduction-sentences in the interests of the wider use of dispositional terms like "is magnetic" opens the door also to objectionable terms of the kind above exemplified, and amounts then to their bald acceptance as primitives. If, moreover, our claim is that for specific terms like "is magnetic" we do, as a matter of common or scientific procedure, have a use wider than that specifiable by such other predicates as are included in E, then we may add such terms to our list of E's primitives. Such piecemeal additions would not simultaneously admit objectionable terms, as would wholesale sanction of the method of reduction sentences.

In support of the above treatment of disposition terms, the following considerations may be adduced:

(a) We have noted the looseness of the general term "is an observational predicate" and the need for supplanting its use by a finite listing in any rigorous treatment. Included in such a listing would be sufficiently clear terms with relatively determinate applications to observable entities. Now it is worth noting that the disposition terms under discussion are all predicates of our initially chosen observable entities, i.e., applicable to nothing outside the range of the variables already specified for E as suitable for its initial, observational predicates. Thus, if the latter are characterized by relatively determinate application to entities within this range, the same may hold for so-called dispositional predicates, especially in view of the fact that determinateness of application is a continuous matter, a question of degree. Furthermore, unless such determinateness did characterize some dispositional terms in contexts wider than those represented by customarily stated test-conditions, we could hardly charge E with inadequacy for omitting them.[21] There is, thus, no theoretical ground for denying to *every* dispositional term the status of observational predicate.

(b) It is also worth remembering that dispositionality is not absolute. Most physical examples recently discussed, e.g., "is soluble at t," "is magnetic at t," are contrasted with ostensible non-dispositional terms, e.g., "dissolves at t," "is placed near iron filings at t," which are themselves physical-object terms. But the latter are often treated as in effect dispositional by phenomenalistically-minded thinkers: consider, for example, Mills' doctrine[22] of matter as permanent possibility of sensation, or Lewis' terminating judgments as expressing the content of objective beliefs.[23] The converse is, also, not unthinkable. If no term is dispositional in an absolute sense, we need not balk at taking some terms customarily labelled "dispositional" as observational predicates in E.

(c) Much of the puzzlement over dispositional terms seems to be a product of platonist semantics, at least in part. Predicates are taken to designate properties, in addition to denoting each entity to which they apply. Properties designated by non-dispositional terms are then said to be themselves observable, while those designated by dispositional terms are said to be not observable, or, at least, not observable to the same degree or in the same sense. To drop platonist semantics is thus to dispose of part of the puzzlement over dispositional terms. For, left only with denotation, we must admit that dispositional as well as nondispositional terms apply to observable entities equally, and without sharp distinction as to determinateness or vagueness. No further question about the relative observability of *properties* remains.

Even had we kept our platonist semantics, it would have been hard to explicate the notion of observability of properties in such a way as to render dispositional properties clearly unobservable in some relevant sense. For, to take just one point, complex properties, each component of which is observable, are presumably to be considered themselves observable. But then we cannot by simple inspection rule out the contingency that a given dispositional property may turn out equivalent to some such complex property, and hence also observable. This and other difficulties arise in analogous forms if, having dropped platonism, we attempt to define some notion of observability or descriptiveness which shall distinguish between certain predicates (as applicable to things in virtue of certain specific sensory qualities) and others, though all apply to observable entities within the same range. This attempt is often made not for the purpose of distinguishing dispositional from non-dispositional predicates, but rather in order to distinguish ethical from non-ethical terms, in which case it seems to me clearly unworkable.[24]

(d) It has been noted under (b) above, that dispositionality is not absolute, but depends on the system chosen. It is now important to see that even within a given system, a predicate P may be dispositional with respect to one predicate or set of predicates Q and not dispositional with respect to some other W, on the basis of which it is fully definable. There is then no point in taking P as ostensibly saying something about possibilities or potentialities quite generally. Thus, we may denote as "played" every record which, at any time, is actually put on the turntable of some record player of familiar type and, with the needle in position, produces recognized music or speech. Aside from these, we may denote as "playable" also certain other objects, e.g., records accidentally shattered before ever reaching the turntable. Asked to explain the latter term, we should naturally try to do so by reference to the former via recourse to notions such as capability, possibility or potentiality, to the use of the subjunctive, or to devices like reduction-sentences. And each of the latter courses seems to point to a queerness inherent in the predicate "is playable"; it talks about something else than what is actually the case in our world. Consider, however, that everything in the above sense playable is a record of music or speech, meeting rather definite specifications as to form, shape, and history, and conversely, that every such record is playable. This means that "is playable" is definable in terms of those predicates by means of which such specifications are stated, without recourse to possibility or the subjunctive. Assuming that such predicates are among the primitive observational terms of a given system S, we now have a situation in which one predicate ("is playable") is dispositional with respect to another ("is played") in S, but is fully definable and non-dispositional with respect to certain others, also in S. The upshot is that, even for a given system, where a predicate P seems to be dispositional in relation to some other predicate Q within the system, we cannot generally attribute some special reference to possibility to the predicate P as such, even as it figures within the given system.

(e) The previous four subsections are all concerned with breaking down the customary sharp dichotomy between dispositional and non-dispositional terms and with pointing out that any predicate having a sufficiently determinate application to chosen observable entities may qual-

Prospects of a Modest Empiricism · 139

ify as a clearly observational predicate, and may even turn out definable by other specified observational terms. In line with this general purpose, it was pointed out that dispositional as well as non-dispositional predicates may apply to observable entities within the same range, as customarily construed, that giving up platonist semantics with its attendant obscurities, we have no clearer way of distinguishing degrees of observationality or descriptiveness for terms with equal applicability to observable entities; and that the attribution of dispositionality or non-dispositionality to any predicate P varies with the chosen system and with the predicates taken as standards. The conclusion as regards introducing so-called dispositional terms into our observational system E seems to be that each term must be judged on its merits, that any such term might be sufficiently clear to qualify as an observational primitive or get itself defined properly, and that if such a term is really needed in E for adequacy it will indeed be of this type.

Is there then no specific problem associated with disposition terms? Such a judgment would be wrong. There is an important problem here, but it is independent of the one with which we have been occupied above, namely, to see whether it is possible to construct an observational E, free from meaningless sentences and adequate for expressing our beliefs in specified domains, e.g., the sciences. This other problem is to define the *relationship between* dispositional terms and what are customarily taken as their respective non-dispositional counterparts in specified contexts. More exactly, it is to define the semantic relative term "is the dispositional counterpart of," or, put otherwise, to define a dispositional operator, say "-ible," attachable to predicates so as to form their dispositional predicate counterparts. Goodman has recently discussed this as "the problem of characterizing a relation such that if the initial manifest predicate 'Q' stands in this relation to another manifest predicate or conjunction of manifest predicates 'A', then 'A' may be equated with the dispositional counterpart – 'Q-able' or 'Q_D' – of the predicate 'Q' " (p. 48).

That this general problem is independent of any given decision on a particular dispositional term is pointed out by Goodman as follows, "Observe first that solution of the general problem will not automatically provide us with a definition for each dispositional predicate; we shall need additional special knowledge in order to find the auxiliary predicate that satisfies the general formula – i.e., that is related in the requisite way to the initial manifest predicate. But on the other hand, discovery of a suitable definition for a given dispositional predicate need not in all cases wait upon solution of the general problem. If luck or abundant special information turns up a manifest predicate 'P' that we are confident coincides in its application with 'flexible,' we can use 'P' as definiens for 'flexible' without inquiring further about the nature of its connection with 'flexes' " (F.F.F., pp. 48-49). It has here been further argued that the decision to regard a term as dispositional varies, and that even where it is positive, it is no bar to observationality in any important sense. In particular, where application of a given term is fairly determinate in contexts wider than those represented by customarily stated test-situations (and hence where definition of a narrower term fails to do our de facto usage justice) no general reason can be given against taking such a term as observational, and pending possible definition in the course of a general attempt at economy, including it as primitive. Since such terms with wide determinate application are the only dispositional terms whose omission might seriously mar E's adequacy, our initial problem can be affirmatively answered for crucial *dispositional* terms: if we add every such term required for E's adequacy to its primitives, we do not in general decrease the observational character of E. This does not mean we should not try to define these terms anyway, but we should try to define as many terms as possible *generally,* to reduce our stock of primitive notions. Nor does this mean that we dispense with the *general problem* of dispositions earlier mentioned, since it is independent. Nor, finally, does this mean that every term *customarily taken* as dispositional will have such a determinate application as to be judged sufficient; for many it will no doubt be advisable to define narrower notions, as Goodman suggests; but these terms are precisely those whose omission represents no threat to E's adequacy, since they have no well-determined, prior uses in wider contexts. It is worth noting that Goodman's example of the possibility of defining "flexible" implies a prior, well-determined application which may control our definition, and hence the correlative possibility of taking this term as observational. Unless, moreover, we were aware of such previous application prior to our attempt at definition, we should have no way of guiding ourselves in our attempt, nor any reason for considering such attempt worthwhile.

13. The problem of theoretical or transcendental terms.

If disposition terms required for descriptive adequacy may be accommodated in E without marring its observational character, as we have above argued, we still face a major obstacle to E's adequacy, i.e., the case of so-called abstract, theoretical, or transcendental terms. These terms, unlike dispositional predicates, do not generally purport to apply to entities within the range of application of our clearly observational terms. They are typically non-observational and beyond the reach even of reduction sentences. It is generally claimed that they are required not because of their usefulness in expressing available observational evidence, but rather because, by their introduction in the context of certain developed theories, comprehensive relationships become expressible in desirable ways on the observable level.

Such theories seem to commit us to a new range of entities as values of

Prospects of a Modest Empiricism · 141

variables to which their transcendental terms may be attached in existentially quantified statements, e.g., "There is an electron," "Something is a positron." For the entities here required are not such as are qualifiable by our hitherto accepted observational predicates. We have, for example, no clearly true sentence such as "x is an electron and x is red," or "x is an electron and x is non-red," for any value of "x." Historically, this is perhaps related to the distinction between primary and secondary qualities, a distinction which, in some form, it is dangerous to overlook in the interpretation of modern scientific theories. Thus, it is by now well known that confusion results both from the popularization of advanced theories through pictorial description in the common language of observation, and from the ascription of exclusive reality to the entities presupposed by such of these theories as we deem true. But to maintain the relevant distinction among ranges seems clearly to mean the abandonment of even our modest, revised empiricism. For the predicates appropriate to one of these ranges are non-observational, and to admit the necessity of theories couched in such terms in order for E to be adequate is to admit that no adequate, purely observational E exists which is even a sufficient condition of cognitive significance. (Thus those who insist on some such distinction by saying that "theoretical entities" have only such properties as are attributed to them by their respective theoretical contexts are, if they let the matter rest here, abandoning even our modest version of empiricism.) Furthermore, aside from empiricism, the perpetuation of a distinction among ranges seems to have generally puzzling aspects, which have troubled philosophers of science recurrently: (a) If adequacy requires that clearly non-observational terms be eligible for admission into the language of science, then, since these include terms ordinarily deemed meaningless, *what* are we believing in committing ourselves to science? (b) If science explains by providing *true* premises from which the relevant problematic data may be derived, how can theories that are *meaningless* in the ordinary sense be said to explain? In short, *even if we are not interested in relating cognitive significance to observationality,* but are concerned with constructing some weakest language adequate for formulating some specified segment of our scientific beliefs, we may be troubled to find ourselves explicitly allowing clearly meaningless units to be built into our language structure, in some ordinary sense of "meaningless."

14. Pragmatism.

The qualification embodied in the final phrase of the last sentence is a clue to one widespread attempt to cope with the troubles discussed: the view which takes the *system* to be the unit of significance, and adopts a wider notion of meaningfulness in deference to scientific practice. This view I shall label 'pragmatism.' The alternative reaction, which I shall discuss in later sections, I call 'fictionalism.'[25] Both views may be considered independently of the issue of empiricism, it seems to me, in view of the final set of remarks in the previous section. For, however we delimit the terms initially admissible (whether by reference to observation or not), we face the problem of interpreting all the others which seem to be required in increasing numbers with the theoretical development of science. Nevertheless, a consideration of this problem in abstraction from empiricism will bear rather directly on it as embodied in our revised form: pragmatism will negate this empiricism, while fictionalism will render it at least possible.

Pragmatism, then (in my terminology), accepts as fact that no initial listing of admissible terms is sufficient for the formulation of our scientific beliefs, and that the admission of any term is conceivable (hence legitimate) on the grounds of its utility in prediction and theoretical simplification. It admits, furthermore, that some such terms are meaningless, *in one usual sense*. But here it takes the bull by the horns, claiming that this sense is irrelevant for analyzing our scientific beliefs and practices. For any sense of "meaningless" which renders what is predictively useful meaningless is inadequate for philosophy of science, however relevant it may be in other contexts. If science finds, e.g., transcendental theories fruitful within whole systematic contexts, our notions of cognitive significance must reflect this fact. If science introduces terms not only by reference to prescientific usage or explicit test-methods, but within the network of whole theoretical frameworks justified by their predictive utility in subsequent inquiry, this is a *bona fide* fact about cognitive significance, not a problem. We must, accordingly, for the pragmatists, admit the *significance* of whole systems with unrestricted vocabularies, provided that they are, at some points, functionally tied to our initially specified language. Since, however, this proviso excludes no term at all, and virtually no system at all (every system meets this requirement by addition of one conjunct in the initial language, and every term is part of some system meeting this requirement), pragmatism supplements it by stressing some simplicity factor which presumably is to eliminate certain systems, but which is not to be so stringent as to eliminate every system which overflows the bounds of the initially specified language. In no account is the treatment of simplicity very precise, but in some accounts it is intended as a matter of degree so that cognitive significance is broadened correspondingly. It is, further, not very clear how considerations of simplicity are to be applied in determination of *confirmedness* or *truth* as distinct from *significance*. Nevertheless, the pragmatist reaction to the problem of interpreting transcendental terms and their theoretical contexts is clearly to accept their fruitfulness and to defend, in consequence, a broader notion of significance applicable to whole systems. This systematic emphasis is supported also by reference to well-known analyses of testing which show the theoretical revisability of every segment of a system when any segment is ostensibly under review. A corollary of this pragmatist treatment is its insistence that questions of ontology are scientific questions, since it takes the range of significant ontological assertion to be solely a function of scientific utility in practice and denies all independent language restrictions based on intuitive clarity or observationality. And rejecting such independent restrictions, it solves the two initial difficulties noted above (at the end of section 13) by (a) denying that we believe meaningless assertions in committing ourselves to science and by (b) affirming the possible truth and explanatory power of transcendental theories.

To what extent is the pragmatist position in favor of a broader notion of significance positively supported by the arguments it presents? Its strong point is obviously its congruence with the *de facto* scientific use of transcendental theories and with the interdependence of parts of a scientific system undergoing test. These facts are, however, not in themselves *conclusive* evidence for significance, inasmuch as many kinds of things are used in science with no implication of cognitive significance, i.e., truth-or-falsity; and many things are interdependent under scientific test without our feeling that they are therefore included within the cognitive system of our assertions. Clearly "is useful," "is fruitful," "is subject to modification under test," etc., are applicable also to non-linguistic entities, e.g., telescopes and electronic computers. On the other hand, even linguistic units judged useful and controllable via empirical test may conceivably be construed as nonsignificant machinery, and such construction is not definitely ruled out by pragmatist arguments. This, even if we accept pragmatism's positive grounds, we *need* not broaden our original notion of literal significance. And it further follows that our revised empiricism is not refuted by pragmatism.

15. Pragmatism and fictionalism.

But if not refuted, our empiricism remains beset with the problem of interpreting transcendental terms and theories. If pragmatism's positive grounds do not, that is, *establish* the literal significance of transcendental theories, it is not thereby demonstrated that they are eliminable or otherwise interpretable as nonbeliefs, i.e., mere instruments. Any view which takes them to be either I call 'fictionalism.' Clearly if fictionalism can show how transcendental terms are eliminable from our corpus of scientific beliefs, it will have removed transcendental theories from the domain of beliefs which need to be encompassed in E, and it will have destroyed a major obstacle to our revised empiricism. Short of showing eliminability, if fictionalism can plausibly construe transcendental theories as mere machinery without literal meaning, it will avoid the need for expressing such theories in E, and again make way for our revised empiricism.

16. Instrumentalistic fictionalism.

Perhaps the easiest, and by far the most popular type of fictionalism is one which simply disavows the belief-character of transcendental theories without claiming their eliminability from scientific discourse. Indeed such a fictionalism often goes with a positive indifference to the question of their eliminability, or even champions their ineliminability; we might aptly label this type 'instrumentalism,' and note in passing that some writers have vacillated between pragmatism and instrumentalism (in our present terminology) or have confused the two. Instrumentalistic fictionalism, then, holds that some scientific theories are not significant, but that they are moreover not intended as formulations of belief or as truths, being employed simply as mechanical devices for coordinating or generating *bona fide* assertions. Hence, again, transcendental theories are said to pose no problem; since they do not represent *beliefs,* we need not worry about including them within any deliberate statement of our beliefs in some restricted language. Our problem, it will be recalled, was that clearly meaningless terms seem required for adequate expression of our scientific beliefs. Whereas pragmatism's answer is to deny that any terms usefully employed in science are meaningless in the relevant sense, fictionalism's answer is to deny that our objectionable terms are required for expression of *beliefs,* though they may be otherwise required. And our instrumentalistic variant supports this denial not by showing how to eliminate such terms from scientific language, but rather by stipulating how 'belief' is to be understood. Correspondingly, certain further stipulations are generally accepted as corollaries, e.g., that transcendental theories be said to *hold* or *fail* rather than to be *true* or *false,* that they are adopted or abandoned rather than believed or denied, etc. Thus instrumentalism takes care of the difficulties mentioned (a) by insisting that we do not strictly *believe* but *hold* or *employ* some statements in science and (b) by generalizing the concept of explanation to allow such held theories to serve as explanatory grounds.

If pragmatism's positive grounds seemed to us unconvincing, instrumentalism's positive grounds seem to consist just in the meaninglessness of transcendental theories. But the point at issue is whether science requires us to believe such theories, and this point is not met but begged by arguing that the answer is negative since the theories are intuitively meaningless. We can, however, be more generous to both pragmatism and instrumentalism by taking them not as arguments but as decisions or resolutions: pragmatism's ascribing meaning to transcendental theories represents a decision to apply to them the ordinary language of truth and falsity, and this, coupled with denial of the need for further interpretation, involves a rejection of even a modified empiricism, as we have above

Prospects of a Modest Empiricism · 145

formulated it. Instrumentalism's denial of the belief-character of transcendental theories represents a decision to talk about such theories in different and special ways without any further changes. Taken as basic decisions, there would seem to be no way of refuting either position, and to this extent at least, ontology is independent of science. There is no way to refute the instrumentalist's denial of the belief-character of various theories which he continues to employ. We may charge his implicit conception of the nature of belief with being tenuous and merely verbal, and we may declare his disavowals of belief to be rather hollow unless he gives up using the sentences which he claims intellectually to disavow. Yet, if he sticks to his guns, and continues to remind us that we all *use* all kinds of objects which we hold meaningless, and feel no guilt upon reflection in continuing to use them, then he is secure. Ontology, then, is relative to the person, and independent of the used language. Just as our common use of available technology does not commit us all equally to the same beliefs, so our common use of scientific language does not dictate that we should all draw the same line between literal sense and nonsense therein.

Coming back to our modified empiricism now, it appears that if pragmatism, in choosing to deny it, does not thereby refute it, instrumentalism renders it trivial. For if the range of our *beliefs* is freely specifiable by intellectual decision independently of the *content of our discourse,* we can always guarantee E's adequacy by simply deciding to exclude recalcitrant sentences from this range. Our judgments of recalcitrance will, of course, vary; but one consistent with our modified empiricism is trivially always possible.

If, however, we interpret our modified empiricist problem more stringently and more objectively, i.e., as not allowing for such a trivial answer, we must require of the empiricist fictionalist not simply that he appropriately adjust his terminology of belief but that he provide a method for eliminating transcendental terms and theories from scientific *discourse,* or of treating them within his discourse otherwise than as significant.

17. Syntactic fictionalism.

One such course open to the fictionalist is to provide a syntax for transcendental theories. Goodman and Quine[26] have in part thus dealt with the problem of treating mathematics nominalistically. Unable, at the time of their study, to translate all of mathematics into a nominalistic language, they developed a nominalistic syntax language enabling them to talk *about* and deal with the untranslated residue, thus *independently* supporting the claim that this residue could be treated as mere machinery without literal significance. Though, in one sense, they did not eliminate this residue, they did go considerably beyond a mere statement that it might be considered as machinery only. For they provided an alternative language without the (to them at that time) objectionable features of the original, such that it was capable of doing much the same job. As they put their view, "our position is that the formulas of platonistic mathematics are, like the beads of an abacus, convenient computational aids which need involve no question of truth. What is meaningful and true in the case of platonistic mathematics as in the case of the abacus is not the apparatus itself, but only the description of it: the rules by which it is constructed and run. These rules we do understand, in the strict sense that we can express them in purely nominalistic language. The idea that classical mathematics can be regarded as mere apparatus is not a novel one among nominalistically minded thinkers; but it can be maintained only if one can produce, as we have attempted to above, a syntax which is itself free from platonistic commitments. At the same time, every advance we can make in finding direct translations for familiar strings of marks will increase the range of the meaningful language at our command (p. 122)."

Such a syntactical approach has relevance far beyond the question of platonistic mathematics. It is in general open to the fictionalist who wishes to disavow the belief-character of some segment of received scientific discourse in more than the trivial sense discussed above in connection with instrumentalism. In the special case of nominalism, it was by no means initially obvious that a syntax could be constructed without platonistic features. Having such a syntax for mathematics, it seems possible to extend it to specified parts of empirical science by addition of predicates applicable to the extra-logical notation contained therein. In particular, such syntax could be developed for transcendental theories which the fictionalist cannot eliminate through translation but which he finds it objectionable to take as significant. For the non-nominalist who objects to taking as significant some particular transcendental theory, the task is, of course, much easier, for he has available all the tools of platonistic syntax. In a less trivial sense than that of instrumentalism, then, our modified empiricism may be feasible through syntactic construction. Note again, incidentally, that ontology turns out independent of our received scientific discourse, through the possibility of variable syntactic reinterpretation, and that only such elements as are needed for the applicability of syntactic predicates may be sufficient in the extreme case.

18. Eliminative fictionalism.

We may, finally, require our modified empiricism to show how transcendental terms may be eliminated from scientific discourse in favor of some other object-language discourse which is equivalent in some appropriate sense. Here it is well to recall that transcendental theories are justified generally as making possible the statement of comprehensive relationships

Prospects of a Modest Empiricism · 147

(in desirable ways) on the observable level. Thus, if a way could be shown of appropriately stating these observational relationships in some theory, S, which otherwise differed from its transcendental counterpart only by lacking sentences with any transcendental term, S would be, in a reasonable sense, equivalent to that counterpart.

One such method is that of Craig, who states as one of his results, "... if K is any recursive set of non-logical (individual, function, predicate) constants, containing at least one predicate constant, then there exists a system whose theorems are exactly those theorems of T in which no constants other than those of K occur. In particular, suppose that T expresses a portion of a natural science, that the constants of K refer to things or events regarded as 'observable,' and that the other constants do not refer to 'observables' and hence may be regarded as 'theoretical' or 'auxiliary.' Then there exists a system which does not employ 'theoretical' or 'auxiliary' constants and whose theorems are the theorems of T concerning 'observables.' "[27]

Professor Hempel, discussing Craig's method, states concisely what is involved and what sense of "equivalence" is here relevant: "Craig's result shows that no matter how we select from the total vocabulary V_T' of an interpreted theory T' a subset V_B of experiential or observational terms, the balance of V_T', constituting the 'theoretical terms,' can always be avoided in sense (c)." This sense, which Hempel distinguishes from definability and translatability, he calls "functional replaceability" and describes as follows, "The terms of T might be said to be avoidable if there exists another theory T_B, couched in terms of V_B, which is 'functionally equivalent' to T in the sense of establishing exactly the same deductive connections between V_B sentences as T."

Professor Hempel offers, however, two reasons against the scientific use of Craig's method, "no matter how welcome the possibility of such replacement may be to the epistemologist." One reason is that the functionally equivalent replacing system constructed by Craig's method "always has an infinite set of postulates, irrespective of whether the postulate set of the original theory is finite or infinite, and that his result cannot be essentially improved in this respect.... This means that the scientist would be able to avoid theoretical terms only at the price of forsaking the comparative simplicity of a theoretical system with a finite postulational basis, and of giving up a system of theoretical concepts and hypotheses which are heuristically fruitful and suggestive – in return for a practically unmanageable system based upon an infinite, though effectively specified, set of postulates in observational terms."[28]

It should be obvious that any proposal like Craig's for meeting our present demand for elimination of transcendental terms will be judged in various ways in accordance with varying approval of its tools and subsidiary concepts. Moreover, such variation may be independent of the question of modified empiricism as such. In particular, a dissatisfaction with systems containing infinite, effectively specified sets of postulates may or may not be justified, but is at any rate independent of modified empiricism as we have formulated it. Further, though the relevant notions of heuristic fruitfulness and suggestiveness, simplicity, and practicality are not very precise, suppose it granted that Craig's functionally equivalent system is indeed inferior to its counterpart in all these respects. This is irrelevant to our modified empiricism. If Craig's replacing system renders such empiricism possible, this represents an intellectual gain no worse for the fact that the system is unwieldy and not likely to be used by the practicing scientist. The case is analogous to ordinary definition, where we try to minimize the complexity of our primitive basis at the cost of replacing short and handy definienda by cumbersome definientia in terms of a simple few primitives. Obviously, no one intends these definientia to be used in practice in place of their definienda, but neither does anyone seriously maintain that their formulation therefore represents less of an intellectual gain.

One may however, with Goodman,[29] suggest the infinity of postulates in Craig's replacing system not as representing a practical difficulty, but rather as indicating that the deductive character of the original system is not sufficiently reflected by its replacement. That is to say, if transcendental theories serve to enable finite postulation, no replacement is equivalent *deductively* in every relevant sense if it fails to serve thus also, even though it does accurately reflect the whole class of relevant postulates-or-theorems (assertions) of the original. If specific empiricist programs are to be interpreted in accord with this point of view, then, even granted Craig's result, they are not proven generally achievable, and continue to represent non-trivial problems in individual cases. It seems to me, however, that if we take these programs as requiring simply the reflection of non-transcendental assertions into replacing systems without transcendental terms, then we do not distort traditional notions of empiricism, and we have to acknowledge that Craig's result does the trick; the further cited problems remain but they are independent of empiricism as above formulated.

Professor Hempel's second reason against the scientific use of Craig's method is that "The application of scientific theories in the prediction and explanation of empirical findings involves not only deductive inference, i.e., the exploitation of whatever deductive connections the theory establishes among statements representing potential empirical data, but it also requires procedures of an inductive character, and some of these would become impossible if the theoretical terms were avoided." He illustrates in terms of the following four sentences, where 'magnet' is taken to be a theoretical, i.e., non-observational term:

Prospects of a Modest Empiricism · 149

(5.1) The parts obtained by breaking a rod-shaped magnet in two are again magnets.

(5.2) If x is a magnet, then whenever a small piece y of iron filing is brought into contact with x, then y clings to x. In symbols:

$$Mx \supset (y) (Fxy \supset Cxy)$$

(5.3) Objects b and c were obtained by breaking object a in two, and a was a magnet and rod-shaped.

(5.4) If d is a piece of iron filing that is brought into contact with b, then d will cling to b.

Now, says Hempel, given (5.3) (and assuming (5.2)), we are able to deduce, with the help of (5.1), such sentences as (5.4). But (5.3) is non-observational, containing '*Ma*', itself not deducible from observational sentences via (5.2) which states only a necessary, but not a sufficient condition for it. Thus, if (5.4) is to be connected by our theory here with other observational sentences, an *inductive* step is necessary, leading to (5.3), i.e., to '*Ma*' specifically, from observational sentences. E.g., '*Ma*' might be *inductively* based on a number of instances of '*Fay* $\supset$ *Cay*,' assuming that we have no instance of '*Fay* · ~ *Cay*'. This is so, since such instances confirm '(y) (*Fay* $\supset$ *Cay*)', which, by (5.2), partially supports '*Ma*'. Thus, our hypothesis (5.1) takes us, in virtue of (5.2), from some observational sentences, i.e., instances of '*Fay* $\supset$ *Cay*', to observational sentences such as (5.4), but the transition requires certain inductive steps along the way. But though Craig's functionally equivalent system retains all the deductive connections among observational sentences of the original system, it does not, in general, retain the inductive connections among such sentences. Hempel concludes, "the transition, by means of the theory, from strictly observational to strictly observational sentences usually requires inductive steps, namely the transition from some set of observational sentences to some non-observational sentence which they support inductively, and which in turn can serve as a premise in the strictly deductive application of the given theory."

With respect to this argument, we might question by what theory of confirmation '(y) (*Fay* $\supset$ *Cay*)' supports '*Ma*'; it surely is not Hempel's satisfaction criterion of confirmation.[30] But this is irrelevant to the important point brought out by Hempel's argument, viz., that since functionally equivalent systems (of Craig's type) are not logically equivalent to their originals, they need not (on *any* likely view of confirmation) sustain the same confirmation relations as these originals, even among purely observational sentences. And this despite the fact that they do preserve the same deductive relations among such sentences by retaining all original theorems couched in purely observational terms. Thus, if we do not attempt an observational reduction of the *whole* of our theoretical discourse in given scientific domains via definition and translation or syntactic construction, but aim merely to isolate the *observational part* of such discourse, we must be careful to construe this part adequately, i.e., as comprising not only a deductive network but also a wider confirmational range. Specific empiricist programs would then seem to be not achievable generally by means of Craig's result, in the light of Hempel's argument. One further line of attack might be to clarify the inductive relation sufficiently to enable agreement on just which sentences confirm which, relative to a given theoretical system, and then to strive for an independent, observational specification of such confirmation-pairs, to supplement the appropriate Craigian equivalent.[31] A second approach would be to try to meet each specific case by strengthening the replacing functional equivalent with hypotheses designed to yield just those inductive relations borne by the original, in which we are particularly interested.

In the case of Hempel's above example, for instance, the inductive relation in question is between instances of '*Fay* ⊃ *Cay*' plus the observational parts of (5.3), (objects *b* and *c* were obtained by breaking object *a* in two [*Bbca*] and *a* was rod-shaped [*Ra*]) and (5.4). Such an inductive relation might also be expressed without theoretical terms by the following statement, which, owing to the universal quantification, can at best establish only an inductive relationship between its instances and the observational sentence derived:

$$(x)\{[(y)(Fxy \supset Cxy) \cdot Rx] \supset (z)[(\exists w)(Bzwx) \supset (u)(Fzu \supset Czu)]\}[32]$$

19. Summary and conclusion.

If our journey has yielded no single, easy solution as a climax, its difficulty has nevertheless earned for us the right to stop and get our bearings. For we have traveled a long way from the conception of empiricism as a shiny, new philosophical doctrine for weeding out obscurantism and cutting down nonsense wherever they crop up. We have, furthermore, seen that even if we take empiricism as the proposal of a general meaning-criterion in terms of translatability into a chosen artificial language, we run into trouble. We have thus come to restrict the empiricist's job to providing merely an adequate sufficient condition of significance on an observational basis, in the form of an observational system capable of housing our scientific beliefs.

Even this restricted task has, however, turned out to have quite difficult obstacles before it. While the inclusion of needed disposition terms seemed to us not as formidable a problem as hitherto thought, we found theoretical terms to be generally resistant to straightforward empiricist interpretation. Considering this difficulty in the light of a number of recent approaches to philosophy of science, we found that the pragmatic rejection of our restricted empiricism does not constitute a refutation, while intrumentalism's easy solution fulfills such empiricism in only the most trivial sense. Taking empiricism's task as the provision of an appropriate modification of scientific discourse itself rather than simply of our notions of belief, we found the possibility of syntactic reinterpretation promising, though less intuitively satisfying then a direct reinterpretation of the object-language of science proper. Our final examination of Craig's method for eliminating theoretical terms wholly from such language while preserving its observational segment intact led to the conclusion that this method is, in itself, incapable of achieving the goal of our restricted empiricism.

It appears, in sum, that even a modest empiricism is presently a hope for clarification and a challenge to constructive investigation rather than a well-grounded doctrine, unless we construe it in a quite trivial way. Empiricists are perhaps best thought of as those who share the hope and accept the challenge – who refuse to take difficulty as a valid reason either for satisfaction with the obscure or for abandonment of effort.

NOTES

I wish to thank Professors C. G. Hempel, N. Goodman, and N. Chomsky for their criticisms of an earlier draft of this paper and for helpful discussions of related problems.

1. "Analytic," as used throughout the present paper, embraces both analytically true and analytically false statements.

2. A. J. Ayer, *Language, Truth, and Logic*, 2d ed., (London and New York, 1947), 15-16.

3. Ibid., 38-39.

4. Ibid., 35.

5. See C. G. Hempel, "Problems and Changes in the Empiricist Criterion of Meaning," *Revue Internationale de Philosophie* 11 (1950); reprinted in L. Linsky, *Semantics and the Philosophy of Language*, (Urbana, 1952), for the classic treatment, with numerous examples, of the difficulties involved in early attempts.

6. I. Berlin, "Verifiability in Principle," *Proceedings of the Aristotelian Society* 39: 225-48, esp. p. 234, and Ayer; op cit., 11-12.

7. A. Church, review of Ayer, op. cit., *Journal of Symbolic Logic* 14 (1949).

8. N. Goodman, *Fact, Fiction, and Forecast* (Cambridge, Mass., 1955), p. 58, n. 1.

9. Op. cit., 50 ff.

10. I mean by "materially adequate" here something both more restricted and vaguer than C. A. (I); the two ought not to be confused.

11. N. Goodman, *The Structure of Appearance* (Cambridge, Mass., 1951), 1-11.

12. Ibid., 11-26.

13. Ibid., 11.

14. Professor Hempel, in correspondence, informs me that, while he thinks I am right in criticizing the translatability criterion, he would perhaps now wish to construe such requirement pragmatically rather than semantically, i.e., not by reference to extensions or truth-values, but rather by reference to the ability of the utterer to restate his utterance within some empiricist language. While favoring the development of a pragmatic concept of translation (see "On Synonymy and Indirect Discourse," in Part 1 above), I am skeptical of a method which asks the utterer to restate, i.e. *translate*, his own utterances, and I wonder if any purely pragmatic notion, without some semantic conditions, is adequate. Professor Hempel indicates doubt, moreover, concerning the sense in which an isolated utterance could be said to be restated as a formula in some system whose interconnections contribute to its meaning, and he suggests therefore that the whole question of empirical significance may concern only the expressions of certain well-specified systems.

15. Hempel, op. cit., 51.

16. M. White, "A Finitistic Approach to Philosophical Theses," *Philosophical Review* 60 (1951): 299-316.

17. N. Goodman, *The Structure of Appearance* (Cambridge, Mass., 1951), 42-55.

18. R. Carnap, "Testability and Meaning," *Philosophy of Science* 3 (1936) and 4 (1937).

19. C. G. Hempel, *Fundamentals of Concept Formation in Empirical Science* (Chicago, 1952), 23-29.

20. N. Goodman, *Fact, Fiction, and Forecast* (Cambridge, Mass., 1955), 50.

21. That is, inadequacy in descriptiveness, or in capacity for formulating evidence-statements. Certain so-called dispositional terms might be defended on much the same grounds as theoretical terms, however, in a manner to be discussed in later sections, and involving no requirement for reduction-sentences. Such defense must, of course, be treated differently, but my aim here is to counter the prevalent idea that partial specification of the meaning of disposition terms by reduction sentences creates a special intermediate class of terms with unique functions in formulating our information about observables, i.e., capable of *continued* adequacy in expressing such information with the *continuing growth* of scientific investigation. Either a term has a clear enough denotative use to be observationally defined or to serve as an observational primitive (in which case the method of reduction-sentences is unnecessary) or else it must be justified on the same grounds as theoretical, non-observational

terms (in which case the method of reduction-sentences is also unnecessary). See Goodman, F.F.F., p. 60, n. 11, for criticism of reduction as a kind of definition.

22. J. S. Mill, *An Examination of Sir William Hamilton's Philosophy*, 6th ed. (London, 1889) p. 225 ff., esp. p. 233.

23. *An Analysis of Knowledge and Valuation* (LaSalle, Ill., 1946).

24. I. Scheffler, "Anti-Naturalist Restrictions in Ethics," reprinted as article 1 in Part 3 below.

25. The labels I introduce here and in following sections are related to, but not intended as names of, specific philosophies associated with familiar historical movements or individual thinkers. They refer rather to characteristic trends, somewhat oversimplified and idealized, perhaps, in comparison to actually held philosophies. Nevertheless, they are influential and salient trends and will be recognizable, I hope, as elements of much of the recent literature in philosophy of science and theory of knowledge. As recent illustrations (in a loose sense) of pragmatism see R. Carnap, "Empiricism, Semantics, and Ontology," *Revue Internationale de Philosophie* 11 (1950), reprinted in L. Linsky, *Semantics and the Philosophy of Language* (Urbana, Il., 1952); and W. V. Quine, "Two Dogmas of Empiricism," *Philosophical Review* (1951), included in W. V. Quine; *From a Logical Point of View* (Cambridge, Mass., 1953); of instrumentalistic fictionalism, see Toulmin, S. E., *The Philosophy of Science* (London, 1953). Regarding vacillation between the latter two trends, see Nagel's discussion of Dewey in *Sovereign Reason* (Glencoe, Ill., 1954), pp. 110-15.

26. N. Goodman and W. V. Quine, "Steps Toward a Constructive Nominalism," *Journal of Symbolic Logic,* 12 (1947): 105-22.

27. W. Craig, "On Axiomatizability Within a System," *Journal of Symbolic Logic* 18 (1953): 31, text and n. 9. See also W. Craig, "Replacement of Auxiliary Expressions," *Philosophical Review* 65 (1956): 38-55.

28. C. G. Hempel, "Implications of Carnap's Work for the Philosophy of Science," to appear in the forthcoming Carnap volume of the *Library of Living Philosophers*.

29. This point was made by Goodman in correspondence.

30. "A Purely Syntactical Definition of Confirmation," *Journal of Symbolic Logic* 8 (1943): 122-43. See also C. G. Hempel, "Studies in the Logic of Confirmation," *Mind,* n. s. 54 (1945): 1-26 and 97-121, especially pp. 107 ff.

31. It would also be desirable to clarify the notion of induction sufficiently to enable critical evaluation of the assumption that where the Craigian equivalent fails to reflect the confirmation relationships of the original theoretical system, it is itself confirmed to a lesser degree by the available evidence on which that original rests. Perhaps even more interesting would be the examination of the analogous assumption for the *strengthened* functional equivalent discussed immediately below.

32. This formula is due to Professor Hempel, who suggested it to me as an improvement over my original version.

4. INDUCTIVE INFERENCE: A NEW APPROACH

ON WHAT GROUNDS do we choose the theories by which we anticipate the future? How do we decide what to predict about cases never before observed? These questions concerning what is traditionally called "induction" are among the most fundamental and most difficult which can be asked about the logic of science. Much reflection has been devoted to these questions in recent years, but no contribution has proved more incisive and challenging than that of Nelson Goodman of the University of Pennsylvania, whose papers on induction and allied problems have activated lively philosophic controversy over the past twelve years.

In 1955, Goodman published *Fact, Fiction, and Forecast*,[1] in which he presented the outlines of a new approach to the understanding of induction. This recent work has also aroused considerable comment by philosophers, both in print and out, and it is safe to say that the discussion is still in its early stages. The scientific public is, however, largely unaware of this new development, just as it was largely unacquainted with the controversies that preceded it. If there is no real boundary between science and the philosophy of science, the consideration of fundamental research in the logic of science ought not to be confined, even at the early stages, to circles of philosophers. The aim of this article is thus to acquaint the scientific reader with the background and the direction of Goodman's investigations, as they bear on the interpretation of induction.

The starting point for all modern thinking about induction is David Hume's denial of necessary connections of matters of fact: between observed cases recorded in the evidence and predicted cases based on the evidence there is a fundamental logical gap, which cannot be bridged by deductive inference. If, then, the truth of our predictions is not guaranteed by logical deduction from available evidence, what can be their rational justification? This challenge, arising out of Hume's analysis, has evoked a variety of replies. Leaving aside the reply of the skeptics, who are willing to admit that all induction is indeed without rational foundation, and that of the deductivists, who strive vainly to show Hume wrong, we find two replies which have

This paper appeared in *Science* 127 (24 Jan. 1958): 177-81.

gained wide popularity, the first primarily among philosophers, the second among scientists as well.

The first reply criticizes the assumption that rational justification can be only a matter of deduction from the evidence, pointing out that the normal use of expressions such as "rational," "reasonable," "based on good reasons," and so forth sanctions their application to statements referring to unexamined cases, and hence not deducible from accumulated evidence. This reply, although true, is, however, woefully inadequate. For not every statement which outstrips available evidence is reasonable, though some are. Outstripping the evidence is, to be sure, no bar to rationality, but neither does it guarantee rationality. If we are to meet the challenge posed, we must go on to formulate the specific criteria by which some inductions are justified as reasonable while others are rejected as unreasonable, though both groups outstrip the available evidence. Now it is likely that at least part of the reason why this further task has been slighted is that the adequacy of the second reply has largely been taken for granted.

This second reply, stated in one form by Hume himself, is that reasonable inductions are those which conform to past regularities. In modern dress, it appears as the popular assertion that predictions are made in accordance with general theories which have worked in the past. What leads us to make one particular prediction rather than its opposite is not its deducibility from evidence but rather its congruence with a generalization thoroughly in accord with all such evidence, and the correlative disconfirmation of the contrary generalization by the same evidence. (I shall refer to this hereafter as the "generalization formula.") Of course, if no relevant evidence is available to decide between a given generalization and its contrary, or if the available evidence is mixed, neither generalization will support a particular inductive conclusion. But it is only to be expected that every limited body of evidence will fail to decide between *some* generalization and its contrary, and hence that we will generally not be able to choose between *every* particular prediction and its opposite. It is sufficient, therefore, for a formulation of the criteria of induction to show how certain bodies of evidence enable us to decide between certain conflicting inductions. This the generalization formula seems to accomplish. For if there is evidence which consistently supports a given generalization, then the contrary generalization is *ipso facto* disconfirmed, and our particular inductive conclusions seem automatically selected for us. There are, of course, details to be taken care of, relating to such matters as the calculation of degrees of support which generalizations derive from past evidence, but, in principle, we have our answer to the challenge of induction.

It is this sanguine estimate which has been thoroughly upset by Goodman's researches. Published in 1946 and 1947, his early papers in the philosophical journals dealt with a variety of interrelated questions: the nature of scientific law, of dispositional properties, of potentiality, of relevant conditions, of counterfactual judgments, of confirmation or induction.[2] They immediately aroused a storm of controversy. What made the papers so disturbing to the philosophic community was the fact that, while all these questions were shown to be intimately connected, Goodman's logically rigorous attempts to answer them without going around in circles ended in a big question mark. Appearing at a time when logicians had been making considerable progress in analyzing other aspects of scientific method, these results came as a shock. Goodman's investigations, it seemed, had sufficed to undermine all the usual formulas concerning the most basic concepts of the logic of science, but his repeated and ingenious efforts to supply a positive alternative had all turned out fruitless. In the philosophic discussions that followed, every attempt was made to skirt Goodman's disheartening results. They were declared unimportant for the practicing scientist. The initial questions were asserted to be insoluble, hence worthless. Many papers, on the other hand, proposed what seemed perfectly obvious solutions that turned out to be question-begging. Only a very few authors fully recognized the seriousness of the situation for the philosophy of science and tried to cope with it directly.[3]

In 1953, with the whole matter still very much unsettled, Goodman delivered a series of three lectures at the University of London, in which he again addressed himself to the problem. These lectures, together with his major 1946 paper, were then published together in his book *Fact, Fiction and Forecast*. Here Goodman essayed a new and positive approach to some of the major questions he had faced earlier. He did not offer his book as a final solution to all the original problems. He did, however, present a fresh approach, worked out with sufficient rigor to put discussion of it on a fruitful basis. But we are getting ahead of our story and must now return to see how Goodman's early work affected the theory of induction.

How did Goodman's early papers upset complacency with respect to the generalization formula (according to which we make these predictions congruent with generalizations thoroughly in accord with past evidence)? We may profitably approach this matter in the light of a passage from J. S. Mill's *Logic*. Although it does seem true that, for every particular induction we make, there is some generalization related to it in the manner described, Mill argues that generalizations which are equally well supported by available evidence vary in the sanction they provide for their respective particular inductions: "Again, there are cases in which we reckon with the most unfailing confidence upon uniformity, and other cases in which we do not count upon it at all. In some we feel complete assurance that the future will resemble the past, the unknown be precisely similar to the known. In others, however invariable may be the result obtained from the instances which have been observed, we draw from them no more than a very feeble presumption that the like result will hold in all other cases.... When a chemist announces the existence and properties of a newly discovered substance, if we confide in his accuracy, we feel assured that the conclusions he has arrived at will hold universally, though the induction be founded but on a single instance.... Now mark another case, and contrast it with this. Not all the instances which have been observed since the beginning of the world in support of the general proposition that all crows are black would be deemed a sufficient presumption of the truth of the proposition, to outweigh the testimony of one unexceptionable witness who should affirm that in some region of the earth not fully explored he had caught and examined a crow, and had found it to be grey. Why is a single instance, in some cases, sufficient for a complete induction, while in others myriads of concurring instances, without a single exception known or presumed, go such a very little way towards establishing an universal proposition?"[4]

And Goodman gives an analogous example when he writes: "That a given piece of copper conducts electricity increases the credibility of statements asserting that other pieces of copper conduct electricity, and thus confirms the hypothesis that all copper conducts electricity. But the fact that a given man now in this room is a third son does not increase the credibility of statements asserting that other men now in this room are third sons, and so does not confirm the hypothesis that all men now in this room are third sons. Yet in both cases our hypothesis is a generalization of the evidence statement. The difference is that in the former case the hypothesis is a *lawlike* statement; while in the latter case, the hypothesis is a merely contingent or accidental generality. Only a statement that is *lawlike* – regardless of its truth or falsity or its scientific importance – is capable of receiving confirmation from an instance of it; accidental statements are not." (F.F.F., p. 73)

But it is Goodman's further formulation of the problem that is crucial. For what has so far been shown is that, in addition to all credible particular inductions, generalization from the evidence also would select certain incredible ones. Now Goodman shows that among these incredible ones lie the very negations of our credible predictions concerning new cases. To apply his previous example, it is not merely that by generalization we selectively establish, in addition to the credible prediction that the next specimen of copper will conduct electricity, also the incredible one that the next present occupant of this room to be examined is a third son. Rather, we do not even establish that the next specimen of copper conducts electricity, for we can produce a generalization equally supported by the evidence and yielding the prediction that it does not. Or, putting this point in the form of a specific example, while the available evidence clearly supports:

(S_1) All specimens of copper conduct electricity.

and clearly disconfirms its contrary:

(S_2) All specimens of copper do not conduct electricity.

this is not sufficient to yield the particular induction concerning a new copper specimen c, to be examined:

(S_3) c conducts electricity.

since the same evidence also and equally supports:

(S_4) All specimens of copper are either such that they have been examined prior to t and conduct electricity or have not been examined prior to t and do not conduct electricity.

while clearly disconfirming *its* contrary:

(S_5) All specimens of copper are either such that they have been examined prior to t and do not conduct electricity or have not been examined prior to t and do conduct electricity.

thus giving rise to the negate of S_3:

(S_6) c does not conduct electricity.

if it is assumed true that:

(S_7) c has not been examined prior to t.

For cases assumed new, then, the generalization formula selects no particular inductions at all. Merely to be told to choose our inductions by reference to theories which work relative to past evidence is hence to be given worthless advice. Nor does this situation improve with the accumulation of relevant data over time. For even if we later find S_6 false and add S_3 to our evidence, leading to a rejection of S_4, we do not thereby eliminate other hypotheses which are exactly like S_4 but which specify times later than t. Accordingly, no matter how much empirical data we have accumulated and no matter how many hypotheses like S_4 we have disconfirmed up to a given point in time, we still have (by the generalization formula) contradictory predictions for every case not yet included in our data. No matter how fast and how long we run, we find we are standing still at the starting line.

This predicament holds, of course, only for cases assumed to be new. Using our previous example, if neither S_7 nor its negate is assumed, then S_4

yields neither S_3 nor S_6, while if S_7 is assumed false, then S_4 coincides with S_1, implying S_3 rather than S_6. This is not surprising, however, since, if S_7 is false, c is identical with one of our original evidence cases, all of which are described by the evidence itself as conducting electricity; S_3 is thus implied deductively by the evidence at hand, given the general understanding that no cases have been omitted.

As soon as we leave the safe territory of examined cases, however, and try to deal with a new one, generalization yields contradictory inductions, deciding for neither. And, further, since the adoption of a generalization constitutes wholesale endorsement of appropriate particular inductions yet to be made, then even if we do not know about some specific case that it is a new one, our unrestricted adoption of generalizations gets us into trouble if we can make the assumption of novelty for at least one case within the appropriate range. Since, moreover, we patently do choose between contradictory inductions covering new cases, as well as between competing generalizations, the generalization formula must be wrong as a definition of our inductive choices. In our previous example, we obviously in practice would *not* hold S_4 equally supported by uniformly positive evidence supporting S_1, nor would we under such conditions have any hesitation in rejecting S_6 in favor of S_3. This clearly indicates that the generalization formula is not adequate to characterize our inductive behavior. We apparently employ additional, nonsyntactic criteria governing the extension of characteristics of our evidence-cases to other cases in induction.

These criteria of what Goodman calls "projectibility" select just those generalizations *capable* of receiving support from their positive instances and in turn sanctioning particular inductions. Projectible hypotheses may, in individual cases, fail to sanction any particular inductions (for example, in cases where we have two such hypotheses which conflict), but no nonprojectible hypothesis sanctions any induction, no matter how much positive support it has in the sense of the generalization formula. Goodman's problem is then to define projectibility, which is, in turn, needed to define induction. Since counterfactual judgments (for example, "If this salt, which has not in fact been put in water, had been put in water, it would have dissolved.") are, moreover, construable as resting upon just such generalizations as are projectible, that is, legitimately used for induction (in this case, "Every sample of salt, when put into water, dissolves."), and, furthermore, are themselves used to explain dispositional predicates, such as "is soluble," the definition of projectibility would throw light on these additional issues as well.

It may be thought that the characterization of projectibility can be accomplished rather easily, simply by ruling out generalizations making reference to time. Recall that, in our above example, the trouble arose because the available evidence equally supported S_1 and S_4. But whereas the predicate "conducts electricity" makes no reference to time, the predicate "has been examined prior to t and conducts electricity or has not been examined prior to t and does not conduct electricity" makes reference to time of examination, and moreover can be explained, given such reference, in terms of the former predicate. It may further be pointed out that, without assumption S_7 (making reference to time of examination), no contradiction arises. It is only when we add S_7 to S_4 that S_6, which contradicts S_3, is derived. Why not use this, then, as a rule for eliminating S_4 – namely, its requiring an additional assumption about time of examination to produce one of our contradictory inductions?

The answer is that the situation is easily reversed. Symbolize the predicate "conducts electricity" by C and the other, more complicated one, of S_4, by K; symbolize "has been examined before t" by E. It is true that, as the present argument maintains, K is then definable as

$$(E \text{ and } C) \text{ or } (\text{not-}E \text{ and not-}C)$$

("has been examined before t and conducts electricity or has not been examined before t and does not conduct electricity"). However, it is also true that, taking K as our primitive idea, C is definable as

$$(E \text{ and } K) \text{ or } (\text{not-}E \text{ and not-}K)$$

Furthermore, in the latter mode of description, S_1 would become:

(S_1') All specimens of copper are either such that they have been examined prior to t and have the property K or have not been examined before t and do not have the property K.

while S_4 would become:

(S_4') All specimens of copper have the property K.

To derive a parallel to S_3, we need to show that a new case c does not have the property K. This we can do if we now supplement S_1' with S_7 getting:

(S_3') c does not have the property K.

And we derive our contradictory particular induction, parallel to S_6, from S_4', without using S_7:

(S_6') c has the property K.

Thus, neither the employment by a hypothesis of a predicate referring to time nor its need of supplementation by S_7 in order to produce contradiction is a reliable clue with which to try to repair the generalization formula. Neither is, strictly speaking, any clue at all.

Inductive Inference: A New Approach

But perhaps the generalization formula is being applied too narrowly. We have, after all, been considering isolated statements in abstraction from other, relevant and well-established, hypotheses. In the above illustration we have, for instance, so far ignored the fact that available evidence also supports (by the generalization formula) a number of hypotheses of the following kind:

(S_8) All specimens of iron conduct electricity.
(S_9) All specimens of wood fail to conduct electricity.

and that these in turn lend credence to the following larger generalization:

(S_{10}) All classes of specimens of the same material are uniform with respect to electrical conductivity.

This larger generalization, having independent warrant and conflicting with S_4, serves thereby to discredit it, thus eliminating the troublesome induction S_6. In this way, it may be argued, the generalization formula can be rendered viable simply by taking account of a wider context of relevant hypotheses.

It takes but a moment of reflection, however, to see the weakness of such an argument. For, by reasoning analogous to that initially employed in introducing S_4, it will be seen that the very same evidence which supports S_8, S_9, and S_{10} also and equally (by the generalization formula itself) supports:

(S_8') All specimens of iron have the property K.
(S_9') All specimens of wood fail to have the property K.
(S_{10}') All classes of specimens of the same material are uniform with respect to possession of the property K.

This latter large generalization, it will be noted, produces just the opposite effect from that of S_{10}. It conflicts with S_1, thereby, by analogous argument, discrediting it and eliminating the induction S_3 rather than S_6. Which of these conflicting large generalizations shall we now choose to take account of, S_{10} or S_{10}'? It is evident that we are again face to face with the very problem with which we started and that the proposal to repair the generalization formula by referring to other relevant hypotheses selected by it serves merely to postpone our perplexity. For these other hypotheses, in conflict themselves, are of no help unless we have some way of deciding which of them are projectible. In the face of difficulties such as these, *it becomes impossible to explain our choice of predictions by reference to whether or not they accord with generalizations which work,* no matter how widely the scope of this principle is construed.

Goodman's new idea is to utilize pragmatic or historical information that may fairly be assumed available at the time of induction, and to define projectibility in terms of such extrasyntactic information. The generalization formula, it will be recalled, rests on the notion of an *accordance* between a predictive generalization and the evidence by which it is supported, an accordance which can be determined solely by an examination of the generalization and its evidence-statements. In this sense, the relation of accordance is formal or syntactic (as the relation of deduction is), making use of no material or historical information. Goodman now suggests that, in order to specify the predictive generalizations we choose on the basis of given evidence, we need not restrict ourselves merely to the syntactic features of the statements before us. Rather, he makes the radical proposal that we use also the historical record of past predictions, and in particular, the *biographies* of the specific terms or predicates employed in previous inductions. Our theories, he suggests, are chosen not merely by virtue of the way they encompass the evidence, but also by virtue of the way the language in which they are couched accords with past linguistic practice.

His basic concept is "entrenchment," applicable to terms or predicates in the degree to which they (or their extensional equivalents, that is, words picking out the same class of elements, like "triangle" and "trilateral") have actually been previously employed in projection: in formulating inductions on the basis of positive, though incomplete evidence. To illustrate with our previous example, the predicate "has been examined prior to t and conducts electricity or has not been examined prior to t and does not conduct electricity" is less well entrenched than the predicate "conducts electricity," *because the class it singles out has been less often mentioned in formulating inductions*. The factor of actual historical employment of constituent predicates or their equivalents can thus be used to distinguish between hypotheses such as S_1 and S_4, which are equal in point of available positive instances. Goodman appeals, then, to "recurrences in the explicit use of terms as well as to recurrent features of what is observed," suggesting that the features which we fasten on in induction are those "for which we have adopted predicates that we have habitually projected" (F.F.F., pp. 96, 97). With this idea as a guide, Goodman first defines presumptively projectible hypotheses. Next, he defines an initial projectibility index for these hypotheses. Finally, he defines degree of projectibility by means of the initial projectibility index as modified by indirect information embodied in what he calls "overhypotheses," *which must themselves qualify as presumptively projectible*. The latter use made of indirect evidence is worked out with great care and detail and is of independent theoretical interest.

Roughly, degree of projectibility is to represent what Goodman earlier called "lawlikeness," (that is, that property which, together with truth, defines scientific laws) and constitutes therefore not only an explanation but also a refinement of the latter. With the explanation of lawlikeness,

Goodman suggests that the general problem of dispositions is solved. For this general problem is to define the *relationship* between "manifest" or observable predicates (for example, "dissolves") and their dispositional counterparts (for example, "is soluble") and manifest predicates may now be construed as related by true lawlike or projectible hypotheses to their dispositional mates. Other problems, such as the nature of "empirical possibility" are also illuminated by this approach, and some light is thrown on the difficult question of counterfactual judgments which, however, still resists full interpretation.

The most natural objection to Goodman's new approach is that it provides no explanation of entrenchment itself. In using this notion to explain induction, however, Goodman does not at all rule out a further explanation of why certain predicates as a matter of fact become entrenched while others do not. His purpose is to formulate clear criteria, in terms of available information, that will single out those generalizations in accordance with which we make predictions. The strong point of his treatment is that his criteria do indeed seem effective in dealing with the numerous cases he considers.

A possible misconception concerning the use of "entrenchment" as a basic idea is that it may lead to the ruling out of unfamiliar predicates, thus stultifying the growth of scientific language. Unfamiliar predicates may, however, be well entrenched if some of their extensionally equivalent mates have been often projected, and they may acquire entrenchment indirectly through "inheritance" from "parent predicates" – that is, other predicates related to them in a special way outlined in detail in Goodman's discussion (p. 105). Furthermore, Goodman's criteria provide methods for evaluating *hypotheses,* not predicates, so that wholesale elimination of new scientific terms is never sanctioned in his treatment.

As remarked previously, the critical discussion of Goodman's new approach is still in its early stages.[5] His formulations will undoubtedly undergo further refinement and revision with continuing study, but even in their present form they will have contributed much toward putting important questions in the philosophy of science on a scientific basis.

NOTES

1. N. Goodman, *Fact, Fiction and Forecast* (Cambridge: Harvard University Press, 1955).

2. ———, "A Query on Confirmation," *J. Philosophy* 43 (1946): 383; "The Problem of Counterfactual Conditionals, *J. Philosophy* 44 (1947): 113.

3. See in particular R. Carnap, "On the Application of Inductive Logic,"

Philosophy and Phenomenological Research 8 (1947): 133; N. Goodman, "On Infirmities of Confirmation Theory," *Philosophy and Phenomenological Research* 8 (1947): 149; R. Carnap, "Reply to Nelson Goodman," *Philosophy and Phenomenological Research* 8 (1947): 461.

4. J. S. Mill, *A System of Logic.* (London: 1843; new impression, 1947), book 3, chap. 3, sect. 3, p. 205.

5. See, in this connection, the long study of *Fact, Fiction and Forecast* by J. C. Cooley [*J. Philosophy* 54 (1957): 293] and Goodman's reply [*J. Philosophy* 54 (1957): 531].

5. REFLECTIONS ON THE JUSTIFICATION OF INDUCTION

THE GENERALIZATION FORMULA, resting on the notion of support by positive instances, has been found inadequate to characterize our normal selection of hypotheses through experience. But such inadequacy as a characterization does not mean that consistent evidential support is irrelevant to the confirmation of hypotheses. It shows only that such support is not in itself sufficient to characterize confirmation, leaving open the possibility that supplementation by additional criteria may yield an adequate account. We have seen, in this connection, the proposal of a criterion of projectibility, which is clearly not intended as a characterization of confirmation, but only of *confirmability* by instances. Assuming, for the present, that we have suitable criteria of confirmability, together with the generalization formula as restricted by such criteria, do we now have a sufficient basis for explaining the *confirmation* of hypotheses?

It does not follow that this must be so, from the arguments we have reviewed. For the generalization formula, in itself, was seen to be radically defective, in providing *no* selection at all among hypotheses, of the sort that would differentiate between contradictory inductions. Supplementation by criteria of confirmability does indeed overcome *this* difficulty; it enables at least *some* selection among hypotheses, stringent enough to provide legitimate differentiations among contradictory inductions. But it may well be that there are *further* selections to be accounted for, that some conflicts among equally projectible hypotheses are in fact normally decided by further criteria, in advance of a direct decision on the basis of additional evidence. For example, it may be that some such further criterion, based on a consideration of the relative strengths of competing hypotheses, enters into certain of our inductive choices. *If* this is indeed the case, then there are some selections we normally make among hypotheses, which are *not* explained by a simple fusion of generalization formula and projectibility

---

This paper is drawn from Part 3 of my *The Anatomy of Inquiry* (New York: Alfred A. Knopf, 1963); new printing Indianapolis: Hackett Publishing, 1981), 314-25.

criteria. The question of further criteria is complex, and surely not yet settled.

Irrespective of the existence of such further criteria, however, it seems clear that the factors so far discussed, namely, consistent evidential support and projectibility via entrenchment, are genuine considerations in induction, in that they account for a wide range of inductive choices. Leaving aside the prospects of extension and refinement which await further studies, and attending simply to the range so far explained, let us now ask if induction has, at least here, been justified.

We have seen the difficulties attending philosophical efforts to answer Hume's challenge. Induction cannot, we saw, be justified through appeal to its alleged presuppositions. Nor can it be justified by reference to the past successes of inductive procedures or predictive policies. Deductive rationales for induction are illegitimate, since inductive judgments are just those whose truth is not demonstrable on the basis of true evidential data.

We noted, further, the recent view that the solution lies in a firm rejection of all demands for a deductive rationale, coupled with an insistence that the concept of justification normally applies to non-deductive inferences. In considering this view, we remarked its failure to specify criteria which differentiate between proper and improper non-deductive inferences, and thus its failure to say under what conditions particular inductions are in fact justified. In the latter respect, we noted that, though it is similar to Hume's own approach, this recent view undertakes less than he did in formulating his theory of habit formation.

We considered, accordingly, the generalization formula as a modern version of Hume's theory, concluding with a review of the theory of projectibility developed by Goodman. At best, we now have a specification that in fact differentiates between proper and improper inductive choices within a significant range. We have thus, in one clear sense, shown certain normal inductive choices to be justified, for we have indicated the rules by which they are singled out from the rest. We have, however, rejected the idea that these rules are immutable features of nature or human nature, and have, in effect, admitted their possible variability.

But in so doing, have we not raised, in an urgent form, the question how the rules themselves are justified? If variety is open to us, what makes our accepted rules preferable? To answer that they are *ours* seems no more satisfying for the theory of induction than for the theory of practice or morals. But, unlike the justification of practical or moral rules, we cannot without circularity justify inductive rules by reference to the consequences they are found to have in experience. Appeal to convention fails to satisfy; appeal to consequences runs in a circle. What is left?

To this problem, Goodman addresses himself in the following words:

Reflections on the Justification of Induction

> The validity of a deduction depends not upon conformity to any purely arbitrary rules we may contrive, but upon conformity to valid rules... But how is the validity of rules to be determined? Here again we encounter philosophers who insist that these rules follow from some self-evident axiom, and others who try to show that the rules are grounded in the very nature of the human mind. I think the answer lies much nearer the surface. Principles of deductive inference are justified by their conformity with accepted deductive practice. Their validity depends upon accordance with the particular deductive inferences we actually make and sanction. If a rule yields inacceptable inferences, we drop it as invalid. Justification of general rules thus derives from judgments rejecting or accepting particular deductive inferences.
>
> This looks flagrantly circular... But this circle is a virtuous one. The point is that rules and particular inferences alike are justified by being brought into agreement with each other... The process of justification is the delicate one of making mutual adjustments between rules and accepted inferences; and in the agreement achieved lies the only justification needed for either.
>
> All this applies equally well to induction. An inductive inference, too, is justified by conformity to general rules, and a general rule by conformity to accepted inductive inferences. Predictions are justified if they conform to valid canons of induction; and the canons are valid if they accurately codify accepted inductive practice.[1]

Goodman thus offers a model of rule-justification that escapes the defects of both conventionalism and utilitarianism (i.e., the appeal to consequences in experience). It may be argued that this model is, moreover, of interest for rules of practice as well as rules of induction. For to justify a practical rule by appeal to the nature of its consequences raises the issue of justification once again, with respect to the principle by which these consequences are selected. And to justify the practical rule by showing it to be derived from another, or stipulated by some convention (actual or hypothetical), analogously raises the question of justification with respect to the governing rule or convention presumed. To construe rule-justification as at some point tying rules back to cases seems thus to provide a way of putting a limit on the upward process by which justified rule leads to more general rule or principle, and the latter leads to a further principle, rule, or convention, etc.

Nonetheless, it is important to see that the advantages offered by Goodman's model of rule-justification do not require that the other modes of justification be surrendered: it is enough if *at some point* justifying principles are tied back to cases; particular rules may then be justified by *connecting* them with the principles in question. It would, indeed, be an oversimplification to suppose that each and every rule is itself adjusted with cases, independently. Rules may, alternatively, be justified through systematic connectedness with other rules or principles, which are themselves adjusted to cases. In some circumstances, indeed, a "systematic" justification of this sort seems clearly preferable to a direct justification by case-adjustment, for the relevant governing principle is much more basic than the rule being judged, and would dislodge it in case of conflict, even if the rule were supported by direct case-adjustment. It is the case-adjustment of the whole set of principles which is in point, rather than that of any given rule that may be in question.

If these considerations are correct, then another way is conceivable in which rules of induction may be justified. They may be shown to fall under or exemplify more general principles, which are themselves supported by overall case-adjustment. There must, of course, be no appeal to experiential consequences in the process, for such appeal would involve the old trouble of circularity through implicit reference to inductive principles. But appeal to consequences is not the only way in which inductive rules may be connected with larger principles. An alternative is to exhibit these rules as exemplifying more comprehensive principles, themselves ultimately supported by case-adjustment. Such a justification does not, of course, preclude direct case-adjustment of the rules themselves, and may rather reinforce it. However, even where direct case-adjustment exists, such justification would be far from trivial, for it would remove the impression of *uniqueness* from the inductive rules in question, showing them to be akin to others, all alike exemplifying certain accepted comprehensive principles.

If such a course is possible, then the challenge to justify induction may be given an additional interpretation: the challenge is *not* to provide a deductive guarantee of inductive success, nor an elaboration of inductive presuppositions, nor an inductive argument for the reliability of inductive policies. Rather, the challenge is *to show that inductive rules exemplify more comprehensive accepted principles, and are thus not idiosyncratic.*

If we look at a basic feature of Goodman's projectibility rules in this light, we may indicate (in a relatively speculative fashion) some directions in which the answer to such a challenge might be sought. This basic feature is (roughly) the preferability of entrenched over non-entrenched predicates, in the formulation of hypotheses which cover the available evidence and go beyond it.

We have here a kind of conceptual inertia or conservatism which has been remarked also in other contexts, under a variety of labels. Quine writes, for example, of "familiarity of principle," noting that it is such familiarity "we are after when we contrive to 'explain' new matters by old

Reflections on the Justification of Induction · 169

laws; e.g., when we devise a molecular hypothesis in order to bring the phenomena of heat, capillary attraction, and surface tension under the familiar old laws of mechanics. Familiarity of principle also figures when 'unexpected observations'... prompt us to revise an old theory; the way in which familiarity of principle then figures is in favoring minimum revision."[2] He thinks of familiarity of principle as a kind of "conservatism, a favoring of the inherited or invented conceptual scheme of one's own previous work."[3]

In another place, he speaks of "the totality of our so-called knowledge or beliefs, from the most casual matters of geography and history to the profoundest laws of atomic physics or even of pure mathematics and logic" as "a man-made fabric which impinges on experience only along the edges." In this fabric, certain statements are more central than others in that they are relatively less likely to be revised in the face of "recalcitrant experience."[4]

> For example, we can imagine recalcitrant experiences to which we would surely be inclined to accommodate our system by reëvaluating just the statement that there are brick houses on Elm Street, together with related statements on the same topic. We can imagine other recalcitrant experiences to which we would be inclined to accommodate our system by reëvaluating just the statement that there are no centaurs, along with kindred statements. A recalcitrant experience can, I have urged, be accommodated by any of various alternative reëvaluations in various alternative quarters of the total system; but, in the cases which we are now imagining, our natural tendency to disturb the total system as little as possible would lead us to focus our revisions upon these specific statements concerning brick houses or centaurs. These statements are felt, therefore, to have a sharper empirical reference than highly theoretical statements of physics or logic or ontology. The latter statements may be thought of as relatively centrally located within the total network, meaning merely that little preferential connection with any particular sense data obtrudes itself.[5]

Several points concerning Quine's views are worth notice. Whereas projectibility involves a certain principle of conservation with respect to predicates (or their extensions), Quine refers, in effect, to a principle of conservation applicable to whole systems of statements. It is clear, too, that he takes the latter principle to be relevant not merely to the so-called empirical sciences, but also to logic and mathematics. By "familiarity of principle" he intends a favoring of the whole "conceptual scheme" of statements rather than a favoring of those single *statements* which are most familiar. For the point is to revise the whole system as little as possible, and to do this may require saving relatively unfamiliar but central *statements* and discarding relatively familiar ones with which they conflict, where the opposite choice would give us a more unfamiliar *system*, as a structured whole. Finally, there are two components in Quine's "familiarity of principle," one counseling minimum revision when the scheme collides with observation or is otherwise found unsatisfactory, the other counseling maximum extension of the scheme to cover new areas that may be brought to light.

Neither of these components of "familiarity of principle" operates alone, on Quine's view. In particular, considerations relating to the "simplicity" of the total scheme enter in, and may, in fact, conflict with the demands of "familiarity of principle." In Quine's opinion, "Whenever simplicity and conservatism are known to counsel opposite courses, the verdict of conscious methodology is on the side of simplicity."[6] Thus, minimum revision under observational challenge is indicated only if no loss in systematic simplicity is anticipated; and maximum extension to new areas is indicated only if it is not accomplished through a sacrifice of simplicity, when compared with alternative ways of incorporating the new material.

We have, then, a suggested principle of conservation of *schemes* as well as a principle of conservation of *predicates* (or extensions) – a principle which is, moreover, not restricted in its application to the empirical sciences. Scheme-conservation, as one principle of scientific strategy, has been frequently noted (under different names), and its justification has not generally been put in terms of demonstrated or probable success. It has more often been held reasonable as effort-saving, or as maximizing intelligibility, where its justification has at all been considered. In the latter form of justification, it has in effect been assimilated to ordinary and philosophical explanations in which the unfamiliar is reduced to the familiar.

The latter idea, moreover, has also been applied to the use of so-called models and analogies in theorizing. A new theoretical scheme seems, in general, to be preferred, in point of intelligibility, to the degree that a familiar model can be provided for it, or an explicit analogy of restricted sort exhibited between it and the inherited scheme. As E. Nagel has written,

> an analogy between an old and a new theory is not simply an aid in exploiting the latter but is a desideratum many scientists tacitly seek to achieve in the construction of explanatory systems. Indeed, some scientists have made the existence of such an analogy an explicit and indispensable requirement for a satisfactory theoretical explanation of experimental laws. And conversely, even when a new theory does organize systematically a vast array of experimental fact, the lack of marked analogies between the theory and some familiar model is sometimes given as the reason why the new theory is said not to offer a 'really satisfactory' explanation of those facts. Lord Kelvin's inordinate fondness for mechanical models is a notorious example of such an

attitude; he never felt entirely at ease with Maxwell's electromagnetic theory of light because he was unable to design a satisfactory mechanical model for it. More recently, a distinguished physicist has argued that a theory for which no visualizable models can be given is just as good as one for which such models are available, provided that both theories enable us to handle experimental problems equally well; and he has made clear that in this latter respect the mathematical formalism of current quantum theory, for which no satisfactory model of this kind is known, is unusually successful. Nevertheless, he has also registered the uncomfortable sense of loss, shared by many physicists, because quantum theory offers no 'explanation' of the experimental facts – a feeling he attributes to the circumstance that we can construct for the theory no physical model in which the 'interplay of elements [is] already so familiar to us that we accept them as not needing explanation.' (P. W. Bridgman, *The Nature of Physical Theory*, Princeton, 1936, p. 63.) It is a matter of historical record that there are fashions in the preferences scientists exhibit for various kinds of models, whether substantive or purely formal ones. Theories based on unfamiliar models frequently encounter strong resistance until the novel ideas have lost their strangeness, so that a new generation will often accept as a matter of course a type of model which to a preceding generation was unsatisfactory because it was unfamiliar. What is nevertheless beyond doubt is that models of some sort, whether substantive or formal, have played and continue to play a capital role in the development of scientific theory.[7]

It would seem that we have here another principle, of *model*-conservation, which operates where scheme-conservation is overridden, perhaps by the demands of systematic simplicity, in meeting an observational challenge or incorporating new data. In such a case, the more "simple," newly preferred schemes are also more deviant, as a whole, relative to the hitherto accepted schemes, than their rejected alternatives. Now the new principle of conservation would seem to come into play, counseling the relative preferability of such of the "simple" schemes as can be provided with familiar models or analogies, of sorts that may be specified or understood from the context. The contribution of model-conservation to intelligibility might be said to consist in its allowing us to continue seeing the old in the new – if not directly by extension of our old scheme, then by extension of certain features of it, as embodied in explanatory models or analogies with roots in our past thinking. (Parallels might be cited with the field of perception, in which antecedent schemas, categories or sets tend to mold our manner of seeing what is new, thus rendering it "intelligible";

the issues here have been illuminatingly treated by E. Gombrich[8] and N. R. Hanson.[9])

We have sought to indicate, in speculative vein, some directions in which an answer to Hume's challenge might be sought – assuming the challenge is to show that inductive rules are not idiosyncratic. We have, in fact, focused particularly on Goodman's projectibility theory, which contains a qualified principle of conservation of predicates or extensions. We have suggested as relevant parallels the principles we referred to as "scheme-conservation" and "model-conservation," which have to do with cases where the inherited theoretical scheme is under threat of forced revision or requires accommodation to new (though compatible) realms of knowledge. The latter principles, we remarked, are neither isolated nor coordinate. Presumably, there is, aside from the demand to avoid violation by the evidence, also a demand for overall, systematic simplicity, which ranks higher than either principle. Furthermore, model-conservation comes into play, it was suggested, only after scheme-conservation has been applied.

There are, to be sure, numerous questions that have been ignored in our speculations. Our above-mentioned principles have themselves been characterized only very sketchily. The notion of systematic simplicity has, furthermore, hardly been characterized at all. (It is perhaps worth noting, in passing, that systematic simplicity must in some way tie in with restrictions on systematic predicates, for, otherwise, any scheme would be transformable into one that is as simple as you like.) But the main purpose of the present discussion, it should be recalled, is *not* to characterize the operative principles of inquiry, nor even to supply a justification of certain inductive rules. Rather, it is to suggest *a manner of conceiving such justification,* and to illustrate this conception by pointing out some plausible directions for the justification of projectibility principles.

We have proposed that one might justify the projectibility principles by pointing out that they are in fact not idiosyncratic. For, like the other two principles discussed, they exemplify a comprehensive principle of conceptual conservation, which counsels the preservation of as much of our intellectual equipment as possible, provided the weightier demands of fact and systematization are satisfied. The latter principle might, in turn, be further justified as effort-saving, and thus tied to principles of practice, or it might be taken to exemplify a principle of intelligibility and thus tied to common as well as philosophical explanations, and to the realm of perception. Either of the latter courses might be preferred to the other, and they might, of course, be combined; alternatively, they might both be thought superfluous. Wherever the appeal to a more comprehensive justifying principle ends, however, the last such principle in the chain is to be justified through case-adjustment. Though the latter mode of justification is, thus, ultimately required, still the fact that it is postponed for one or two steps may enable us to view the framework of induction as something which is not unique, as something which has a general point, or serves a broader purpose.

We return now to the question of scope. We noted, toward the beginning of the present section, that Goodman's proposal (following the main tradition of discussions of induction) is ostensibly limited to simple universal hypotheses of observational sort: the extra-logical predicates are presumed to denote, and elements are presupposed which may be determined to satisfy these predicates, and so to support or violate the hypotheses under consideration. Perhaps some proposal for extending projectibility theory beyond this scope will be forthcoming. Nonetheless, it is well to note the actual limitation of this scope by reference to *de facto* scientific theories. We have seen, in the last Part, the important place of theoretical constructions which are non-observational, and we have noted several of the issues of interpretation to which they give rise.

More recently, we have noted the principles of scheme-conservation and model-conservation, which are not limited to systems of an observational sort. If it be granted that theoretical notions play an important role in science (e.g., in integrating observational hypotheses into compact systems with inductive import and suggestive value for further inquiry), it will probably also be granted that the principles of scheme-conservation and model-conservation in fact enter into our choices among alternative theoretical constructions. We may, more generally, wish to construe such choices as reasonable in principle, though different in detail from the selection of observational hypotheses on the basis of instances recorded in the evidence.

NOTES

1. *Fact, Fiction and Forecast* (Cambridge: Harvard University Press, 1955), 66-67.

2. Willard Van Orman Quine, *Word and Object* (New York and London: published jointly by The Technology Press of The Massachusetts Institute of Technology and John Wiley, 1960), p. 20.

3. Ibid.

4. Willard Van Orman Quine, *From a Logical Point of View*, 2d ed. (Cambridge: Harvard University Press, 1961), chap. 2, section 6, pp. 42, 43.

5. Ibid., 43-44.

6. Quine, *Word and Object,* 20-21.

7. Ernest Nagel, *The Structure of Science* (New York: Harcourt, Brace & World, 1961), 114-15. By permission of the publisher and the author.

8. E. H. Gombrich, *Art and Illusion* (Bollingen Series 35 · 5 [New York: Published for Bollingen Foundation by Pantheon Books, 1960]).

9. Norwood Russell Hanson, *Patterns of Discovery* (Cambridge: Cambridge University Press, 1958).

6. THE PARADOXES OF CONFIRMATION.

WE NOTE certain surprising consequences of joining the equivalence condition with Nicod's rule, taken as a sufficient condition of confirmation. Let us review the matter by recalling the following sentences:

(1) $(x)(\text{Raven } x \supset \text{Black } x)$
(2) $(x)(\sim \text{Black } x \supset \sim \text{Raven } x)$
(3) $(x)((\text{Raven } x \lor \sim \text{Raven } x) \supset (\sim \text{Raven } x \lor \text{Black } x))$

An object *b*, satisfying both antecedent and consequent of (2) is, by Nicod's (sufficient) condition, a confirming instance of (2). Sentence (2) is, however, logically equivalent to (1). Therefore, by the equivalence condition, *b* is a confirming instance of (1), though it seems paradoxical to suppose a non-black non-raven to confirm the hypothesis that all ravens are black. (The above argument can, of course, be put with equal force in terms of the *report* '$\sim$ Black *b* · $\sim$ Raven *b*', rather than the *object b*.)

Furthermore, (1) is also logically equivalent to (3), and should thus, by the equivalence condition, be confirmed by any object, *c*, that confirms (3). By Nicod's condition, however, *c* confirms (3) if it is a black thing, or else a non-raven, black or otherwise. Thus any non-raven confirms the hypothesis that all ravens are black, and any black object confirms the same hypothesis. By analogous argument, we get the surprising result that any non-raven or black thing confirms the hypothesis that all non-black things are non-ravens.

Now these "paradoxical" results flow from conditions which seem, in isolation, perfectly reasonable, even obvious. That an object satisfying both antecedent and consequent of a universal conditional such as (1) *confirms* it seems the most elementary truth about confirmation. That logically equivalent statements have exactly the same weight, as elements of scientific argument, and, in particular, are identically related to instances, seems equally plain. The obviousness of these conditions accounts for the widespread interest accorded the paradoxes of confirmation. For, quite irrespective of concurrence with Hempel's own "satisfaction criterion," there is

---

This selection is drawn from Part 3 of my *The Anatomy of Inquiry* (New York: Alfred A. Knopf, 1963; new printing Indianapolis: Hackett Publishing, 1981), 258-91.

broad consensus as to these conditions themselves. Consequently, there has been general concern over their engendering problematic results.

Of interest to our main theme is the way in which the apparently innocent idea of *a generalization's positive instances* turns out to have unforeseen complexities; the generalization formula, using this idea, is thus not as straightforward as it seems. In the present and the following sections, we shall, however, isolate the question of the paradoxes themselves and consider some of the discussions to which they gave rise, attending first to Hempel's treatment and then to further proposals. Following these considerations, we shall pick up the main thread once more.

It will be recalled that Hempel rejects Nicod's criterion (i.e., as being both a necessary and sufficient condition of confirmation) in part because it violates the equivalence condition. He, further, proposes an alternative criterion which does satisfy this condition. Since his criterion, moreover, gives the same results as Nicod's rule, taken *as a sufficient condition* only, it is clear that the paradoxes of confirmation arise within Hempel's construction. We may illustrate this fact by referring again to (1), (2), and (3) above. The developments of these sentences, for the object b, are, respectively (using 'R' for 'Raven', and 'B' for 'Black'):

(4) $\quad\quad\quad \sim Rb \vee Bb$
(5) $\quad\quad\quad Bb \vee \sim Rb$
(6) $\quad\quad\quad (\sim Rb \cdot Rb) \vee (\sim Rb \vee Bb)$

Clearly, the report:

(7) $\quad\quad\quad\quad \sim Bb \cdot \sim Rb$

implies each of the above developments (4), (5), and (6), thus directly confirming, and therefore confirming, each of (1), (2), and (3).
Clearly, too, the report:

(8) $\quad\quad\quad\quad \sim Rb$

alone, also confirms (1), and (2) and (3) as well, as does the report 'Bb' (which we shall not further discuss, for brevity).
Finally, the report:

(9) $\quad\quad\quad\quad Rb \cdot Bb$

confirms not only (1) and (3), but (2) also.

It is apparent that Hempel himself needs to give some account of the paradoxes in order to make his construction plausible though, as we have suggested above, these paradoxes are by no means peculiar to his construction. Now these "paradoxes" are not formal contradictions; they do not render their containing theories inconsistent. Rather, they represent a violation of

The Paradoxes of Confirmation · 177

our initial sense of the range of positive instances. A construction containing them thus collides with our intuitions in the matter. Intuitively, for example, we take reports (7) and (9) to be quite unequal in their force relative to (1), the latter report clearly positive, the former not. Conversely for (2), we take (7) intuitively as representing a positive case, unlike (9). Intuitively also, we judge (8) as a positive case of neither (1) nor (2). Yet, a construction yielding the paradoxes equalizes the force of (7), (8), and (9) with respect to (1) and (2). What is needed, then, is an explicit treatment of the conflict between construction and intuition: we may modify the construction so as to bring it into line with our intuition, or we may maintain our construction, offering an explanation of the intuitive divergence which shows why this divergence may be disregarded for theoretical purposes.

Hempel takes the latter course. Before presenting his own explanation of the intuitive divergence, however, he criticizes two alternative proposals, based on the idea that scientific hypotheses should not be represented as universal conditionals, after the manner of (1). The first proposal suggests that the hypothesis 'All ravens are black' should rather be represented as:

(10) $\qquad (x)(Rx \supset Bx) \cdot (\exists x)(Rx)$

while the hypothesis that no non-black thing is a raven should be represented as:

(11) $\qquad (x)(\sim Bx \supset \sim Rx) \cdot (\exists x)(\sim Bx)$

each case with an attached premise asserting the existence of an object satisfying the antecedent-predicate. The effect of this proposal is to break the equivalence between 'All ravens are black' and 'No non-black thing is a raven', for, clearly, (10) is not logically equivalent to (11).

Now, for the object b, the development of (10) is:

(12) $\qquad (\sim Rb \lor Bb) \cdot (Rb)$

while the analogous development of (11) is:

(13) $\qquad (Bb \lor \sim Rb) \cdot (\sim Bb)$

The report (9), i.e., '$Rb \cdot Bb$', clearly implies (12) but not (13), which it contradicts. On the other hand, the report (7), i.e., '$\sim Bb \cdot \sim Rb$', does just the reverse, implying (13) but contradicting (12). The report (8), i.e., '$\sim Rb$', implies neither, though it contradicts only (12). Thus (9) but not (7) confirms (10) (which replaces (1)), while (7) but not (9) confirms (11) (which replaces (2)), and (8) confirms neither. We have, it seems, explained the intuitive inequalities between (7), (8), and (9) while maintaining our construction intact. For the construction itself yields these inequalities with respect to (10) and (11), and while it is perhaps natural to err in using

(1) and (2) rather than the former pair to represent scientific hypotheses, our intuition is satisfied as soon as this error is noted and duly corrected.

The foregoing proposal is, however, rejected by Hempel for the following reasons: First, the suggested representation of general hypotheses conflicts with accepted scientific practice, which construes as equivalent such paired statements as 'All sodium salts burn yellow' and 'Whatever does not burn yellow is no sodium salt'. To represent them, in the usual way, as universal conditionals preserves their equivalence, while to represent them, according to the proposal, as conjunctions of such conditionals with disparate existential sentences destroys their equivalence.

Secondly, the proposal is ambiguous as respects alternative conditional formulations. Hempel's example is: 'If a person after receiving an injection of a certain test substance has a positive skin reaction, he has diphtheria'. Using obvious abbreviations, we have at least the following symbolic alternatives:

(14) $(x)((Px \cdot Ix \cdot Rx) \supset Dx)$
(15) $(x)((Px \cdot Ix) \supset (Rx \supset Dx))$
(16) $(x)(Px \supset ((Ix \cdot Rx) \supset Dx))$

The proposal gives us no way of telling which alternative to pick, and hence no way of deciding which existential statement to construe as implied by the original hypothesis, whether '$(\exists x)(Px \cdot Ix \cdot Rx)$', or '$(\exists x)(Px \cdot Ix)$', or '$(\exists x)(Px)$'.

Finally, many scientific hypotheses clearly do not make an existential claim at all. We may here include, perhaps, cases where the denial of such a claim is implicit in the context, as well as cases where the attribution of such a claim appears highly implausible, the theory having rather an "ideal," "hypothetical," or "contrary-to-fact" interpretation. Hempel writes,

> it may happen that from a certain astrophysical theory a universal hypothesis is deduced concerning the character of the phenomena which would take place under certain specified extreme conditions. A hypothesis of this kind need not (and, as a rule, does not) imply that such extreme conditions ever were or will be realized; it has no existential import. Or consider a biological hypothesis to the effect that whenever man and ape are crossed, the offspring will have such and such characteristics. This is a general hypothesis; it might be contemplated as a mere conjecture, or as a consequence of a broader genetic theory, other implications of which may already have been tested with positive results; but unquestionably the hypothesis does not imply an existential clause asserting that the contemplated kind of cross-breeding referred to will, at some time, actually take place.[1]

The Paradoxes of Confirmation · 179

The second proposal rejected by Hempel is the following: The representation of a scientific hypothesis should include, along with the universal conditional, a specification of some class, taken as the *field of application* of the hypothesis in question. The paradoxes of confirmation arise only when such specification is omitted, the specification serving to limit confirming instances to the field indicated. Thus (1) and (2) would be supplanted, respectively, by:

(17) $\quad\quad\quad\quad (x)(Rx \supset Bx)$ $\quad\quad\quad\quad$ [Class: Ravens]

and

(18) $\quad\quad\quad\quad (x)(\sim Bx \supset \sim R)$ $\quad\quad\quad\quad$ [Class: Non-black things]

The report (9), i.e., '$Rb \cdot Bb$', would then confirm the sentence (17) but not (18), since b falls within the class of ravens but not non-black things. On the other hand, the report (7), i.e., '$\sim Bb \cdot \sim Rb$', would confirm the sentence (18) but not (17), since b here falls within the class of non-black things but not ravens. The report (8), i.e., '$\sim Rb$', would confirm neither, since b here does not fall within the class of ravens, and is not described by the report or any of its consequences as belonging to the class of non-black things.

Or, on another variant of the same general idea, the class of ravens might be specified as field of application for (1), '$(x)(Rx \supset Bx)$', *and* its logical equivalents, thus allowing the report (9) to confirm (1) and all its equivalents, but disallowing both (7) and (8). The latter variant, unlike the first, preserves the equivalence condition. For while, on the first variant, for example, (9) confirms (1) but not (2), the latter variant has (9) confirming both (1) and (2). However, the latter variant (unlike the first) violates Nicod's sufficient condition, since, though (7) satisfies both antecedent and consequent predicates of (2), it is not considered a confirming instance.

Both variants of this second proposal thus deal with the paradoxes by modifying the construction so as to bring it into line with intuition. They yield intuitive inequalities between (7), (8), and (9) only by giving up either the equivalence condition or Nicod's sufficient condition. Indeed, the latter variant, in giving up Nicod's condition, also is at odds with our intuitive judgment of (7) as representing a positive case of (2), unlike (9). The second proposal, as a whole, seems inadequate for it provides no special arguments against those considerations supporting the equivalence condition and Nicod's sufficient condition. What recommends it is the fact that specification of some *field of application* is a weaker device than the conjoining of an existential sentence to a hypothesis, and so it escapes some of the criticism leveled against the proposal previously considered; in particular, it avoids the dubious attribution of existential import to scientific hypotheses.

Nonetheless, as Hempel points out, this second proposal is still subject to other objections made against the first one. The field of application to be specified is not clearly determined. Further, the proposal has "no counterpart in the theoretical procedure of science, where hypotheses are subjected to various kinds of logical transformation and inference without any consideration that might be regarded as referring to changes in the fields of application."[2]

Having rejected the two proposals just considered, involving a change in our representation of scientific hypotheses, Hempel proceeds to his own treatment of the paradoxes of confirmation. As indicated earlier, his course is to maintain the construction incorporating the equivalence condition and Nicod's sufficient condition, *and* to offer an explanation of the intuitive divergence which shows why it may be disregarded theoretically. As he puts it, "The impression of a paradoxical situation is not objectively founded; it is a psychological illusion."[3]

To support the latter judgment, he offers the following considerations: First, there is a prevalent, though mistaken, view that hypotheses of the form 'All A's are B's' are about A's only; perhaps the idea is that such hypotheses assert *something* (i.e., being B) about every element of the indicated *subject class* (i.e., the class of A's). But though the sentence form may, in fact, reflect some special interest in A's, this is a practical or psychological point which does not bear on the logical issue. From a logical point of view, such hypotheses have to be taken as imposing restrictions upon, and thus saying something about, all objects whatever within the logical type of the quantifier expression 'all'. Such hypotheses may be rendered:

(19) $$(x)(Ax \supset Bx)$$

and the latter expression says explicitly that, *no matter what object x* is chosen, *if x is A, then x is B* – that is to say, *no x is both A and not B, every x whatsoever* is not A, or is B. (19) thus says about *everything* (within the logical type of the variable 'x') that it conforms to the prescription 'not-A or B'. There is, in short, no reason to separate black ravens, non-black non-ravens, and black non-ravens, as instances of 'All ravens are black'.

Secondly, underlying the intuitive inequalities among positive instances of a hypothesis, there is frequently the illegitimate introduction of extra information. Clearly, in judging the relevance of given instances to a hypothesis h according to any construction, we must assume, for each instance, that we have no other information bearing on h; otherwise we may, in actuality, be judging the relevance of *instance + extra information* rather than that of *instance* alone. If we are not careful to exclude disguised references to extra information, we may suppose real inequalities to obtain among instances

The Paradoxes of Confirmation

when in fact there are only apparent inequalities, due to extra information we have attached to these instances.

Hempel suggests, as an illustration, the sentence: 'All sodium salts burn yellow',

(20) $\qquad (x)(Sx \supset Yx).$

One might feel that yellow-burning sodium constitutes a positive instance of (20), but that neither yellow-burning nor non-yellow-burning non-sodium constitutes such a positive instance. One might attempt to support this feeling, moreover, by arguing as follows: Surely, we should not be adding strength to the hypothesis (20) if we held a bit of pure ice in a colorless flame and found that it did not turn this flame yellow, even though, admittedly, such ice is non-yellow-burning non-sodium, and thus satisfies the equivalent of (20), i.e.:

(21) $\qquad (x)(\sim Yx \supset \sim Sx).$

Hempel admits that the latter experiment would indeed fail to produce additional strength for (20), but denies that a general inequality has thereby been shown between instances that are both S and Y and those that are neither. For compare this ice experiment with another, in which an *unspecified* object a is held in a flame and determined to lack Y, and then analyzed as containing no sodium salt. In the latter case, it is no longer paradoxical to hold that the experiment *has* brought forth an instance, a, constituting strengthening evidence for (21), and therefore (20) as well. But the only significant difference between the two experiments is that, whereas a was initially unspecified, the *first* test substance was initially specified as pure ice, which we *independently* know to be free of sodium salt. As a result, we *independently* knew *this* test substance to constitute a positive instance of (20), and the experiment therefore could produce no *new* positive instance. Let us call this first test substance 'b'. Then, the development of (20) for b is:

(22) $\qquad \sim Sb \vee Yb$

(since (20) literally says, of everything x, that x is not S or Y). Now, if we already have the information:

(23) $\qquad \sim Sb$

we have a premise, concerning b, that implies (22), and thus confirms (20); we know b to constitute a positive instance, in short. We shall thus not be strengthening (20) by finding experimentally either 'Yb' or '$\sim Yb$', for to conjoin either finding to (23) will still, at most, give us a report on b implying (22); it will not give us a *new* positive instance of, i.e., a new object satisfying, (20). No wonder, then, that we felt our first experiment to add no support to (20). The trouble was not, however, that *b* lacked both *S* and *Y*, but rather that *b* was an *old* instance, in virtue of independent information. By contrast, if we are given the object *a*, which we do *not* initially know to be a positive instance of (20), and if we find subsequent reason to accept the report '$\sim Ya \cdot \sim Sa$' on the basis of experiment (such a report implying the relevant development '$\sim Sa \lor Ya$'), we have indeed added support to (20), though *a*, like *b*, lacks both *S* and *Y*.

Analogously, an unspecified item *d*, found to be both *S* and *Y*, strengthens (20); on the other hand, an experimental finding of '*S*' for object *e*, initially known to be yellow-burning, adds no strength to (20), since *e* is already known to represent a positive instance of this hypothesis, which asserts that everything is not *S* or *Y*. Thus, whether a given finding will be judged to add strength to a hypothesis does not depend on its producing a special *sort* of instance, among all those considered to be positive by Hempel's (or an analogous) construction. It does, however, depend upon the extra information assumed prior to the finding in question. To preclude all such extra information in judging the relevance of an instance to a hypothesis is thus to remove certain felt inequalities among instances construed as positive by the construction; it helps, accordingly, to dissolve the paradoxes themselves. The exclusion of extra assumptions requires that the question be put properly from the start. As Hempel suggests, in reference to our recent discussion, "... we have to ask: Given some object *a* (it happens to be a piece of ice, but this fact is not included in the evidence), and given the fact that *a* does not turn the flame yellow and is no sodium salt – does *a* then constitute confirming evidence for the hypothesis? And now – no matter whether *a* is ice or some other substance – it is clear that the answer has to be in the affirmative; and the paradoxes vanish."[4]

As we discussed the case of (20), trouble arose because of an initial extra assumption that the test item *i* was not-*S*, or was *Y*, and therefore already known as a positive instance of (20). But is there not, already at this point, sufficient paradox in supposing *i* to be a positive instance simply because not sodium, or yellow-burning? Yet this supposition is required for the earlier explanation to be at all convincing, for it is this supposition which renders further experimental inspection of *i* superfluous. Here again, argues Hempel, intrusion of extra information is at the root of our feeling of strangeness in taking *i* as positive evidence for (20). As a matter of fact, (20) says that everything is not-*S* or *Y*, and if *i* is not-*S*, or if *i* is *Y*, it conforms to the condition laid down as universal by (20). Suppose we rigorously observe what Hempel calls "the methodological fiction" that we have no other evidence under consideration but the single report:

(24) $\sim Si$

Then (24) supports the hypothesis:

(25) $\qquad\qquad\qquad (x)(\sim Sx)$

and therefore, surely, the weaker:

(20) $\qquad\qquad\qquad (x)(Sx \supset Yx).$

Or, suppose our sole report to be:

(26) $\qquad\qquad\qquad Yi.$

Then (26) supports also:

(27) $\qquad\qquad\qquad (x)(Yx)$

and surely, *a fortiori*, the weaker (20). Now, we do not normally think of (24) or of (26) as confirming, respectively, the strong hypotheses (25) and (27) (nor of these hypotheses as confirmed relative to available evidence), because we have independent information that the latter are false, i.e., that there *is* some sodium salt and that there *is* something not yellow-burning. Once recognizing the need to exclude such extra information, however, we no longer find it strange to take our sole report (24) or (26) as providing positive evidence for (20). The conclusion drawn by Hempel is that the paradoxes "are due to a misguided intuition in the matter rather than to a logical flaw in the two stipulations from which the 'paradoxes' were derived."[5]

FURTHER DISCUSSIONS OF THE PARADOXES OF CONFIRMATION

Despite Hempel's attempt to explain the paradoxes as illusions stemming from (i) a faulty view concerning the reference of universal conditionals, and (ii) the improper intrusion of extra information, there has been considerable further discussion of the paradoxes, and several other proposals have been put forward for dealing with them. We will make no attempt, in this section, to give an exhaustive review of this discussion, nor to evaluate every alternative proposal. Rather, we shall attempt to provide some notion of the range of diverse approaches, and to indicate relevant points of general interest for the theory of confirmation.

We turn first to a claim, put forward by J. W. N. Watkins, that the paradoxes are wholly avoidable in a "Popperian theory of confirmation," i.e., a theory construing confirmation not in terms of conforming instances simply, but rather in terms of those instances determined by unsuccessful attempts at falsification.[6] With respect to the hypothesis:

(1) $\qquad\qquad\qquad$ All ravens are black,

Watkins writes, "On a Popperian theory of confirmation, this hypothesis is confirmed by an observation-report of a black raven, not because this reports an instance of the hypothesis – a white swan is also an instance of it – but because it reports a satisfactory test of the hypothesis: a raven has been examined unsuccessfully for non-blackness. On this view, statements about non-ravens which do not report tests of our hypothesis cannot confirm it."[7]

Presumably the intuitive inequality as between:

(2) $\qquad\qquad Ra \cdot Ba$

and

(3) $\qquad\qquad \sim Ba \cdot \sim Ra$

is explainable by reference to an inequality with respect to *testing*: (2) represents a satisfactory test, whereas (3) does not. Such differences with respect to testing yield just the intuitive inequalities which diverge from Hempel's construction, producing the paradoxes of confirmation. The solution is, then, to alter the construction by suitably incorporating some relevant notion of testing, in order to bring confirmation theory into line with our intuitions in the matter.

Watkins grants, however, that (1) is equivalent to:

(4) $\qquad\qquad$ All non-black things are non-ravens

and agrees, moreover, that "if observations confirm one formulation of a hypothesis they confirm any logically equivalent formulation."[8] If the idea is, then, to consider (2) as confirming (1) because it reports a satisfactory test of the hypothesis in that a raven has been examined unsuccessfully for non-blackness, we must similarly, it would seem, consider (3) as confirming (4) in that a non-black thing has been examined unsuccessfully for ravenness. But then, since Watkins accepts the equivalence of (4) and (1), as well as the equivalence condition, he must agree that (3) also confirms (1), on his "Popperian theory." Thus, the intuitive inequality as between (2) and (3) is not explainable by reference to an inequality with respect to testing. For no inequality with respect to testing has been produced. If it is a defect of any construction such as Hempel's to accord equal confirmatory weight to (2) and (3) relative to (1), the same defect arises for Watkins' "Popperian theory." If Hempel needs to account for this "paradoxical" divergence of his construction from intuition, then so does the proponent of Watkins' "Popperian theory," for mere adoption of the latter does not avoid the paradoxes.[9]

Responding to arguments of the sort just outlined, Watkins has granted that, in allowing what he calls "flatly conflicting observation reports" (e.g., (2) and (3)) to confirm the same hypothesis, his "Popperian theory" is exactly similar to Hempel's. But he has argued, further, that his "Popperian theory" is less paradoxical than Hempel's, in that it refuses to consider *any* instance as confirming unless it represents a *test* of the hypothesis in question.[10]

This idea is elaborated by Watkins as follows: *First*, as he writes,

> Whether or not a certain experimental situation would provide a test for a theory depends on whether or not existing knowledge of that situation indicates that further investigation of it might lead to the falsification of the theory. If we already know that object a is a black raven, that b is black, and that c is non-raven, we *know* that further investigation of a, b, and c will *not* falsify 'All ravens are black'.... If, on the other hand, we know that d is a grey object, nature not yet ascertained, that e is a raven, color not yet ascertained, that f is a grey bird, species not yet ascertained, that g is a bird in the Hamburg Zoo reputed to be a green raven, then we know that further investigation of d, e, f, and g *might* falsify our theory.[11]

Secondly, assuming that we have in every case the requisite conditions for testing the theory, distinctions as to *degree* of confirmation, varying with the severity of the test, still apply. In Watkins' words:

> If various experiments are performed all of whose outcomes turn out to be favorable to the hypothesis, the experiment which best confirms the hypothesis is the one whose outcome, given existing knowledge *minus* the hypothesis in question, was most likely to be unfavorable to the hypothesis. Thus if further investigation of g, the alleged green raven in Hamburg, revealed that g is not, after all, a non-black raven, that would confirm 'All ravens are black' better than would a further investigation of the grey object d which revealed it to be no raven.[12]

Watkins' new argument is thus that Hempel's construction is paradoxical in failing to acknowledge *testing* as a general condition of confirmation.

The way in which Watkins introduces his new argument is worth noting. He writes, "Hempel argued that the 'paradoxes', though counterintuitive, are harmless and do not constitute an objection to his theory of confirmation. I claimed that they are not harmless. I still consider that claim correct; but... my main argument in support of it is incorrect... Hempel's criterion allows flatly conflicting observation-reports to confirm the same hypothesis only if they are over-specified... but on a Popperian criterion two flatly conflicting observation-reports may also confirm the same hypothesis if they are over-specified."[13] (Over-specification is here meant to rule out their being contradictory.) He then goes on to present his new argument that Hempel's theory is paradoxical in failing to acknowledge the notion of testing in appropriate fashion.

Now Hempel's argument for the harmlessness of the paradoxes clearly involved just the *paradoxes of confirmation* as we have earlier discussed them, for example, the construing of reports (2) and (3) as both confirming (1). Watkins' initial argument, that Popperian theory avoids allowing such "flatly conflicting" reports to confirm the same hypothesis, having now been withdrawn, Popperian theory is presumably also faced with the *paradoxes of confirmation*. Now the question is whether or not *these* paradoxes are harmless. Watkins (in the passage just noted) continues to say they are *not* harmless. But he offers no proposal for dealing with them; we have earlier seen, indeed, that the concept of *testing* provides no explanation of the intuitive inequality between (2) and (3). Furthermore, Watkins apparently supposes that his *new* argument supports the superiority of "Popperian theory" with respect to the paradoxes we have all along been discussing. For it is apparently the substitution of the new for the withdrawn argument which supports the continuing claim that these paradoxes are peculiarly harmful to Hempel's confirmation theory. The new argument, however, does *not* address itself to the *paradoxes of confirmation;* it purports rather to reveal a new "paradoxical" consequence of Hempel's theory, which Popperian theory lacks: the consequence that instances satisfying a hypothesis are confirming irrespective of whether they represent tests of the hypothesis in question.

There seems to be a serious confusion here, based on the word 'paradox.' If Watkins holds the original *paradoxes of confirmation* to be harmful, then they constitute a defect in the Popperian theory as well as Hempel's theory, and he offers no proposal for explaining or eliminating them in "Popperian" terms. If, on the other hand, Watkins is rather concerned to show his new *paradoxical consequence* to be harmful to the one but not the other theory, his argument is simply irrelevant to *the paradoxes of confirmation,* even though it purports to point out a paradox which concerns confirmation. In either event, the claim that the *paradoxes of confirmation* constitute an objection to Hempel's theory, though they are avoidable in a "Popperian theory of confirmation," has not been sustained.

What Watkins' *new* argument does, in effect, is to introduce another independent factor, that of *testing,* into the situation, but this factor is perfectly impartial as between those *intuitively unequal* instances or reports (e.g., (2) and (3)) whose *theoretical equalization* gives rise to the paradoxes of confirmation. That the testing factor is indeed impartial in the manner indicated is explicitly stated by Watkins, in his comparison of Hempel's and Popper's theories:

Hempel's "instantiation" or "satisfaction" theory of confirmation has the consequence that 'All ravens are black', being instantiated or satisfied by all black ravens, all black things and all non-ravens, is *automatically* confirmed by *any* observation-report that an object is a black raven, or black, or no raven. Popper's testability theory has the very different consequence that 'All ravens are black' is confirmed by an observation-report that an object is a black raven, or black, or no raven *only* if this reports a *test* of the hypothesis.... If I want to confirm the hypothesis I can do so, on Hempel's theory, by sitting in my study and listing its familiar contents. Popper's theory prescribes a more arduous course – I must go out searching for, or devising, test situations and then investigate them more closely to see if I cannot falsify the hypothesis.[14]

The idea is apparently that, provided the evidence represents a test of the hypothesis, it confirms 'All ravens are black' whether it reports a black raven, a non-raven, or a black object, and, no matter which of these it reports, it will *not* confirm the above hypothesis unless it represents a test of the latter. The testing factor thus represents an extra, impartial condition placed on instances in order for them to be confirming.

Watkins might suggest that the *paradoxes of confirmation* have been at least *limited* in scope, in that pairs such as (2) and (3) now confirm (1) only if they are possible outcomes of test situations. But *such* limitation is hardly to the point: the *paradoxes* are here limited only in the sense that the *scope of confirmation* as such has been narrowed. Furthermore, even were there some *interesting* limitation on the paradoxes, such limitation would not constitute an *explanation* of them, i.e., an account of the gap between intuition and construction, with respect to the confirmatory force of instances.

We shall conclude our consideration of Watkins' view with some remarks concerning the new paradoxical consequence he attributes to Hempel's theory: the consequence that instances satisfying a hypothesis are confirming irrespective of whether or not they represent tests of the hypothesis in question. Although this consequence is, as we have argued, independent of the *paradoxes of confirmation,* it will be of interest to consider it on its own account. What is it that is paradoxical about this consequence? Watkins apparently interprets it to mean that we may confirm, i.e., strengthen the support of, a hypothesis at a given time by experimental determinations which are superfluous in the light of knowledge available at that time. Such interpretation is, however, mistaken. It confuses the notion of evidence which is *positive,* with the notion of evidence which *strengthens the support of a hypothesis at a given time*. Watkins, in stressing the latter notion, indicates his concern with a conception of confirmation quite different from that to which Hempel addresses himself. This may be seen if we consider again Watkins' two elaborations of the testing factor. One is clearly irrelevant to Hempel's objectives in that it involves a distinction of *degrees:* "If various experiments are performed all of whose outcomes turn out to be favorable to the hypothesis, the experiment which *best* confirms the hypothesis is the one whose outcome, given existing knowledge *minus* the hypothesis in question, was most likely to be unfavorable to the hypothesis."[15] As we saw at the very outset, Hempel aims at a *qualitative* characterization of confirming instances. Distinctions of degree among such instances are irrelevant to, but compatible with, this aim.

The other elaboration of the testing factor is not as obviously irrelevant, but equally so in fact. "Whether or not a certain experimental situation would provide a test for a theory depends on whether or not existing knowledge of that situation indicates that further investigation of it might lead to the falsification of the theory. If we already know that object *a* is a black raven . . . we *know* that further investigation of *a* . . . will *not* falsify 'All ravens are black'. . . . If, on the other hand, we know that *d* is a grey object, nature not yet ascertained . . . then we know that further investigation of *d* . . . *might* falsify our theory."[16]

The first point to notice here is the intrusion of the time factor, as indicated by the phrases "existing knowledge," "further investigation," "already know," "not yet ascertained," as well as reference to "experimental situations." Accordingly, an instance does not simply confirm or fail to confirm a hypothesis; it does so at a given *time* (taken as determining a particular state of knowledge, presumably variable also with persons). The gray object *d* is at time *t* not known to be a non-raven (by someone *J*). It is therefore, at *t*, for *J*, capable of being further investigated with negative results. At t_1, *J* determines *d* to be a non-raven; *d* therefore confirms 'All ravens are black' for *J* at t_1. But *d* is thereafter known by *J* to be a non-raven and is thus no longer capable of being further investigated with negative results. Thus, at t_2, *d* does not, for *J*, confirm 'All ravens are black'.

Clearly, the notion involved here does not lend itself to many of the typical uses to which talk of confirmation is put. *J* may, for example, insist, at t_2, that *d* is one item of confirming, i.e., positive, evidence for his hypothesis; he may be bewildered to be told that no sooner does an instance test positively, than it ceases to confirm that for which it provided a positive test. He may be equally bewildered to learn that, as a result, his "best established" beliefs are, *at the moment* (since he is performing no experiments) without confirmation, and hence no better confirmed than his worst. He may, further, maintain that *d* confirms the hypothesis quite independently of whether anyone else is, at the moment, determining that it is no raven.

Nonetheless, there *is some* notion which seems to operate in the

The Paradoxes of Confirmation · 189

indicated fashion. This is the notion of a given report's *strengthening a hypothesis at a given time (for a given person)*. The experimental report, '*d* is a non-black non-raven' may, in fact, at t_1, serve to strengthen the hypothesis, 'All ravens are black', for *J*. Further, at t_2, it will not, typically, *again* strengthen the above hypothesis for *J*. This is *not* to say, however, that *J* will no longer regard it as relevant and favorable, nor that he will fail to regard it as constituting positive evidence irrespective of anyone else's state of knowledge. In sum, an instance which *strengthens* must not only be positive but *new;* whether an instance or experiment confirms a hypothesis is ambiguous as between its being merely *positive or favorable,* and its being *strengthening at the time in question.*

We have encountered the notion of *strengthening a hypothesis at a given time,* in the last section, in Hempel's discussion of the ice experiment: if we know initially that *a* is pure ice, and hence contains no sodium salt, we already know *a* to be a positive instance of 'All sodium salts burn yellow'. Thus to put *a* in the flame and find it does not turn the flame yellow is indeed to fail to strengthen the hypothesis, for it can provide no *new* instance of it. It does *not* follow that *a* is not a favorable instance at all.

Hempel, however, is not concerned to explicate the *strengthening of a hypothesis at a time.* Nor does he offer methodological prescriptions as to how to go about strengthening hypotheses. Thus, he excludes reference to extra information, time, and persons, asking, "Given some object *a* (it happens to be a piece of ice, but this fact is not included in the evidence), and given the fact that *a* does not turn the flame yellow and is no sodium salt – does *a* then constitute confirming evidence for the hypothesis?"[17] Obviously, even if a given item *does* constitute confirming evidence, in this sense, it will not necessarily *confirm,* in the sense of *adding strength* to the hypothesis at a specified time for a given person. (Clearly, Watkins, in the passage last cited, operates in a context different from that of Hempel.)

If we have indeed two distinguishable notions, then we need independent analyses and an account of their relationships. But it will hardly be possible to charge Hempel's theory with a paradoxical consequence on the assumption that he aims to explain *the strengthening of hypotheses.* For, given his objective of accounting for the non-temporal non-personal notion of a *confirming instance* of a hypothesis, it is not at all paradoxical that such instances turn out independent of the assumed knowledge of persons at given times. Analogously, given the objective of accounting for the *strengthening of hypotheses,* it is not paradoxical that a given item strengthens a specified hypothesis for *J* at t_1 but not at t_2.

It is thus misleading to say, as Watkins does, "If I want to confirm the hypothesis I can do so, on Hempel's theory, by sitting in my study and listing its familiar contents."[18] (For such contents, though positive, *confirming* instances are, normally, old and do not (for Hempel) strengthen the hypothesis under such circumstances.) As well say that Watkins counsels us, if we are asked for the evidence in support of our views, to disclaim all the accumulated support gathered from tests until five minutes ago, and to devise methods of testing these views in fresh situations. The constructive task, in sum, is to analyze and relate the several ideas associated with 'confirmation', as found in scientific practice, rather than to pit them against one another.

We return again to the paradoxes of confirmation, and address ourselves to the proposal to explain them by reference to the *size of classes* related in certain ways to the hypothesis in question. The proposal is due to J. Hosiasson-Lindenbaum,[19] but it has figured in a variety of treatments by other writers. The general approach of Hosiasson-Lindenbaum involves a theory of *degrees* of confirmation, but some of her ideas have relevance for the paradoxes, as we have encountered them, involving *qualitative* confirmation. These ideas are discussed, in the latter connection, by Hempel, in the course of his study.

Of the proposal emerging from these ideas, Hempel writes:

> Stated in reference to the raven-hypothesis, it consists in the suggestion that the finding of one non-black object which is no raven, while constituting confirming evidence for the hypothesis, would increase the degree of confirmation of the hypothesis by a smaller amount than the finding of one raven which is black. This is said to be so because the class of all ravens is much less numerous than that of all non-black objects, so that – to put the idea in suggestive though somewhat misleading terms – the finding of one black raven confirms a larger portion of the total content of the hypothesis than the finding of one non-black non-raven. In fact, from the basic assumptions of her theory, Miss Hosiasson is able to derive a theorem according to which the above statement about the relative increase in degree of confirmation will hold provided that actually the number of all ravens is small compared with the number of all non-black objects.[20]

Essentially the same proposal, though interpreted in terms of the risk of falsification, is presented in a paper by D. Pears.[21] "People," he writes, "are too myopic to verify general hypotheticals, and so want to make the fullest use of the limited evidence which they do get. They therefore consider that a general hypothetical does not merely escape falsification; but is confirmed to the extent that it ran the risk of being falsified." But, he argues, the raven-hypothesis runs less risk of being falsified when the search for counterexamples (i.e., non-black ravens) is conducted among non-black things than when it is conducted among ravens. If there is, in fact, a counterexample, it is more likely to turn up in the latter case, "since the class of things which are ravens is smaller than the class of things which are not black."[22] This statement about the relative size of the classes mentioned represents an assumption which people make, and their making it explains why black ravens are thought to provide more confirmation than non-black non-ravens. For a black raven is a member of a class, i.e., *ravens,* in which counterexamples have a greater relative frequency than they have in the class of *non-black* things, to which non-black non-ravens belong – *if* there *are* counterexamples at all. Thus, to have eliminated a given member of the raven class as a counterexample is to have done more to show there *are* no counterexamples than is shown when a member of the non-black class is eliminated as a counterexample. To show there are no counterexamples is, of course, just to show that the raven-hypothesis is true, for this hypothesis *is* the denial that there exists anything which is both a raven and non-black.

Finally (paraphrasing Pears' further argument), to determine an object as being no raven or black is to determine simply that it fails of being a counterexample, without also locating it more narrowly within some class (smaller than the universe) where counterexamples must be found if they exist at all. It is, in effect, to have eliminated an object at large as a counterexample to the hypothesis. But, on the assumption that neither the raven class (surely) nor even the non-black class exhausts the universe of objects, the relative frequency of counterexamples, if any, among objects at large in the universe will be smaller than their frequency among either ravens or non-black things. The raven-hypothesis thus runs least risk of being falsified when the search for counterexamples is conducted among objects at large. Hence to show that such an object is eliminated as a counterexample is to do least in showing that there are in fact no counterexamples, i.e., that the raven-hypothesis is true. We now can explain why the determination of something as *non-raven or black* is thought to provide less confirmation for our hypothesis than its determination as *black raven* or *non-black non-raven.*

A hasty reading of the foregoing proposal for explaining the paradoxes might suggest that it is simply irrelevant, since it concerns itself with *degrees* of confirmation. Such a suggestion would be wrong. The proposal does not simply *introduce* distinctions of degree, but purports to show how such distinctions, coupled with natural assumptions about the relative size of classes, *explain* the intuitive inequality among reports, all of which represent *qualitatively* positive instances. The proposal, in effect, maintains as legitimate the equalization of the reports in question, from the standpoint of a qualitative confirmation theory, but explains the resultant divergence from intuition by appeal to assumptions as to class size, allegedly affecting our judgment of degree in just the manner required by intuition.

After presenting an account of Hosiasson-Lindenbaum's version, Hempel criticizes the proposal on the following counts: First, the requisite assumption as to relative class size is not warranted in every case in which the paradoxes arise. It is not simply that ravens happen to be fewer than non-black things whereas, for some other hypothesis than the raven-hypothesis, the antecedent class might not be smaller than the complement of the consequent class. Rather, argues Hempel, even for the raven-hypothesis itself, the required numerical assumption depends in part upon the choice of language within which hypotheses are to be expressed.

If we choose a "thing-language," with physical things of finite size as our individual elements, then we are probably justified in assuming ravens to be fewer than non-black things. On the other hand, if we choose a "co-ordinate-language," in which finite space-time regions are individuals, the raven-hypothesis needs to be expressed in some such manner as, 'Every space-time region which contains a raven, contains something black'. Now, "even if the total number of ravens ever to exist is finite, the class of space-time regions containing a raven has the power of the continuum and so does the class of space-time regions containing something non-black; thus, for a co-ordinate language of the type under consideration, the above numerical assumption is not warranted. Now the use of a co-ordinate language may appear quite artificial in this particular illustration; but it will seem very appropriate in many other contexts, such as, e.g., that of physical field theories."[23]

Secondly, even choosing a thing-language, we must recall that there is no *logical* guarantee that the required numerical assumption is true. If ravens are fewer than non-black things, this is a fact of nature, and "it remains an empirical question, for every hypothesis of the form 'All P's are Q's', whether actually the class of non-Q's is much more numerous than the class of P's; and in many cases this question will be very difficult to decide."[24]

Now Hempel's first criticism, regarding relativity of the numerical assumption to language choice, might be countered as follows: At most what this criticism shows is that the assumption (in the case of the raven-hypothesis, for example) is *unwarranted* for some choice of language. But warrant is not at all to the point. What is needed is an explanation of the source of our intuitive inequalities among evidential reports. For this purpose, it is sufficient to show that the numerical assumption is actually held; it is not further required that it be shown to have warrant. Conversely, to rebut the proposed explanation, it is not sufficient to show the assumption to lack warrant; it must further be shown that it is not actually held. In fact, Pears puts his version of the proposal in just this way: the numerical assumption is one that is generally held, and the fact that it is generally held explains why it is also generally thought that black ravens provide more confirmation than non-black non-ravens. Furthermore, Hempel's own explanation of the paradoxes traces the intuitive inequalities to a prevalent view concerning the reference of universal conditionals, which is nonetheless *mistaken,* and to the intrusion of extra information, which is *improper.* Surely, it is no rebuttal of his explanation merely to point out the mistake and the impropriety in question.

The foregoing argument is not, however, as sound as it may seem. For consider Hempel's own explanation of the paradoxes. It is true that he traces the intuitive inequalities to a certain view which is held in fact, though mistaken, and to a prevalent intrusion of extra information, which is nonetheless improper. But his *explanatory claim* rests on the assertion that, when the mistaken view is recognized as such and given up, and when the intrusion in question is acknowledged as illegitimate and abandoned, the paradoxes no longer arise, for the intuitive inequalities vanish.

In the present case, however, the situation seems to be quite different. The intuitive inequalities are traced to a numerical assumption which is in fact held; so far the analogy holds. But if the explanatory efficacy of this proposal is to be upheld, it must further be the case that the intuitive inequalities vanish in those cases where the numerical assumption is recognized as unwarranted and accordingly given up. This, however, does not appear to be claimed nor does it appear to happen. Non-black non-ravens still appear intuitively unequal to black ravens with respect to 'All ravens are black' when translation from a thing-language to a coordinate language has been effected (or when the relativity to choice of language has been made clear). The report 'This is a space-time region neither containing anything black nor containing a raven' retains its "paradoxical" character when construed as representing a positive instance of 'Every space-time region which contains a raven, contains something black'. In general, it may be remarked that the direction of this proposal differs from that of Hempel's own answer. Hempel suggested that the paradoxes are *psychological illusions* resting, in every case, on mistaken conceptions which need to be given up. The present proposal suggests that, at least in typical cases, the "paradoxical" inequalities *arise from a real factor in the situation* (i.e., the described class-size relationships), and are thus *not* illusory. This real factor is, however, as we have seen, not pervasive enough to do the job demanded of it.

Thus, the numerical assumption in question cannot be supposed to explain the paradoxes of confirmation. Even though there are cases where the assumption does in fact have warrant, and even granted that our concept of confirmatory *degree* may reflect such warrant in the manner suggested by the proposal, the paradoxes require explanation in other terms.

Hempel's second criticism reinforces this view, by pointing out that the paradoxes are invariant though the numerical assumption cannot be supposed invariably to hold. Where the assumption is doubtful, or clearly false, paradoxicality nonetheless persists. We may illustrate by suggesting the hypotheses 'All molecules are inanimate' and 'All invertebrates lack kidneys'. Whether we take as doubtful or false the required numerical assumptions (i.e., that molecules are fewer than animate things, and that invertebrates are fewer than things not lacking kidneys) it remains paradoxical to take the family cat as a positive instance of either hypothesis.

There is, however, a further criticism, beyond the two presented by Hempel, that may here be offered against the proposal under consideration: it does not fully address itself to the paradoxes of confirmation. These paradoxes involve intuitive inequalities (among reports) *which themselves vary* with respect to logically equivalent versions of "the same hypothesis." The present proposal abstracts from the question of equivalent versions altogether by speaking only of *the raven-hypothesis*. It thus *cannot* explain the paradoxes of confirmation fully.

Consider, for example, the equivalent variants of *the* raven-hypothesis:

(1) $\qquad (x)(Rx \supset Bx)$
(4) $\qquad (x)(\sim Bx \supset \sim Rx)$

and the reports:

(2) $\qquad Ra \cdot Ba$
(3) $\qquad \sim Ba \cdot \sim Ra$
(5) $\qquad \sim Ra \lor Ba$

Paradoxes arise from the fact that the equivalence condition and Nicod's sufficient condition, together, equalize the reports (2), (3), and (5), with respect to (1), and with respect to (4). Such equalization is contrary to intuition, which treats these reports differently. But the intuitive differences *vary* with the formulation of the hypothesis chosen. With respect to (1), (2) appears positive, but (3) and (5) do not. With respect to (4), on the other hand, (3) appears positive, but (2) and (5) do not. Those answers to the paradoxes earlier considered, resting on the attachment of different existential statements to (1) and (4), or the specification of different fields of application for them, or indication of the prevalent but mistaken construal of (1) and (4) as referentially different – all attempt to account for the *variation* of intuitive inequalities among reports, with choice of (1) or (4) to represent *the* (abstract) raven-hypothesis.

The present proposal makes no mention of such variation at all, yielding simply inequalities among (2), (3), and (5). Relative to (1), the main inequality yielded is indeed the one wanted, i.e., (2) comes out better than (3). (The further inequality yielded by Pears' version, that (3) is better than (5), is also welcome, since intuitively (5) appears positive for neither (1) nor (4), whereas (3) appears positive for at least (4).)

But the main inequality yielded is the *reverse* of what is wanted relative to (4). For here, intuition asks that (3) come out better than (2), and the proposal persists in saying that (2) is better than (3). To put it differently, the proposal simply ranks (2), (3), and (5) in degree of confirmation, granting a positive weight to each relative to the raven-hypothesis, *no matter how formulated.* It follows that (2) equally confirms (1) and (4), though intuition denies it confirms (4). It follows also that (3) equally confirms (1) and (4), though intuition denies it confirms (1). (It follows, finally, that (5) equally confirms (1) and (4), though intuition denies it confirms either.)

Though, relative to (1), the qualitative difference between (2) and (3) *is* appropriately correlated with a difference in degree, this latter difference does *not* properly reflect, but runs counter to, the qualitative difference relative to (4). Thus, given the single report (2), there is no explanation of the qualitative difference when we shift from (1) to (4), and, analogously, given the report (3), there is no explanation of the qualitative difference when we shift from (4) to (1).

At best, then, we have in the present proposal only a partial answer to the paradoxes, and one which violates our troublesome intuition as much as it tends to support it. Coupled with the objections discussed above, the last criticism indicates the weakness of the appeal to class size as an explanation of the paradoxes of confirmation. We may, of course, independently wish to acknowledge class size as a relevant factor in determining *degrees* of confirmation, but that is another story.

We turn now to another idea for explaining the paradoxes, suggested by N. Goodman.[25] This idea rests on the observation that logically equivalent statements do not generally have the same or logically equivalent contraries, so that reports equally satisfying logically equivalent statements may yet differ in the way they eliminate alternative hypotheses.

Hempel's answer to the paradoxes, it will be recalled, suggested that the intuitive inequalities are due to (i) faulty views concerning the reference of universal conditionals, and (ii) improper intrusion of extra information (the basic idea of the latter point attributed by him to Goodman). Since the paradoxes are, on this account, due to *faulty and improper* conceptions, Hempel argued that "the impression of a paradoxical situation is not objectively founded; it is a psychological illusion."[26] We have seen how alternative attempts to ground the paradoxes on objective features of the situation (e.g., the testing factor, and class size relationships) failed to yield the desired inequalities in every case, thus reinforcing Hempel's view of the paradoxes as illusory.

Goodman now suggests that even when faulty and improper conceptions are removed, there remains an objective, logical feature of the situation that yields just the inequalities required by intuition. Discussing the confirmation of (1) by (3), he writes:

> We arrive at the unexpected conclusion that the statement that a given object is neither black nor a raven confirms the hypothesis that all ravens are black. The prospect of being able to investigate ornithological theories without going out in the rain is so attractive that we know there must be a catch in it. The trouble this time, however, lies not in faulty definition, but in tacit and illicit reference to evidence not stated in our example. Taken by itself, the statement that the given object is neither black nor a raven confirms the hypothesis that everything that is not a raven is not black as well as the hypothesis that everything that is not black is not a raven. We tend to ignore the former hypothesis because we know it to be false from abundant other evidence – from all the familiar things that are not ravens but are black. But we are required to assume that no such evidence is available. Under this circumstance, even a much stronger hypothesis is also obviously confirmed: that nothing is either black or a raven. In the light of this confirmation of the hypothesis that there are no ravens, it is no longer surprising that under the articifial restrictions of the example, the hypothesis that all ravens are black is also confirmed. And the prospects for indoor ornithology vanish when we notice that under these same conditions, the contrary hypothesis that no ravens are black is equally well confirmed.[27]

Everything in this passage until the last sentence is familiar, suggesting that the paradoxicality of supposing (3) to confirm (1) is wholly due to improper reference to extra information, and that removal of such reference leaves (2) and (3) on a par with respect to their confirmation of (1). But the last sentence introduces the idea that (3) is still, in a special sense, worse than (2). For whereas (3) confirms (in the "satisfaction" sense) both (1) *and* its contrary:

(6) $(x)(Rx \supset \sim Bx)$

(2) confirms (1) but disconfirms (6). Report (2) is therefore not, like (3), merely a bit of indoor ornithology.

The situation is, moreover, fittingly reversed when we consider (4). For the contrary of (4) is *not* (6), but rather:

(7) $(x)(\sim Bx \supset Rx)$

and here we find that, although (2) and (3) both confirm (4), only (3)

The Paradoxes of Confirmation · 197

disconfirms (7), while (2) confirms it. Finally, (5) confirms (1) and (4), but fails to disconfirm either (6) or (7). The report:

(8) $\sim Ra \cdot Ba$

is even worse, confirming each of (1), (4), (6), and (7). We thus find the required inequalities in their requisite variation with choice of (1) or (4) to represent *the* raven-hypothesis.

The basic datum upon which the present idea rests is the fact that logically equivalent statements may have contraries which are not logically equivalent to one another. Thus, although (1) and (4) are logically equivalent and thus "have the same content," or "impose the same restrictions upon objects," their contraries, respectively (6) and (7), are not themselves logically equivalent. The suggestion is, then, that reports which alike represent positive instances of the same (abstract) hypothesis may yet differ in their confirming-or-disconfirming relationships to different statements which are contraries of equivalent formulations of the hypothesis in question.

The construction which takes both (2) and (3) as positive instances of (1) is thus correct; the intuitive inequality between (2) and (3) does not require revision of their equal status *with respect to* (1). Rather, the construction itself may show this inequality by yielding differential status respecting (6), for (2) disconfirms (6) while (3) confirms it. An analogous argument gives the desired reverse conclusion where (4) is taken rather than (1).

The upshot of this idea is thus that the paradoxes are not wholly illusory but arise out of intuitions marking objective distinctions of a logical sort. We have, furthermore, a new complication to cope with, respecting the notion of confirming instances. For to say that a report represents a confirming instance of a given (abstract) hypothesis is not in itself sufficiently informative with regard to statements that may be taken as contraries. Where a given contrast is intended, say between (1) and (6), the fact that (3) confirms (1) will not be of much interest, for it also confirms (6). (It is worth noting here that the generalization formula wrongly assumes that a hypothesis has but one contrary, for it speaks of *the* contrary of a hypothesis being ruled out by positive instances of the latter.)

In ordinary parlance, the intent to make a particular contrast is indicated by choice of a particular *formulation* of the hypothesis, e.g., (1) rather than (4). The air of paradox results from (i) such a choice, with its attendant expectation that a confirming report will be one that *selectively* confirms (1), relative to its contrary, (6); and (ii) the construal of confirming reports as satisfying the equivalence condition, and therefore *not* guaranteed to fulfill the above expectation. (For we have seen that, though (3) confirms (4) and *selectively* confirms it relative to its contrary (7), it does *not* selectively confirm (1) relative to *its* contrary, (6).)

Thus, we have some such notion as *selective confirmation* which does *not* fulfill the equivalence condition – *not* because a report may satisfy some statement but fail to satisfy its equivalent, but rather because it may satisfy also the contrary of one but fail to satisfy, and indeed violate, the contrary of the other. Thus some apology is due to Nicod, for aside from other criticisms leveled against his original criterion, that resting on the equivalence condition seems to have been too strong. In fact, the notion of *selective* confirmation *does* go beyond the invariant content of a hypothesis and does *not* carry across equivalent formulations. Logically equivalent statements do bear the same *satisfaction* relations to all instances but not the same *"confirmation"* relationships generally, unless these are initially reduced to satisfaction relationships. We thus need to reconsider the status of the equivalence condition.

Hempel argues that the equivalence condition is necessary because it would be absurd to suppose that it was "sound scientific procedure to base a prediction on a given hypothesis if formulated in a sentence S_1, because a good deal of confirming evidence had been found for S_1; but that it was altogether inadmissible to base the prediction ... on an equivalent formulation S_2, because no confirming evidence for S_2 was available."[28] But take (4) as our case of S_1, and imagine all the evidence to consist of statements such as (3). True, (3) satisfies (4) and also (1). But it also satisfies the contrary of (1), i.e., (6).

Do we have any reason, so far, for predicting that a new-found raven will be black rather than not? Since (1) and (6) together imply that there are no ravens, our new-found raven forces us to give up at least one of these statements. If we give up (6) and predict 'Black', we can retain (4). If we give up (1) and predict 'Not black', we have to give up (4) as well, for (4) and (6) are incompatible, given the existence of our raven. We might suppose we have here a reason for retaining (1) and predicting 'Black'. But, on the contrary, if we predict 'Black', thus saving (4), we shall need to give up another hypothesis hitherto confirmed, i.e., that nothing is black, whereas if we yield (4) and predict 'Not black', we can save the latter hypothesis. Here, it seems, is a case where basing a prediction directly on (4) (i.e., predicting 'Non-raven' for a new instance of 'Non-black') is beyond suspicion, while basing a prediction directly on its equivalent, (1), is a matter of balanced decision. The reason, furthermore, is *not* that "no confirming evidence" (in the sense of satisfaction) is available for (1), but that whatever is available also supports its contrary, (6).

The foregoing discussion is no argument *for* Nicod's original criterion, which suffers from other defects mentioned earlier. Nor is it an argument *against* constructions such as Hempel's, which satisfy the equivalence condition. For in terms of Hempel's definitions of confirmation, disconfirmation, and neutrality, requisite notions of *selective confirmation,* of various sorts, can be further explained, in (roughly) the manner in which they were introduced in our discussion above. What *has* been suggested is the *importance* of such notions of selective confirmation. We need to *distinguish,* in any event, between *evidence which simply accords with a statement,* and evidence which *accords with it but not also with its contrary* (in the extreme case violating its contrary). It may be further suggested that the *confirming* of a hypothesis perhaps typically involves the *favoring* of it in this way as against a contrary one. Thus, if the concept of a *positive instance* is construed simply in terms of *according with* a given statement, we shall have to provide also for the stronger notion (or notions) of a *positive instance which also favors* the statement in question.[29]

NOTES

1. Carl G. Hempel, "Studies in the Logic of Confirmation," *Mind,* n.s. 54 (1945); (I) pp. 1-26; (II) pp. 97-121. The passage cited here is from (I), p. 16.

2. Ibid., 18.

3. Ibid.

4. Ibid., 20.

5. Ibid., 20-21.

6. J. W. N. Watkins, "Between Analytic and Empirical," *Philosophy* 32 (1957): 112-31; and "A Rejoinder to Professor Hempel's Reply," *Philosophy* 33 (1958): 349-55. By permission of the author. The latter was a response to Carl G. Hempel's article in the same volume, "Empirical Statements and Falsifiability," 342-48.

7. Watkins, "A Rejoinder to Professor Hempel's Reply," 351.

8. Watkins, "Between Analytic and Empirical," 116.

9. This paragraph draws on my "A Note on Confirmation," *Philosophical Studies* 11 (1960): 21-23. A similar criticism had earlier been offered by H. G. Alexander, "The Paradoxes of Confirmation," *British Journal for the Philosophy of Science,* 9 (1958): 227-33, in a paper that came to my notice only after my note had been written in spring 1959.

10. J. W. N. Watkins, "Professor Scheffler's Note," *Philosophical Studies* 12 (1961): 16-19. By permission of *Philosophical Studies* and University of Minnesota Press. On the points at issue, there was, aside from published papers, a correspondence between Watkins and myself, in response to my "A Note on Confirmation." My published reply to "Professor Scheffler's Note" is "A Rejoinder on Confirmation," *Philosophical Studies* 12 (1961): 19-20. Watkins has set forth an extended presentation of his recent views on these

matters, in "Confirmation, Paradox, and Positivism," included in the forthcoming book: *The Critical Approach: Essays in Honor of Karl Popper,* ed. M. Bunge.

11. Watkins, "Professor Scheffler's Note," 18.

12. Ibid.

13. Ibid., 17.

14. Ibid., 18.

15. Ibid.

16. Ibid.

17. Hempel, "Studies in the Logic of Confirmation," 20.

18. Watkins, "Professor Scheffler's Note," 18.

19. Janina Hosiasson-Lindenbaum, "On Confirmation," *Journal of Symbolic Logic* 5 (1940): 133-48.

20. Hempel, "Studies in the Logic of Confirmation," pp. 21-22, fn. 2.

21. David Pears, "Hypotheticals," *Analysis* 10 (1950): 49-63. The passage cited in the text is from p. 50. By permission of the publisher.

22. Ibid.

23. Hempel, "Studies in the Logic of Confirmation," p. 21, fn. 2.

24. Ibid.

25. I am indebted to Goodman for discussion of his idea, which amplified his published treatment in *Fact, Fiction, and Forecast.*

26. Hempel, "Studies in the Logic of Confirmation," (I) 18.

27. Goodman, *Fact, Fiction, and Forecast,* 71-72.

28. Hempel, "Studies in the Logic of Confirmation," 12-13.

29. These notions must, furthermore, be distinguished from that of *strengthening instances,* i.e., those which are novel at t, and accord with, or selectively favor, the hypothesis at stake.

For another sort of development of the general notion of favoring instances, see S. Morgenbesser, "Goodman on the Ravens," *Journal of Philosophy* 59 (1962): 493-95.

7. SELECTIVE CONFIRMATION AND THE RAVENS

(WITH NELSON GOODMAN)

GIVEN HEMPEL'S "satisfaction" criterion of confirmation,[1] a notion of *selective confirmation* may be introduced,[2] which allows us to characterize evidence not simply as according with a hypothesis, but, further, as favoring the hypothesis rather than its contrary. Thus, the statement that all ravens are black is not merely *satisfied* by evidence of a black raven but is *favored* by such evidence, since a black raven disconfirms the contrary statement that all ravens are not black, i.e., satisfies its denial. A black raven, in other words, satisfies the hypothesis *that all ravens are black rather than not:* it thus selectively confirms *that all ravens are black*.

By contrast, a nonblack nonraven satisfies the hypothesis that all ravens are black, but it also satisfies the contrary hypothesis that all are not black; nor is this surprising in view of the fact that it satisfies still a third hypothesis, i.e., that there are no ravens at all. Thus, a nonblack nonraven may be said to satisfy or accord with the hypothesis that all ravens are black, but it can hardly be said to favor it as against its contrary, since it equally satisfies this contrary. It cannot, in other words, be held to satisfy the hypothesis *that all ravens are black rather than not*, i.e., to provide selective confirmation *that all ravens are black*. Insofar as, in ordinary circumstances, we often have in mind the question whether available evidence *favors* a hypothesis rather than simply according with it, it is of general interest to examine the notion of *selective confirmation*.

SELECTIVE CONFIRMATION AND THE PARADOXES OF CONFIRMATION

The notion of selective confirmation takes on special interest in connection with the "paradoxes of confirmation": If we suppose (1) that a simple universal conditional is confirmed by joint satisfaction of its anteced-

This paper appeared in the *Journal of Philosophy* 69 (10 Feb. 1972): 78-83.

ent and consequent, and if we also adopt (2) the so-called *equivalence condition,* requiring that whatever confirms any statement confirms also all its logical equivalents, we must conclude that a blue chair, for example, confirms the hypothesis that all ravens are black. For it satisfies both antecedent and consequent of the contrapositive equivalent of the latter hypothesis, i.e., "All nonblack things are nonravens." Similarly, the hypothesis that every nonblack thing is a nonraven will be confirmed not only by nonblack nonravens but also by objects satisfying *its* contrapositive equivalent, i.e., by ravens that are black. Hempel's "satisfaction criterion" of confirmation, in particular, yields such results since it incorporates both assumptions (1) and (2) above.

Now *selective confirmation* turns out to bear on the paradoxes in that, unlike the "satisfaction criterion," it fails to fulfill the equivalence condition; that is to say, it is free of the above-mentioned assumption (2). For it rests on the idea of favoring a conditional hypothesis as against its contrary, and logically equivalent hypotheses as a matter of fact have contraries that are not logically equivalent to one another. So, for example, the hypothesis (i) *that all ravens are black rather than not,* differs logically from the hypothesis (ii) *that all nonblack things are nonravens rather than not,* since the excluded contrary in the first case is "All ravens are not black," but the excluded contrary in the second case is "All nonblack things are ravens." It is thus not surprising that what satisfies (i) fails to satisfy (ii), and vice versa. Correspondingly, what *selectively confirms* "All ravens are black," for example, may not selectively confirm its logical equivalent, "All nonblack things are nonravens." And indeed, a black raven which, as we earlier saw, selectively confirms "All ravens are black," fails to selectively confirm "All nonblack things are nonravens," since it satisfies also the latter's contrary, i.e., "All nonblack things are ravens." And contrariwise, too, a nonblack nonraven which, as we saw earlier, fails to provide selective confirmation for "All ravens are black," *does* favor the hypothesis that all nonblack things are nonravens. For it satisfies the latter but also the denial of its contrary; that is, it negates the contrary "All nonblack things are ravens".

So selective confirmation, explained in terms of "satisfaction", which does fulfill the equivalence condition, itself violates this condition. It thus affords a way of interpreting the paradoxes of confirmation. For though a blue chair satisfies the hypothesis that all ravens are black, it does not satisfy it any more than it does the hypothesis that all ravens are not black; it does not *favor* or *selectively confirm* it as against its contrary. And though a black raven satisfies the hypothesis that all nonblack things are nonravens, it equally satisfies the contrary hypothesis that they *are* ravens, and so cannot be said to favor either. If we now take into account the particular contrast indicated by assertion of a *given* hypothesis rather than some equivalent, and assume that our aim is to find evidence *favoring,* and not merely *satisfying* the hypothesis in question, we have objective grounds for honoring the feeling that a statement may be "confirmed" by evidence that fails to "confirm" its equivalent.

MR. FOSTER'S CRITICISM

In a recent paper,[3] Lawrence Foster suggests that the above response to the paradoxes[4] involves a rejection of "the equivalence condition in order to have a theory of confirmation that allegedly accords with our intuitions," and Foster contrasts this approach with that of Hempel, who keeps the equivalence condition, arguing that our contrary intuitions are misguided. Foster also suggests that Scheffler offers an independent argument against the equivalence condition, upon which argument the plausibility of the *selective confirmation* approach depends, at least in part.

In contrast to these suggestions which we consider rather misleading, it should be pointed out that AI does not independently attack the equivalence condition "in order to" develop a satisfactory theory of confirmation. Rather, the notion of selective confirmation, found to provide an objective logical feature underlying the paradoxes, occasions a reconsideration of the equivalence condition, since "selective confirmation does go beyond the invariant content of a hypothesis and does not carry across equivalent formulations." Nor is there even a general rejection of the condition as such. As AI remarks, there is here no "argument *against* constructions such as Hempel's, which satisfy the equivalence condition. For in terms of Hempel's definitions of confirmation, disconfirmation, and neutrality, requisite notions of *selective confirmation,* of various sorts, can be further explained, in (roughly) the manner in which they were introduced in our discussion above. What *has* been suggested is the *importance* of such notions of selective confirmation. We need to distinguish in any event, between *evidence which simply accords with a statement,* and evidence which *accords with it but not also with its contrary*... It may be further suggested that the *confirming* of a hypothesis perhaps typically involves the *favoring* of it in this way as against a contrary one" (290/1).

There is in short, in AI, no absolute choice of a single idea to represent pre-analytic conceptions of confirmation, but rather a plurality of notions. There is, moreover, no incompatibility between Hempel's concept of satisfaction and that of selective confirmation; in fact, since the latter is defined in terms of the former, it is equally available to Hempel. Surely there is no difficulty in the fact that the former, though not the latter, harbors an equivalence condition. Finally, *that* the latter does not incorporate such a condition follows from the way it is defined; the matter does not rest on independent argument.

Foster argues that to connect confirmation theory with a theory of rationality requires fulfillment of the equivalence condition, for "it would be odd to maintain that it is rational to believe a hypothesis S_1 on the basis of evidence E but not rational to believe S_2 on the basis of E even though S_1 and S_2 are logically equivalent." We do not find this argument at all persuasive. For, on the one hand, it does not follow, from the fact that E selectively confirms S_1 but not S_2, that it is rational to believe S_1 but not S_2. And, on the other hand, it is, we suggest, not odd at all to hold it rational to believe that all ravens are black *rather than not*, given solely a black raven, while denying the rationality of believing, on the same evidence, that all nonblack things are nonravens *rather than not*. Foster further states that according to his "confirmation intuitions, it seems clear that finding a black raven does increase the credibility of the hypothesis that all nonblack things are nonravens." But since a black raven does not *favor* the latter hypothesis as against its contrary, is it intuitively obvious that the hypothesis achieves any *net* advantage?

Foster says that "Scheffler's argument against the equivalence condition is mistaken." He suggests that if the argument of AI (290) is refuted, the barriers to reinstatement of the equivalence condition for confirmation have been removed. But even if AI (290) is eliminated utterly from consideration, it remains clear that selective confirmation, as defined, fails to fulfill the equivalence condition.

The argument in AI (290) is, as a matter of fact, intended as an auxiliary consideration, against Hempel's contention that surrender of the equivalence condition is at odds with the deductive uses of hypotheses in scientific reasoning. Hempel rightly argues that a deductive argument will remain valid if any of its premises are replaced by logical equivalents. In particular, a prediction got from one hypothesis will be equally derivable from any logically equivalent hypothesis. He claims, however, that this principle is violated if a concept of confirmation is adopted that fails to satisfy the equivalence condition.

Instead of directly (and properly) attacking the latter claim, AI (290) attempts rather to show that in certain circumstances, a prediction "based directly" on one form of a theory is beyond suspicion, while another "based directly" on an equivalent form is not. Specifically, it is argued that, (in the circumstances in question) where "All nonblack things are nonravens" is selectively confirmed but its contrapositive is not, to predict that a new nonblack thing will be a nonraven is clearly reasonable, while to predict that a new-found raven will be black is not; instantiating the antecedent of the former conditional deductively yields an acceptable prediction, whereas parallel instantiation of the latter's antecedent deductively yields a prediction that is a matter of "balanced decision."

Foster does not question the example, but he rightly criticizes the use made of it to suggest that it bears directly on Hempel's point. The example, that is, does *not* show that the *same* prediction is derivable from one but not another logical equivalent, and in stressing this point Foster is perfectly correct. Instead it shows that parallel predictions, derivable by identical deductive processes (e.g., modus ponens) from equivalent hypotheses in combination with parallel minor premises, nevertheless differ in point of scientific reasonableness. So the adoption of the concept of *selective confirmation* does not, it seems, have the stronger (and unwelcome) consequences stated by Hempel, and the argument in AI does not show it does. What *is* established is the weaker (though still surprising) consequence that certain deductively parallel features of equivalent hypotheses may vary in their predictive significance when some but not all of these hypotheses are selectively confirmed. In sum, the stronger consequence, which *is* a defect, does not follow from rejection of the equivalence condition, whereas the weaker consequence (mitigating the predictive parallelism of equivalent hypotheses) follows but is no defect at all.

MR. FOSTER'S PROPOSAL

Foster proposes one more solution of the paradox of the ravens, dismissing (on grounds that are unclear) Hempel's explanation that the paradoxical flavor results from equivocation between the stipulated evidence for a given case and the total body of evidence actually available.

Foster adopts the satisfaction criterion of confirmation, modified by the requirement of projectibility of hypotheses, and takes all equivalents of a confirmed hypothesis to be confirmed. He holds that "All nonblack things are nonravens" is, by virtue of its ill-entrenched predicates, not projectible and hence not confirmed by a nonblack nonraven. However, this hypothesis is for him confirmed by a black raven, since the black raven confirms the equivalent hypothesis "All ravens are black."

Since on this account, a black raven but not a thing that is neither black nor a raven would count as evidence that whatever is not black is not a raven, one wonders at Foster's feeling that he has in any way mitigated the paradox. Furthermore, Foster's claim that "All nonblack things are nonravens" is not projectible needs a closer look for many reasons: (1) Even granting that the predicates here are ill entrenched, this seems to illustrate no general principle. Surely 'nonmetallic', 'noncombustible', 'invisible', 'colorless', and many other privative predicates are well entrenched. Furthermore, it should be noted that a privative predicate will be as entrenched as any of its coextensive predicates. (2) Ill-entrenchment of predicates does not make a hypothesis unprojectible; it is projectible if no conflicting hypotheses are at least as well entrenched, and it is neither projectible nor unprojectible if some conflicting hypothesis is equally entrenched while none is better entrenched. (3) One may question whether the predicate to be compared for entrenchment should or should not include the logical constant '$\sim$'.

NOTES

1. Carl G. Hempel, "A Purely Syntactical Definition of Confirmation," *Journal of Symbolic Logic* 8 (1943): 122-43; and "Studies in the Logic of Confirmation," *Mind* n.s. 54 (1945), I, 213 (January): 1-26; II, 214 (April): 97-121.

2. See Israel Scheffler, *The Anatomy of Inquiry* (New York: Knopf, 1963), p. 289. (This book will be referred to in the text as AI). See also Nelson Goodman, *Fact, Fiction, and Forecast* (Cambridge: Harvard, 1955), pp. 71-72. (2d ed., Indianapolis: Bobbs-Merrill, 1965, pp. 70-71.)

3. "Hempel, Scheffler, and the Ravens," *Journal of Philosophy* 68 (25 Feb. 1971): 107-14.

4. Discussed in *The Anatomy of Inquiry,* 286-91, and the passages in *Fact, Fiction, and Forecast* mentioned in fn 2 above.

8. AN IMPROVEMENT IN THE THEORY OF PROJECTIBILITY

(WITH ROBERT SCHWARTZ AND NELSON GOODMAN)

DISCOVERY of a long-overlooked discrepancy in *Fact, Fiction, and Forecast*[1] has pointed the way to an important improvement in the theory of projection. The rules of unprojectibility can now be reduced to one, so that elementary projectibility can be defined in a simple and straightforward way.

On page 102 of the second edition of FFF the hypothesis:

> All emerubies are green.

where the evidence consists of green emeralds and red rubies, all examined before t, is said to be eliminated according to the first rule as a result of conflict with the better entrenched:

> All rubies are red.

Actually, the first rule reads as if it applied only where the *consequent*-predicates differ in entrenchment.

The trouble lies not in the statement on page 102, but in the formulation of the first rule. It was meant, as was often assumed later in the text, to cover all cases of genuine conflict between two hypotheses of unequal entrenchment, i.e., between hypotheses with equally well entrenched antecedent-predicates and unequally well entrenched consequent-predicates, or with equally well entrenched consequent-predicates and unequally well entrenched antecedent-predicates, or with both the antecedent- and consequent-predicates of one better entrenched than the corresponding predicates of the other. If we state the first rule correctly and modify it somewhat, no further rules are needed.

What decides between two hypotheses that conflict by ascribing incompatible predicates to some future cases? We may have to await more evidence.

This paper appeared in the *Journal of Philosophy* 67 (17 Sept. 1970): 605-608.

But if the choice always depended on that, no hypothesis could ever be projected; for, given any supported, unviolated, and unexhausted hypothesis, we can always concoct another such hypothesis that thus conflicts with it. Another consideration must be brought to bear: the relative entrenchment of the conflicting hypotheses.

Since only supported, unviolated, and unexhausted hypotheses are projectible, we may confine our attention to these for the present. Among such hypotheses, *H overrides H'* if the two conflict and if *H* is the better entrenched and conflicts with no still better entrenched hypothesis.[2] Our revised rule then reads:

> A hypothesis is *projectible* if all conflicting hypotheses are overridden, *unprojectible* if overridden, and *nonprojectible* if in conflict with another hypothesis and neither is overridden.

In each of the following cases, the hypothesis in question is overridden and hence is unprojectible:

H_1 "All emeralds are grue", when all emeralds examined before t are found to be grue and hence green,[3] is overridden by "All emeralds are green."

H_2 "All emeralds are grund", when all emeralds examined before t are green, and all are also square, succumbs to "All emeralds are square."

H_3 "All emerubies are green", when all emeralds examined before t are green and all rubies examined before t are red, gives way to "All rubies are red."

H_4 "All emerubies are gred",[4] when all emeralds examined before t are gred and hence green, is overridden by "All emerubies are green."

The evidence for H_1 carries with it, so to speak, the evidence for the overriding hypothesis: that is, whatever the evidence gathered before t, if H_1 is supported and unviolated, the cited conflicting hypothesis will also be supported, unviolated, and not overridden. The same cannot be said for the other cases; and we must consider the effect of different evidence. For example, if all emeralds examined before t are found to be green but either none has been examined for shape or some have been found to be square and others not square, then "All emeralds are square" is either unsupported or violated and so cannot override H_2. Here, however, H_2 conflicts with the equally well entrenched hypothesis "All emeralds are grare [green if examined before t, or not so examined and square]", so that both hypotheses are nonprojectible.[5] And, indeed, if we have found anything, say the Eiffel tower, to be of some shape other than round, say pointed, H_2 will conflict with some such hypothesis as "All Eifferalds are pointed."

Suppose, though, all emeralds examined before t have been both green and round. Now, since all such conflicting hypotheses as "All emeralds are grare" are overridden by "All emeralds are round", H_2 qualifies as projectible. Plainly, projection of H_2 is harmless where the evidence makes projectible two well-entrenched hypotheses, "All emeralds are green" and "All emeralds are round", such that H_2 follows from their conjunction. This is *not* to say that consequences of projectible hypotheses are always projectible; for some such consequences are unsupported or exhausted. But a consequence of a projectible hypothesis meets two of the requirements for projectibility: it is unviolated, and all conflicting hypotheses are overridden. And thus H_2, since also supported and unexhausted by the evidence given, is projectible.

Still, are we content to say that H_2 is projectible in this case? Lingering reluctance to do so arises, it seems, from confusing two senses of "projectible." In one sense, a hypothesis is projectible if support normally makes it credible. In another sense, a hypothesis is projectible only when the actual evidence supports and makes it credible.[6] In the first sense, "All emeralds are green" is projectible. In the second sense, it is not projectible when deprived, by evidence that violates or exhausts it or leaves it in conflict with hypotheses that are not overridden, of its normal capacity to derive credibility from support. On the other hand, "All emeralds are grund" is normally not projectible but may be relieved, by evidence that neither violates nor exhausts it but overrides all conflicting hypotheses, of its normal incapacity to derive credibility from support. In sum, just as a normally projectible hypothesis may lose projectibility under unfavorable evidence, so a hypothesis not normally projectible may gain projectibility under sufficiently favorable evidence.

If all emeralds examined before t have been green, H_3 is projectible only under weird circumstances: either (i) that all rubies examined have been green, so that H_3 follows from the projectible hypotheses "All emeralds are green" and "All rubies are green", or (ii) that nothing of any other color has been found, so that H_3 follows from the projectible "All things are green." If all sapphires examined are blue, H_3 will be nonprojectible because of conflict with the no-less-well entrenched hypothesis "All sapphirubies are blue." And if we have found anything, say the Eiffel tower, of some color other than green, say black, H_3 will conflict with some such hypothesis as "All Eifferubies are black."

Even H_4 becomes projectible when all emeralds examined before t are green and all rubies red; for the formerly overriding hypothesis "All emerubies are green" is now itself overridden by "All rubies are red." Again, while the hypothesis "All emerubies are grund" is not projectible if before t neither emeralds nor rubies have been examined for shape, or some emeralds or rubies found to be of other shapes than round, it is projectible if all examined emeralds are green and all examined rubies round.

The rules set forth in *Fact, Fiction, and Forecast* (chapter 4, section 4) are thus to be replaced by a single one that makes no use of any 'variety of disagreement' between hypotheses other than incompatibility. Indeed, subject to the further considerations[7] under the heading "Comparative Projectibility" in that book, we now have the following definitions:

A hypothesis is *projectible* if and only if it is supported, unviolated, and unexhausted, and all such hypotheses that conflict with it are overridden.

A hypothesis is *unprojectible* if and only if it is unsupported or violated or exhausted or overridden.

A hypothesis is *nonprojectible* if it and a conflicting hypothesis are supported, unviolated, unexhausted, and not overridden.

NOTES

1. Nelson Goodman, *Fact, Fiction, and Forecast,* 2d ed. (Indianapolis: Bobbs Merrill, 1965).

2. So stated, this covers only hierarchies of at most three supported, unviolated, unexhausted, and successively better entrenched and conflicting hypotheses. Hierarchies of more such hypotheses can be covered if necessary by making the definition more general so that a hypothesis is overridden if it is the bottom member of a hierarchy that cannot be extended upward and has an even number of members.

3. Specifications of the available evidence are often elliptical in this paper. In the present case, for example, we tacitly assume also that some emeralds have been examined before t, while some things other than emeralds may or may not have been found to be green or of some other color.

4. Donald Davidson first introduced a case of this sort in "Emeroses by Other Names," *Journal of Philosophy* 63 (1966): 778-80.

5. When some emeralds have been found to be square and others round, we can retreat from these two hypotheses to the weaker hypothesis "All emeralds are square or round", which does not conflict with them but is projectible whereas they are not. If statistical hypotheses are taken into account, H_3 may be *un*projectible, being overridden by some hypothesis concerning shape distribution among emeralds; but the treatment of statistical hypotheses is a complicated matter requiring redefinition of support, violation, conflict, and so on.

6. In a third sense, a hypothesis is projectible only if projectible in both these senses. Of the three senses, only the second is studied in FFF and this paper; and we are concerned only with whether or not a hypothesis is

made credible to some degree, not with degrees of increase or decrease in credibility. Robert Schwartz is planning a further paper on some senses of projectibility.

7. With such modifications as may be made necessary by the present paper.

9. PROJECTIBILITY: A POSTSCRIPT

THE THEORY of projectibility depends upon the comparative assessment of entrenchment among conflicting hypotheses to determine which, if any, may be overridden. Among supported, unviolated, and unexhausted hypotheses, "a hypothesis is *projectible* if all conflicting hypotheses are overridden, *unprojectible* if overridden, and *nonprojectible* if in conflict with another hypothesis and neither is overridden."[1]

If we compare (h) "All emeralds are green" with (h_1) "All emeralds are grue," then, assuming they are not exhausted by time t, some emerald remains to be examined thereafter which is green but not grue or grue but not green. Such conflict between h and h_1 is resolved for us by the fact that 'green' is better entrenched than 'grue' – and hence that h itself is better entrenched than h_1 – while conflicting with no still better entrenched hypothesis available to us. h_1 is, then, overridden by h and, hence, unprojectible (*ibid.*).

The problem of choosing between h and h_1 depends on our assuming conflict between them. Without such assumption, we do not need to choose, but may project them both. Now if we have positive reason to suppose that all emeralds will have been examined by t, then – since every such emerald is green if and only if grue – we not only avoid this assumption of conflict but also assume the contrary.

If, however, we lack positive reason to suppose h and h_1 to be exhausted by t, we presume they are not thus exhausted; that is, we presume conflict between them. For, if our presumption is correct, relative entrenchment enables us to anticipate cases in conflict by the choice of h over h_1, and, if our presumption is in error, we lose nothing by such choice since these hypotheses agree, given exhaustion by t. On the other hand, the contrary presumption of exhaustion by t, though harmless in failing to differentiate h from h_1 if they are in fact exhausted, leaves us – if it is in error – unprepared for cases examined after t, until they are actually confronted. Since there is, moreover, always a "grue" predicate whose critical t is the moment just after the last case examined, the presumption of exhaustion by t is the denial of novelty; h and h_1 simply describe our past

---

This paper appeared in the *Journal of Philosophy* 79 (June 1982): 334-36.

evidence and do not project at all, much less prepare us for differential anticipations of new cases.

We have seen that h_1 is unprojectible. Does its unprojectibility show the projectibility of h? No; for despite the fact that h overrides h_1 and is therefore not itself *eliminated* through conflict with a still better entrenched hypothesis, it may be *blocked* through conflict with an equally well-entrenched hypothesis – one which it neither overrides nor which overrides it.

Consider the hypothesis (h_2) "All stones are fusible,"[2] which is to be supposed supported, unviolated, unexhausted and of equal (maximal) entrenchment with h. If h_2 conflicts with h, i.e., if some emerald stone is green but not fusible or fusible but not green, h is nonprojectible. If, on the other hand, h_2 does not conflict with h, it poses no threat of nonprojectibility for h.

On the assumption of conflict, we must, then, hold h nonprojectible. Without such positive assumption, however, and lacking also a prior assumption to the contrary, we do not need to choose between h and h_2; we can project them both, treating them as not in conflict unless and until positive reason emerges to persuade us otherwise. This policy of projecting both unless we have positive reason for assuming conflict is preferable to suspension of projection altogether. For such suspension deprives us of genuinely projectible hypotheses if there is in fact no conflict between them, whereas, if we project both h and h_2 when in fact they conflict, future experience, guided by such projection, may itself correct our mistaken presumption of projectibility.

The "no-conflict" presumption here of course need not be supposed epistemologically prior to h or requiring antecedent adoption.[3] Rather it is to be considered part of our preferred policy of facilitating projection: unless positive reason indicates otherwise, we presume h projectible, rather than deem it nonprojectible on the score of its relation with h_2.

The role of conflict operates differently in the two comparisons we have considered. Without the assumption of conflict between h and h_1, there is no problem of choosing between them. If no cases remain to be examined after t, none is to be anticipated in projection. The assumption that there will be new cases to be examined after t is thus an assumption that both underlies and requires discriminating projection. It is hardly an assumption we are likely to dispense with unless we have specific reason to assume the contrary for the particular evidence class in question. By contrast, conflict between h and h_2 blocks projection altogether; hence our rationale for minimizing such assumption, allowing projection fuller scope.

NOTES

I thank Joseph Ullian for critical comments and discussion.

1. See the preceding paper.

2. The example is that of Andrej Zabludowski, "Good or Bad, but Deserved: A Reply to Ullian and Goodman," *Journal of Philosophy* 72 (4 Dec. 1975): 779-84, see p. 779.

3. Thus, Haim Gaifman's claim that, for Goodman and Ullian, h "can be projected *only after we have already accepted* another ... hypothesis, namely 'every emerald is either green and fusible or not green and not fusible' [etc.] ... and so on ad infinitum" is *totally groundless*. See Gaifman, "Subjective Probability, Natural Predicates, and Hempel's Ravens." *Erkenntnis* 14 (No. 2, 1979): 105-147, p. 138 (italics added).

10. WHAT IS SAID TO BE

(WITH NOAM CHOMSKY)

1. PROBLEM

A

PROFESSOR QUINE has recently put his ontological criterion[1] thus:

(1) An entity is assumed by a theory if and only if it must be counted among the values of the variables in order that the statements affirmed in the theory be true.[2]

The point of this criterion is not to say what there is, but rather to determine what any theory[3] says there is, independently of whether we share its ontological assumptions or not. Presumably the criterion is thus quite general in scope, holding for every theory T and for every entity x, assumed by T.

Let us now inquire into the ontological assumptions of (1). Turning the criterion back on itself, we ask what entities must be counted among the values of the variables 'T' and 'x' in order that (1) be true as a statement universally quantified with respect to these variables.

Strictly speaking, (1) taken by itself would be true if the universe were empty and it would also be true if the universe were merely empty of theories. In either case, however, it would be silly to ask for, and pointless to give, a criterion for the ontological assumptions of theories. Analogously, there must also be some entity assumed by a theory if the criterion is to have point. Thus, even if (1) does not, in itself, presuppose the existence of theories and entities assumed by theories, its intended use makes sense only when such presuppositions are taken for granted.

Now, however, if (1) is indeed to be universally true with respect to actual theories and the entities they assume, it must in particular be true of the theory:

(2) $(\exists x)(x$ is phlogiston$)$.

We should, that is, presumably be able to say (with the help of (1)) that (2)

This paper appeared in *Proceedings of the Aristotelian Society,* n.s. 69 (1958-59): 71-82.

assumes something that is phlogiston, since some such entity must be counted among the values of the variables of (2) in order that (2) be true. To apply (1) in this way is, however, impossible unless something that is phlogiston is counted among the values of the variables of (1) itself. Under this interpretation of (1), we indeed infer:

(3) $(\exists x)(x$ is assumed by (2) $\cdot$ x is phlogiston),

which clearly assumes phlogiston no less than does (2) itself. We seem to be in the predicament of having to accept the ontological assumptions of every theory, no matter how ridiculous, just by virtue of adopting (1) as a general criterion. Quine has argued against the notion that "Pegasus cannot be said not to be without presupposing that in some sense Pegasus is."[4] Our present problem is rather that Pegasus cannot be said to be assumed by any theory without presupposing that Pegasus is.

Professor Quine has also argued,[5] with respect to controversies over *what there in fact is,* that retreat to a semantical plane allows consistent formulation of such differences, which involve more than just words. I cannot very well say, "Something is Pegasus and assumed by you, but not by me", though I might say, with respect to the statement

(4) $\sim (\exists x)(x$ is Pegasus),

that whereas you deny it, affirming its denial, I deny the latter and affirm (4) itself. If this procedure is to represent any advantage, however, it must be independently determined that (4) does not assume Pegasus, while its denial does. To make such an independent determination is to affirm or deny neither (4) nor its denial; it is to remain above the strife of ontology. This is the task of Quine's criterion. "We look to bound variables in connexion with ontology not in order to know what there is, but in order to know what a given remark or doctrine, ours or someone else's, *says* there is; and this much is quite properly a problem involving language. But what there is is another question."[6] Yet to apply the criterion seems hardly neutral since it results in saying that whereas (4) assumes nothing, there is some entity such that it is both Pegasus and assumed by the denial of (4). If semantical reformulation is needed for disputes over what there is, semantic (or other) reformulation is no less urgent for the criterion itself, at least in the form it takes in (1).

B

There are other reasons why (1) cannot be taken as providing a criterion for the relation-term 'is assumed by' construed as applying between entities and the theories which assume them to exist. What has so far been argued suggests indeed that 'is assumed by' should not be interpreted as

What is Said to Be

such a relation-term at all, since every such term (embedded in appropriate statements) warrants certain existential inferences which, in the case of 'is assumed by', we do not welcome. If London is west of Paris, there exists something that is west of Paris, whereas if phlogiston is assumed by some theory, we should not want it to follow that there exists something such that it is assumed by that theory.

The phlogiston theory is, of course, one that we do not accept. Consider, however the quite reasonable theory:

(5) $(\exists x)(x$ is a table$)$.

(5) says that there is a table and, unlike the phlogiston theory, its truth is not seriously in question. Nevertheless, there is no particular table of which it affirms existence. What entity is assumed by (5) then? To say there is some entity that is both a table and assumed by (5) would be to say too much. Nor would it be correct to say that (5) assumes the existence of the class of tables or of the attribute of tablehood. Though we are quite prepared to agree that there are tables, we cannot readily interpret the ontological assumption of (5) as a relation between it and some entity whose existence is assumed by it.

Professor Quine's criterion must rather be taken as describing a certain relationship between theories and classes of entities. A theory posits the non-emptiness of a class K if and only if the values of its variables must include some member of K in order for it to be true. In such circumstances it may be said to assume (the class of) K's. Thus (5) assumes (the class of) tables since the range of values of its single variable must include at least some table if it is true. It must be especially noted, however, that (5) does not thereby assert the existence of the class of tables. It is we who describe its ontology by reference to this class. To assume a class is thus not to assume it to exist but to suppose it non-empty.

Such an interpretation is elaborated by Quine in the following statement of his criterion:

(6) To say that a given existential quantification presupposes objects of a given kind is to say simply that the open sentence which follows the quantifier is true of some objects of that kind and none not of that kind.[7]

Thus (5) presupposes tables since the open sentence 'x is a table' is true of some members of the class of tables and not true of anything not a member of this class. It is in this sense that the values of the variable in (5) may be said to require the inclusion of some table in order for it to be true.

Note, however, how our previous difficulty arises to plague us again. In order to judge that (5) posits the non-emptiness of the class of tables, we have had to judge that there is some member of this class to which its open sentence truly applies. We have, in short, had to agree with (5) that the class of tables is indeed non-empty.

In the case of (5), this consequence is, of course, perfectly acceptable. But let us now revert to (2). If we are to say that (2) presupposes the non-emptiness of the class of things that are phlogiston, we should need to say that there is some member of this class to which its open sentence truly applies; we should, in fact, need to assert that the class is non-empty.

If, on the other hand, we shrink from this assertion, we must then face the fact that the criterial condition of (6) fails and we are then precluded from judging that (2) assumes phlogiston. But there is no class such that the open sentence of (2) truly applies to some member of it. We need thus to conclude that (2) presupposes no class, assumes objects of no kind whatsoever. If we compare (2) with:

(7) $(\exists x)(x \text{ is a centaur})$

we must, by similar reasoning conclude that (7) also assumes objects of no kind. The two theories (2) and (7) turn out to have the identical (null) ontological assumption. Thus, the upshot is that either we give up trying to distinguish between (2) and (7) or we do distinguish between them at the cost of agreeing that something is phlogiston or a centaur.

The difficulties just considered stem from our interpreting (6), literally, as requiring the non-emptiness of the kind or class which a theory may be said to presuppose. What happens if we delete this requirement, leaving only the one specifying that the open sentence is not true of anything not of that kind? There is some basis in Professor Quine's use of the criterion for such an interpretation. He says, for example:

> (8) That classical mathematics treats of universals, or affirms that there are universals, means simply that classical mathematics requires universals as values of its bound variables. When we say, for example,
>
> $(\exists x)(x \text{ is prime } \cdot x \text{ is greater than } 1,000,000)$,
>
> we are saying that *there is* something which is prime and exceeds a million; and any such entity is a number, hence a universal.[8]

This passage does not seem to require that there must indeed be some universal that is prime and greater than a million. It requires only that if anything is prime and greater than a million, then it is a universal. If we follow the treatment of this passage, we can now say that the theory in question presupposes the class of universals or assumes its non-emptiness, without ourselves having to agree that it is non-empty. We do indeed have to

What is Said to Be · 219

refer to the class, but we may believe it to be null. (Nominalists will, of course, be prevented from applying the criterion, by their general denial of classes, but this is minor compared to other restrictions that flow from this denial.)

We previously posed the problem of not being able to say that (2) assumes phlogiston without ourselves agreeing that something is phlogiston. Now we *can* say that (2) assumes phlogiston, without embarrassment. We admit only that *if* the open sentence of (2) is true of anything, *then* everything of which it is true is phlogiston. We find, however, that our advantage is gained at great cost. For since the open sentence of (2) is true of nothing, everything of which it is true is also a horse. Hence (2) presupposes horses, as well as every other class. Analogously, (7) presupposes not merely centaurs but also unicorns, since it says there are centaurs, and any such entity is a unicorn. (7) also, in fact, presupposes every class and is thus indistinguishable from (2) in ontological assumption. Indeed, were we to deny the existence of numbers, while still admitting classes, we could say that classical mathematics treats of rabbits on the ground that it says there are numbers and every number is a rabbit, there being none.

This difficulty is matched by another, in cases where the appropriate open sentence *does* apply to something. Suppose the theory before us says:

(9) $(\exists x)$ (x is Cicero),
and (10) $\sim (\exists x)$ (x is Tully).

Now, whatever the open sentence of (9) truly applies to is in fact Tully as well as Cicero. Yet, we cannot, in the face of (10), say that the theory as a whole presupposes the unit class of Tully to be non-empty. We cannot well say that the theory makes inconsistent ontological assumptions, though they may in fact be false. But since the unit class of Tully is identical with the unit class of Cicero we cannot say either that the theory presupposes the latter. Suppose, even, that two predicates are assumed by us to be synonymous, say 'is a brother' and 'is a male sibling'. If the theory asserts both:

(11) $(\exists x)$ (x is a brother)
and (12) $\sim (\exists x)$ (x is a male sibling),

we cannot say the theory presupposes the class of male siblings, though in fact all brothers are male siblings. But since the classes are identical, we cannot say either that the theory presupposes the class of brothers. Or, to put it in the earlier terminology, what class must in fact be non-empty in order for the theory (9)-(10) to be true? The unit class of Cicero? But then (10) would be false. What class must be non-empty in order for (11)-(12) to be true? The class of brothers? But then (12) would be false. Since, surely, no other class must be non-empty in order for either theory to be true, we cannot well say that either (9)-(10) or (11)-(12) presupposes any class of objects unless we are prepared to say that they make inconsistent ontological assumptions, presupposing both the emptiness and the non-emptiness of the same class. Barring the latter interpretation (as unnatural or unfair), we should then need to conclude that the two theories are identical in point of ontological assumption.

2. DIAGNOSIS

The source of all these difficulties seems to be the nonextensionality or referential opacity[9] of the relational locution 'x is assumed by T', with respect to 'x', whether 'x' is construed as an individual or a class variable. Referential opacity, as Quine has shown, has restrictive effects not only on interchangeability of components but on allowable quantification. To switch from individual to class variables, in the present instance, allows us to avoid some of the restrictive consequences of such opacity, *i.e.*, we can infer the existence of the class assumed by a theory even where we think it has no members or deny that any particular member is referred to by this theory. We can, moreover, avoid agreeing that any such class is nonempty, by judicious formulation of our criterion. But the basic difficulty remains that we cannot make the distinctions that need to be made with respect to ontological assumptions of theories, as this is ordinarily understood. We cannot, for example, distinguish between a theory affirming centaurs and one affirming unicorns, between a theory affirming brothers and another affirming male siblings. We cannot, further, distinguish between a theory that affirms Cicero, denying Tully, and a theory that affirms brothers denying male siblings, nor can we distinguish either of the latter from the theory that says merely that there are no griffins. The problem here is analogous to that of interpreting belief statements and statements of indirect discourse generally, in that the distinctions that need making are finer than those that can be made on the basis of extensional logic combined with an extensional, non-semantic ontology. Nor, upon reflection, is this analogy surprising. For how far is the description of what a *theory* says from the description of what a *man* says?

3. PROSPECT

A

There are three attitudes that might reasonably be taken toward the problem of a criterion of ontological assumption if our preceding arguments are correct, and these attitudes are not incompatible, though they will have

What is Said to Be

varying appeal to different people. The first is to keep Professor Quine's criterion and to restrict its use to cases in which the recalcitrant distinctions are not at stake. This requires no further comment here.

B

The second consists in trying to formulate an alternative criterion that is able to make all the distinctions that are wanted. One approach might be to replace the criterion we have been considering by a skeleton or schematic framework for constructing an indefinite number of specific ontological criteria. Such a skeleton might look something like this:

(13) *T* makes a ——— -assumption if and only if it yields a statement of the form '($\exists x$) (x is (a) ———)'.

This skeleton would help us to construct specific criteria of an indefinite variety, for example:

(14) *T* makes a phlogiston-assumption if and only if it yields a statement of the form '($\exists x$) (x is phlogiston)',

(15) *T* makes a table-assumption if and only if it yields a statement of the form '($\exists x$) (x is a table)',

and so forth.

The hyphenated term has the advantage of being immune to quantification with respect to its components and hence to unwanted existential inferences as well as to unwanted interchanges of coextensive components. A unicorn-assumption is not a centaur-assumption, and is thus distinguishable from it though the class of unicorns and the class of centaurs are identical: obviously, too, the statements 'something is a centaur' and 'something is a unicorn' are distinct despite this identity of classes. Quantification with respect to '*T*' remains, of course, but is harmless, at least with respect to the difficulties we have discussed.

There are, however, restrictions on such an approach. It might well be required, for instance, that the language *L* in which (15) is phrased share the predicate 'is a table' with the language in which *T* is phrased, at least for every *T* to which (15) is applied. Suppose, to concentrate on (15) for the moment, that *L* does not contain the predicate-phrase 'is a table' at all. To learn that a given *T* makes a table-assumption would in such a case be unenlightening to a speaker of *L* who was not also a speaker of *T*. In particular, he would not thereby be helped to evaluate *T*, by deciding whether or not this assumption were reasonable. Suppose, secondly, that *L* contains the predicate-phrase 'is a table', as does *T*, but that they differ in extension: in *T*, this predicate-phrase applies only to unicorns, that is, to nothing. Here, the information that T makes a table-assumption would mislead the speaker of L in evaluating the reasonableness of T; he would suppose this assumption to be quite well-grounded whereas in fact it is not.

It might be thought that the present restriction to shared predicates is too strong, that it would perhaps be sufficient to require only intertranslatability of statements of T and L. We should, in line with this idea, relax (13) to demand only that a statement be yielded in T that translates into '$(\exists x)$ x is (a) ———)', taken as a statement in L itself. In the previous example, the result would be that instead of attributing to T a table-assumption or excluding T from the range of the criterion (15) altogether, we should be able to say that T makes a unicorn-assumption. This would be warranted by the new, relaxed, criterion since T yields the statement '$(\exists x)$ x is a table)' in T, which translates into '$(\exists x)$ $(x$ is a unicorn)' in L.

Such a relaxation of the criterion would, however, raise familiar difficulties. Suppose that T also contains the identical predicate "is a unicorn", contained in L. Suppose also that T, besides the original statement we have mentioned, also yields the statement '$\sim (\exists x)$ $(x$ is a unicorn)', identical in both T and L. For us to say that T makes a unicorn-assumption in the face of its explicit denial that there are unicorns would again be to impute an inconsistency of ontological assumption to T which, from T's point of view, is a fiction. The two statements are formally compatible in T, – the extensional equivalent of the two predicates, to which we appeal in making the translation, is something that *we* assume, but something that T excludes. T is quite different from a theory that says explicitly '$(\exists x)$ $(x$ is a unicorn)' *and* '$\sim (\exists x)$ $(x$ is a unicorn)'. This difference is something to which the relaxed criterion is insensitive. Nor would a narrower notion of translatability remove all such insensitivities.

The stricter form of the criterion in general avoids analogous insensitivities and in fact requires L to embrace all the languages of those T's to which it is to be applied. In fact this means not merely overlap of vocabulary but identity of logic and syntax. But assuming all these and other necessary qualifications to have been made, what have we? Not a very exciting result: a theory's ontological assumptions are given in its existential statements in all their original, untranslated variety, – a theory, that is, whose syntax and logic are ours. If you want these assumptions, they can be enumerated with enlightening effect only if you understand the vocabulary of the original existential statements. This amounts to compiling a master list of shared predicates, and checking off for each theory those it declares non-empty, by reference to its existential statements.

C

In view of this sort of result, the question arises whether an adequate ontological criterion serves any philosophical purpose that cannot as economically be served by reference to the theories themselves. The third attitude is to incline towards a negative answer to this question and to forgo reference to such a criterion in all the various contexts in which it has hitherto been familiar. Two illustrations must suffice.

One context in which the use of Professor Quine's criterion has been important is in philosophical debate. A says, "There is a blue sky." B says, "Aha, you must now admit that there is blueness as well as skyness." A says, "Ontological assumptions are revealed by reference to the values of variables alone, not by looking to the predicates. These values for me include a sky but neither blueness or skyness." A might, however, reply to the same effect, "The inference you say I must make is one that simply is invalid by my rules. I have no way of going from '$(\exists x)(Px)$' to '$(\exists y)(y = P)$'.[10] You may favour rules that warrant such a step but I commit no error in rejecting such rules."

Another context in which Professor Quine's criterion has figured is in the comparison and evaluation of rival theories. A says, "T_1 assumes universals and T_2 does not; I should prefer to avoid assuming universals, hence I reject T_1 and accept T_2." A's point may, however, be put in reference to the statements of T_1 and T_2 directly. Some statement S of T_1 is an assertion, either of the existence of universals explicitly, or of something which, were A to agree on its existence, would wring from him the further admission of universals in his own system of beliefs. It is this *statement* S that he finds unacceptable. He in fact may evaluate T_1 and compare it with T_2 by reference to any of their statements, not merely their existential ones, trying to gauge in each instance how likely the statement is to be true. The concept of ontological assumption is absorbed by the notion of the likely truth of existential statements, and this notion in turn absorbed by the likely truth of all the statements of a theory.

NOTES

We are indebted to Professor Quine for a helpful discussion of an earlier draft of this paper.

1. After we had submitted this paper, our attention was called to Richard L. Cartwright's article, "Ontology and the Theory of Meaning" (*Philosophy of Science* 21 (1954): 316-25), by Alan Ross Anderson's review in *The Journal of Symbolic Logic* 22 (1957): 393-4. Cartwright's paper is devoted to an examination of Quine's criterion, and anticipates our arguments in

Section I, regarding the universal and null commitments assigned to false existential statements by variants of the criterion, and the unwanted commitments that follow from certain of its applications. Cartwright concludes that any adequate formulation of an ontological criterion will be intensional, and he proposes such a formulation. Our interest, on the other hand, has been to show that extensional alternatives are trivial if adequate, while the philosophical purposes motivating the original ontological criterion are equally served by arguments making no reference to ontological commitment at all. Consequently, though not disagreeing with Cartwright that any adequate, non-trivial criterion is likely to be intensional, we do not see any philosophical point in developing such a criterion, and we feel, further, that any such criterion will be obscure, for reasons similar to those advanced by Quine.

2. *From A Logical Point of View* (Cambridge: Harvard University Press, 1953), 103. All references to Quine in the present paper concern passages in this book.

3. Strictly, any theory in quantificational form, or assumed translatable into quantificational form; *cp.* Quine, 104-105.

4. Quine, 8.

5. Quine, 16.

6. Quine, 15-16, italics in original.

7. Quine, 131.

8. Quine, 103, italics in original.

9. In a somewhat extended sense as compared with Quine's usage, p. 142.

10. From the statement 'Pa', where 'P' and 'a' are constants, our rules will allow the inference to '$(\exists x)(x = a \cdot Px)$' but not to '$(\exists F)(F = P \cdot Fa)$'.

11. REFLECTIONS ON THE RAMSEY METHOD

ONE FLAWLESS METHOD for eliminating theoretical terms is to embrace mysticism, forswearing speech. Another is simply to renounce theory. Less categorical devices are, however, in greater demand – devices which purport not simply to eliminate theoretical talk, but to exchange it for observational talk of no less value. The Ramsey method, replacing all primitive theoretical constants by existentially quantified variables, is such a device, and the problem is to determine whether or not it is indeed value-preserving in the desired manner.[1]

What sorts of value are, however, in point? Clearly, to require the observational substitute to preserve literally all the values of the theoretical original is to require too much, while to count it successful if it preserves just any is to ask far too little. The philosophical aim is, presumably, to preserve specifically cognitive values during the exchange, so that any further appeal of the original may be ascribed altogether to pragmatic features such as convenience or suggestiveness. Indeed, having satisfied ourselves that the original is not, at any rate, cognitively richer than our better understood observational substitute, we may safely retain it for the sake of its pragmatic virtues, its very eliminability serving thus, ironically, to reinstate it. As in the case of definition generally, we here eat our cake and have it too.[2]

Does the Ramsey method, then, preserve the specifically cognitive values of theory in an observational medium? A Ramseyan substitute certainly preserves all purely observational consequences of its theoretical original, yielding no others of its own.[3] Further, and unlike the substitute yielded by Craig's method,[4] the Ramseyan substitute also retains the deductive coherence of its original, merely changing its theoretical constants to variables. Unlike Craig's method, which simply collects all observational theorems of the original, taking as axiomatic some equivalent of each, the Ramsey method does not squander compactness in the process of preserving content.

There are, however, further factors to consider in assessing the cognitive values of theory. The Ramsey method has, for example, expansive effects

---

This paper appeared in the *Journal of Philosophy* 65 (16 May 1968): 269-74.

on ontological commitment, since it replaces noncommittal theoretical-predicate constants with existentially quantified predicate variables ranging over non-individuals. Such a process platonizes, if it does not intensionalize, nominalistic systems, and may swell the ontology of certain nonnominalistic ones. In either case, it has the concomitant tendency to stimulate increases in a system's complexity count, as Goodman reckons it.[5] One need not be a die-hard anti-liberal to be concerned about such inflationary effects, nor does one need the conscience of a conservative to conclude that the Ramsey method will therefore not generally preserve sound cognitive values.

Hempel (in "The Theoretician's Dilemma") and Maxwell[6] have, however, brought ontological considerations to bear in a quite different way against the Ramsey method. Assume a theory T to postulate an unobservable individual, characterizing it through the application of some theoretical predicate P; then the Ramseyan substitute for T will postulate an equally unobservable individual, asserting of it that it belongs to some class (or has some property) $\emptyset$. Since an unobservable entity by any other name is, however, just as unobservable, the Ramseyan shift from a definite to an indefinite referential mode continues to countenance individual theoretical entities despite its elimination of theoretical constants. Thus, Hempel writes:

> ... the Ramsey-sentence associated with an interpreted theory T' avoids reference to hypothetical entities only in letter – replacing Latin constants by Greek variables – rather than in spirit. For it still asserts the existence of certain entities of the kind postulated by T', without guaranteeing any more than does T' that those entities are observable or at least fully characterizable in terms of observables (81).

And, as Maxwell points out, if 'is observable' is itself an observational predicate, then any theorem of the original to the effect that some entity is not observable will itself be preserved by the Ramseyan substitute.[7] Heretofore, we rather took for granted the nontheoretical character of the Ramseyan substitute and asked whether or not it preserved the cognitive values of the original. The Hempel-Maxwell line, granting for argument's sake that cognitive values are preserved by the substitute, questions rather whether it is indeed nontheoretical.

Now I think the *observationality* of a system's primitive (extralogical) vocabulary needs to be distinguished from the *observability* of its ontology, and I agree that the elimination of primitive theoretical terms does not guarantee observability. I do not, however, see in this fact alone a fatal criticism of the Ramsey method. *Thorough* observability, it should first be noted, is in any event an extravagant ideal, since even a system whose uniformly observational primitives denote observable atomic elements will, in general, need to acknowledge also nonobservable elements – for instance, certain complexes of its observable atoms.[8] To complain that the Ramsey method fails to guarantee thorough observability would thus be to quarrel not with Ramsey but with life.

It must, however, be admitted that an attainable sort of observability characterizes systems whose elements are all composed of observable atomic parts.[9] Moreover, it is true that a given system may fail of this attainable sort of observability even where it lacks primitive theoretical vocabulary and is thus throughout *observational;* it may, for example, harbor a principle of dissection which cuts observable elements into sub-observable parts. It follows, consequently, that the Ramsey method, which merely removes the primitive theoretical vocabulary of a system, is of no avail in guaranteeing observability, even of an attainable sort.

Must it then be concluded that the Ramseyan substitute is no less theoretical than its original, even if we set aside the inflationary ontological effects flowing from its stronger logical machinery? To say "yes" to this question is to show oneself insensitive to the variant uses of the word 'theoretical'. For this word is ambiguous in being opposed both to *observationality* and to *observability.* If a system with purely observational (primitive) vocabulary may be considered theoretical in lacking an observable ontology, as the Hempel-Maxwell argument points out, it is equally true that a system with observable ontology may be considered theoretical in lacking a purely observational primitive vocabulary. The Ramseyan substitute has, accordingly, a perfect right to be contrasted with its original as nontheoretical, just in that it is throughout observational in its primitive (extralogical) vocabulary.

In this respect, it seems to me, incidentally, superior to R. M. Martin's recently proposed Ramsey-constant substitute.[10] For the latter seems to guarantee neither observability nor observationality, replacing original theoretical primitives with so-called *"Ramsey constants"* which are both nonobservational and primitive. To illustrate one variant of Martin's approach, consider

$$Ax_1(T_1, \ldots, T_n, O_1 \ldots, O_k)$$

to be the conjunction of all axioms of the system L containing the primitive theoretical constant 'T_1'. In place of the statement that the individual x has the theoretical property T_1, we form the Ramsey sentence

$$(\exists t_1)(\exists t_2)\ldots(\exists t_n)(t_1 x \cdot Ax_1(t_1, \ldots, t_n, O_1, \ldots, O_k))$$

with 't_1', 't_2', etc. taken as variables ranging over theoretical properties, and 'O_1', ..., 'O_k' taken as primitive observational constants. This Ramsey sentence thus says that the individual x has some theoretical property t_1

characterized by the axioms holding for 'T_1'. This latter Ramsey sentence may now be definitionally abbreviated as

$$R_1(O_1,\ldots,O_k, x)$$

within L. Analogous constructions are to be assumed for all primitive theoretical predicates of L. Martin's suggestion is now "to take 'R_1', 'R_2',... 'R_n', for there will be just n of them, as *primitives* of a new system L'''." Such a primitive Ramsey constant as 'R_1', he says, "enables us to gain the effect of saying in L''' that x has the theoretical property T_1" (4). The catch in this proposal is, however, that the Ramsey constants are nonobservational primitives, so that the total vocabulary is again theoretical. By contrast, the Ramseyan substitute proper is throughout observational in its primitive extralogical vocabulary. Observationality may not in the end be enough, but that does not mean it isn't something, and something important, too.

That observationality in itself represents a significant systematic restriction has already been suggested by the effects of the Ramsey method on ontological commitment and complexity count. In these effects, the Ramsey method betrays its inability to preserve ontological economy and conceptual simplicity. I have elsewhere produced examples to show that it further fails to preserve confirmatory relationships.[11] These examples were of the following sort: A nonanalytic theory T, lacking nonanalytic observational consequences, establishes, however, a certain inductive relation R between the observational statements H and K, but not also between H and $\sim K$. T may, in other words, be said to establish a confirmatory link between H and K. The Ramseyan substitute of T, on the other hand, since it turns out analytic and, hence, insensitive to reinterpretation of its constants, links H by the relation R not only to K, but to $\sim K$ as well. It thus loses altogether the confirmatory relation between H and K.[12]

This argument of mine has been countered in the following way:[13] By suitable alteration of the constants of T, an alternative theory T' can be formulated, which establishes a confirmatory link between H and $\sim K$. But, then, H can be said to confirm K, by appeal to T, only if we indeed have reason to choose T over T'. Now the grounds for such a choice cannot be purely observational since, by hypothesis, T has no nonanalytic observational consequences. T must, then, be preferred because it rests on evidence which is not purely observational. Such evidence must, however, be included within the scope of the Ramsey substitute under consideration, in accordance with the "principle of maximum inclusion" required by the Ramsey method. Inclusion of this sort would be likely, however, to destroy the analyticity of the Ramseyan substitute, thus leading *it also* to establish a confirmatory link between H and K.

The latter counterargument seems to me, however, to be wholly ineffective as a defense of the inductive capabilities of the Ramsey substitute. The counterargument rightly claims that we can assert a confirmatory link between H and K, appealing to T, only if we have reason to choose T rather than T'. But the *assertion* of such a confirmatory link is not the point of primary significance. Let it be granted, for the sake of argument, that we cannot now choose between T and T', and so can neither assert that H confirms K nor that it confirms $\sim K$. Still, it remains true that, relatively to T, H confirms K, and, relatively to T', H confirms $\sim K$, whereas, relatively to the Ramseyan substitute of either theory in this example, H confirms neither K nor $\sim K$. Replacing T, in particular, by its Ramseyan substitute thus clearly involves the *loss* of a confirmatory link between certain observation sentences.

The matter of choice between T and T' is an important, but nevertheless an independent question. According to the counterargument we are considering, the differential grounds for T cannot be purely observational, since T has no nonanalytic consequences. To assume, however, that confirmation of T depends on its having such (deductive) *consequences*, which would obviously be preserved by its Ramseyan substitute, begs the very question whether, aside from its deductive yield, a theory can establish inductive links among observation statements, *not* necessarily preserved by its Ramseyan substitute, but capable of providing relevant support for it if determined to hold in fact.

Since my argument rests on examples, I do not claim to have shown that the Ramsey method is generally or irreparably at fault in failing to preserve confirmatory relations. The ramifications certainly deserve further study. I do think, however, that the examples establish that such failure occurs, and that they therefore lend force to the conclusion that the Ramsey method is not generally value-preserving in requisite respects.

NOTES

A preliminary version of this paper was presented to the Philosophy Club at Brandeis University in December 1964, as part of a symposium with Herbert G. Bohnert on the Ramsey method.

1. The original idea is set forth in Frank Plumpton Ramsey, *The Foundations of Mathematics*, ed. R. B. Braithwaite (New York: Humanities Press, 1931), chap. 9, "Theories," pp. 212-36.

Further discussions may be found in Herbert G. Bohnert, *The Interpretation of Theory* (doctoral dissertation, University of Pennsylvania, 1961); Ernest Nagel, *The Structure of Science* (New York: Harcourt, Brace & World, 1961), pp. 141-42; and Israel Scheffler, *The Anatomy of Inquiry* (New York: Alfred A. Knopf, 1963), pp. 203-22. Other relevant discussions will be noted below.

2. The "virtues of definition as a method of eating one's cake and having it" are discussed in the section "Definition and the Double Life" of W. V. Quine's *Word and Object* (New York: MIT Press, 1960), pp. 186-90.

3. See Bohnert, *op. cit.*, and Scheffler, *op. cit.*, pp. 207-208.

4. For Craig's method see William Craig, "On Axiomatizability within a System," *Journal of Symbolic Logic* 18 (March 1953): 30-32; and "Replacement of Auxiliary Expressions," *Philosophical Review*, 65 (January 1956): 38-55.

Discussions may be found in Carl G. Hempel, "The Theoretician's Dilemma," in Herbert Feigl, Michael Scriven, and Grover Maxwell, eds., *Minnesota Studies in the Philosophy of Science*, vol. 2, pp. 37-98, and "Implications of Carnap's Work for the Philosophy of Science," in Paul Arthur Schilpp, ed., *The Philosophy of Rudolf Carnap* (La Salle, Ill.: Open Court, 1963), pp. 685-707; also in Nagel, *op. cit.*, pp. 134-137; and Scheffler, *op. cit.*, pp. 193-203.

5. He finds, for example, that any predicate of classes of individuals, not equivalent to a basis consisting solely of predicates of individuals, is more complex than any such basis; analogous remarks hold for class predicates of higher type not equivalent to bases of the next lower type level. Moreover, relations similar to epsilon itself have complicating effect. See Nelson Goodman, *The Structure of Appearance*, 2d ed. (Indianapolis: Bobbs-Merrill, 1966), pp. 107-17, 214-17, esp. fn. 7 on p. 214; and "Condensation versus Simplification," *Theoria*, 27 (1961): 47-48.

6. Grover Maxwell, "The Ontological Status of Theoretical Entities," in Herbert Feigl and Grover Maxwell, eds., *Minnesota Studies in the Philosophy of Science*, 3:3-27.

7. Maxwell, *op. cit.*, p. 17 and fn. 14. Remarking that Hempel was too charitable, Maxwell says the Ramsey sentence "cannot, even in letter, much less in spirit ... eliminate reference to unobservable (theoretical) entities."

8. A discussion of the contrast between observationality of terms and observability of elements is given in my *The Anatomy of Inquiry*, part 2, sec. 8, pp. 162-67. The extravagance of thorough observability as a systematic condition is treated on pp. 165-66.

9. Even this condition is idealized in certain respects, as noted on p. 166, *ibid.*

10. See R. M. Martin, "On Theoretical Constructs and Ramsey Constants," *Philosophy of Science* 33 (March 1966): 1-13.

11. *The Anatomy of Inquiry*, 218-22.

12. For a simple illustration of these points, cited *ibid.*, take T as '$(x)((Mx \supset Px) \cdot (Mx \supset Rx))$'; take H as 'Ra'; and take K as 'Pa'. H, taken as *inductive* sign (given T) that 'Ma' is true, leads us then to K, since T plus 'Ma' yield K deductively. The Ramseyan substitute of T, i.e., '$(\exists \emptyset)(x)((\emptyset x \supset Px) \cdot (\emptyset x \supset Rx))$' is analytic and, hence, remains true when for 'P' we put '$\sim P$'.

13. This counterargument was suggested by Herbert Bohnert in connection with the symposium mentioned above. The "principle of maximum

inclusion" mentioned in this paragraph is discussed in Bohnert's *The Interpretation of Theory, op. cit.*, and other references noted in footnote 1 above. Its point is to ensure that a quantified variable replacing a given theoretical constant will be properly linked throughout all items where the constant had earlier occurred.

12. EPISTEMOLOGY OF OBJECTIVITY

THE SO-CALLED CERTAINTY of the given cannot protect its purported descriptions from mistake; the given can therefore not provide a fixed control over conceptualization. If we attempt to picture all our beliefs as somehow controlled by our reports of the given, we shall have to concede that these reports are themselves not rigidly constrained by what is given in fact, since they are themselves subject to error. It does no good, then, to suppose that they constitute points of direct and self-evident contact between our belief systems and reality – firm touchstones by which all our other beliefs are to be judged but which are themselves beyond criticism. Observation reports, in short, cannot be construed as isolated certainties. They must survive a continuous process of accommodation with our other beliefs, a process in the course of which they may themselves be overridden. The control they exercise lies not in an *infallibility* which is beyond their reach; it consists rather in an *independence* of other beliefs, an ability to clash with the rest in such a way as to force a systematic review threatening to all.

Such independence has, in fact, been the primary concern of our earlier discussions. We sought to show that the observational testing of a hypothesis is not necessarily a question-begging procedure, that observation need not be fatally contaminated by theoretical categories, that observation reports may, in consequence, perfectly well conflict with cherished hypotheses. But can such a conception of independence be sufficient for a theory of objective control over belief? Does it provide an adequate restriction of arbitrariness in the choice of hypotheses? Conflict provides at best, after all, a motivation for restoring consistency. However, if this is the only motivation I am bound to honor, I am free to choose at will among equally coherent bodies of belief at variance with one another; I need not prefer the consistent factual account to the consistent distortion nor, indeed, to the coherent fairy tale. Faced with a conflict between my observation reports and my theory, I may freely alter or discard the former or the latter or both, so long as I replace my initial inconsistent set of beliefs with one that is coherent. Clearly, this much freedom is too much freedom. Constraints beyond that of consistency must be acknowledged.

This selection is taken from Chapter 5 of my *Science and Subjectivity*, 2d ed. (Indianapolis: Hackett Publishing, 1982), 91-124.

Yet, in denying the doctrine of certainty, have we not made it impossible to do just that? If all our beliefs are infected with the possibility of error, if none of our descriptions is guaranteed to be true, none can provide us with an absolutely reliable link to reality. None can serve, through an immediately transparent correspondence with fact, as an additional, referential constraint upon our choices of belief. Our beliefs float free of fact, and the best we can do is to ensure consistency among them. The dilemma is severe and uncomfortable: swallow the myth of certainty or concede that we cannot tell fact from fancy.

This dilemma lies at the root of much controversy among scientifically minded philosophers in recent decades. A review of certain elements of the controversy will enrich our grasp of the problem and help to elucidate the approach of these lectures. We take as the primary object of such review the debate within the Vienna Circle in the nineteen-thirties concerning the status of so-called protocol sentences in science. Two chief protagonists in this debate were Otto Neurath and Moritz Schlick, the former rejecting the doctrine of certainty and insisting "that science keeps within the domain of propositions, that propositions are its starting point and terminus,"[1] and the latter urging rather that science is "a means of finding one's way among the facts," its confirmation-statements constituting "absolutely fixed points of contact" between "knowledge and reality."[2]

Let us turn first to Neurath who, in his anti-metaphysical zeal, proposes not only that science be purged of phenomenalism and unified through expression in physicalistic language, but also that scientific operations be understood as wholly confined to the realm of statements:

> It is always science as a system of statements which is at issue. *Statements are compared with statements,* not with "experiences," "the world," or anything else. All these meaningless *duplications* belong to a more or less refined metaphysics and are, for that reason, to be rejected. Each new statement is compared with the totality of existing statements previously coordinated. To say that a statement is correct, therefore, means that it can be incorporated in this totality. What cannot be incorporated is rejected as incorrect. The alternative to rejection of the new statement is, in general, one accepted only with great reluctance: the whole previous system of statements can be modified up to the point where it becomes possible to incorporate the new statement.... The definition of "correct" and "incorrect" proposed here departs from that customary among the "Vienna Circle," which appeals to "meaning" and "verification." In our presentation we confine ourselves always to the sphere of linguistic thought [SP, 291].

Against the notion of a primitive and incorrigible set of so-called protocol statements as the basis of science, Neurath is adamant. "There is no way of taking conclusively established pure protocol sentences as the starting point of the sciences," he writes.[3] Aside from tautologies, the protocol as well as the non-protocol sentences of unified science share the same physicalistic form and are subject to the same treatment. The protocol statements are distinguished by the fact that "in them, a personal noun always occurs several times in a specific association with other terms. A complete protocol sentence might, for instance, read: 'Otto's protocol at 3:17 o'clock: [At 3:16 o'clock Otto said to himself: (at 3:15 o'clock there was a table in the room perceived by Otto)]'" (PS, 202). However, the main point to be stressed is not that protocol sentences are distinct but rather that *"Every law and every physicalistic sentence of unified-science or of one of its sub-sciences is subject to ... change. And the same holds for protocol sentences"* (PS, 203).

The motivation for change is the wish to maintain consistency, for "In unified science we try to construct a non-contradictory system of protocol sentences and non-protocol sentences (including laws)" (PS, 203). Thus it is that a new sentence in conflict with the accepted system may dislodge a systematic sentence or may itself be rejected, and "The fate of being discarded may befall even a protocol sentence" (PS, 203).

The notion that protocol sentences are primitive and beyond criticism because they are free of interpretation must be abandoned, for "The above formulation of a complete protocol sentence shows that, insofar as personal nouns occur in a protocol, interpretation must *always* already have taken place" (PS, 205). Furthermore, there is, within the innermost brackets, an inescapable reference to some person's "act of perception" (PS, 205). The conclusion is that no sentence of science is to be regarded as more primitive than any other:

> All are of equal primitiveness. Personal nouns, words denoting perceptions, and other words of little primitiveness occur in all factual sentences, or, at least, in the hypotheses from which they derive. All of which means that *there are neither primitive protocol sentences nor sentences which are not subject to verification* [PS, 205].

Further, since "*every* language *as such,* is inter-subjective" (PS, 205), it is meaningless to talk of private languages, or to regard protocol languages as initially disparate, requiring ultimately to be brought together in some special manner. On the contrary, "The protocol languages of the Crusoe of yesterday and of the Crusoe of today are as close and as far apart from one another as are the protocol languages of Crusoe and of Friday" (PS, 206).

Basically, it makes no difference at all whether Kalon works with Kalon's or with Neurath's protocols, or whether Neurath occupies himself with Neurath's or with Kalon's protocols. In order to make this quite clear, we could conceive of a sorting-machine into which protocol sentences are thrown. The laws and other factual sentences (including protocol sentences) serving to mesh the machine's gears sort the protocol sentences which are thrown into the machine and cause a bell to ring if a contradiction ensues. At this point one must either replace the protocol sentence whose introduction into the machine has led to the contradiction by some other protocol sentence, or rebuild the entire machine. *Who* rebuilds the machine, or *whose* protocol sentences are thrown into the machine is of no consequence whatsoever. Anyone may test his own protocol sentences as well as those of others [PS, 207].

Neurath stresses the place of prediction in science. He argues against phenomenal language that it "does not even seem to be usable for 'prediction' – the essence of science..." (SP, 290), and urges in favor of physicalism that it enables us to "achieve successful predictions" (SP, 286). He hopes that the fruitfulness of social behaviorism will be shown by the "successful predictions" it will yield (SP, 317), and looks forward to the day when a physicalistic sociology will "formulate valid predictions on a large scale" (SP, 317). Yet, true to his self-imposed restriction to the realm of statements alone, he does not construe the success of a prediction as consisting in its agreement with fact. Rather, he declares: "A prediction is a statement which it is assumed will agree with a future statement" (SP, 317).

Despite his refusal, however, to contrast the "thinking personality" with "experience" (SP, 290), to compare statements with " 'experiences,' 'the world,' or anything else" (SP, 291), and to ask such " 'dangerous' questions ... as how 'observation' and 'statement' are connected; or, further, how 'sense data' and 'mind,' the 'external world' and the 'internal world' are connected,"[4] he slips into what he ought surely to have regarded, in a more careful moment, as dangerous metaphysics:

Ignoring all meaningless statements, the unified science proper to a given historical period proceeds from proposition to proposition, blending them into a self-consistent system which is an instrument for successful prediction, and, consequently, for life [SP, 286].

To speak rashly in this way of the relation between science and life is clearly to leave the pure realm of statements and to admit, after all, that science cannot be adequately characterized in terms of consistency alone, that its very point, indeed, is to refer to what lies beyond itself.

Surely, not all self-consistent systems are "instruments for life," in the intended sense. The supposition that unified science issues in such practically useful instruments goes beyond the range of consistency in a manner that is not satisfactorily explained by Neurath's general account. He implies, of course, that practical usefulness accrues to science in virtue of its yielding successful predictions. This explanation is hardly adequate, however, for Neurath understands the success of a prediction to consist simply in its agreement with a later statement; on this criterion all predictions succeed which are followed by reiterations of themselves or by other statements coherent with them.

In the first of the passages by Neurath earlier quoted, he speaks of comparing each new statement with "the totality of existing statements previously coordinated" (SP, 291), to determine whether or not the statement can be incorporated in the totality. Perhaps the idea is that there is one presumably coherent totality which is to be singled out as a standard on each occasion of comparison, namely, that totality last ratified by acceptance and still in force on that occasion. The factor of *acceptance* may thus be thought to constitute a relevant selective consideration beyond consistency – a consideration that, moreover, escapes the dangers of metaphysics by avoiding appeal to a reality to which statements refer.

Now the *acceptance* of a statement is indeed relevantly independent of its *reference,* but acceptance also fails to differentiate between beliefs that are critically accepted on the basis of factual evidence and those that are not. The method of comparison recommended in the passage under consideration thus applies as well to entrenched myths and indoctrinated distortions as to scientific systems. What is of crucial significance, however, is that this method provides no incentive to *revise* the accepted totality of beliefs. For the assumed coherence of this totality can always be preserved by rejecting *all* new conflicting sentences. Neurath concedes, in fact, that the alternative to such rejection, consisting in revision of the accepted totality, is adopted "only with great reluctance" (SP, 291). The mystery, on his account, is why it should ever be adopted at all. Any coherent totality is, so far as his method is concerned, capable of being established as forever safe from revision, and thereby warranted as correct, to boot.

It is, moreover, pertinent to question the assumed interpretation of acceptance: acceptance by whom? The assumption that acceptance singles out one presumably coherent totality on each occasion of comparison is perhaps plausible if we consider just one individual. It is groundless if we take into account the acceptances of the whole "inter-subjective" community in line with Neurath's general attitude. For he deplores the "emphasis on the 'I' familiar to us from idealistic philosophy" (PS, 206), and considers it meaningless to talk of personal protocol languages. "One can," he writes, "distinguish an *Otto-protocol* from a *Karl-protocol,* but not a protocol of one's own from a protocol of others" (PS, 206). A general rather than an

Epistemology of Objectivity

individual appeal to the factor of acceptance, however, yields a multiplicity of conflicting totalities of belief: Which of these is to serve as a standard?

There are passages, indeed, in which Neurath seems to be making no appeal to anything even approximating acceptance. He does not speak, in these passages, of "*the* totality of existing statements previously coordinated," but acknowledges rather a plurality of mutually conflicting totalities simply as abstract choices open to the investigator. To be consistent, the investigator may not choose more than one of these, but there is no further constraint on his choice beyond convenience. Thus, Neurath writes:

> A social scientist who, after careful analysis, rejects certain reports and hypotheses, reaches a state, finally, in which he has to face comprehensive sets of statements which compete with other comprehensive sets of statements. All these sets may be composed of statements which seem to him plausible and acceptable. There is no place for an empiricist question: Which is the "true" set? but only whether the social scientist has sufficient time and energy to try more than one set or to decide that he, in regard to his lack of time and energy – and this is the important point – should work with one of these comprehensive sets only.[5]

We find here, to be sure, a passing reference to plausibility and acceptability, but it is wholly unexplained, and can, moreover, have no point unless we move outside the "sphere of linguistic thought" (SP, 291) in a manner for which Neurath has altogether failed to prepare us. As to the choice among incompatible systems, any one is as good as any other; within the limits of time and energy, the decision between them is arbitrary. The machine analogy earlier quoted does indeed, as Neurath says, make the point "quite clear." The machine detects contradictions but, aside from a general restriction to physicalistic language which may be assumed, no principle of selection is supplied for determining its input. Protocol sentences, distinguished solely by their form, may be chosen arbitrarily for insertion. Nor is there any restriction on the structure of the machine beyond its requiring the inclusion of at least some laws, presumably also distinguished by their form alone. So long as no contradiction has been detected among its virtually arbitrary elements, moreover, the machine is to be taken as the very embodiment and standard of correctness. The picture is one of unrelieved coherence free of any taint of fact. Since any consistent statement or system whatever can be accommodated by some such machine, any such statement or system can be fastened upon, held to be correct, and thenceforth protected forever from revision. The dogmatism of certainty has given way to the dogmatism of coherence. We have here not a picture of science but a desperate philosophical caricature.

What impels Neurath to construct this caricature? To appreciate his philosophical motivation is to gain a deeper understanding of the basic dilemma we face between coherence and certainty. He is, as we have seen, opposed to the idea that protocol sentences are above criticism because totally free of interpretation, serving simply to register the raw facts as given. On the contrary, he upholds the view that all statements in science are subject to change, insisting that observation reports may themselves be discarded under pressure of conflict with other scientific statements. Accordingly he emphasizes the "unified" nature of science, that is to say, the fact that no statement is an island – that each can survive only within a systematically harmonious community of statements. In place of the doctrine that selected statements provide an infallible contact with reality and are thus privileged to exercise unilateral control over the rest, Neurath urges a fluid and egalitarian conception: control is provisional, mutual, and diffused throughout the community of statements, resting in no case upon infallible access to fact.

Indeed, the very notion of such access seems to require the supposition that statement and reality might, through direct comparison, be determined to correspond with each other. But such a supposition is meaningless from Neurath's point of view. One can certainly compare statements with statements, but to imagine that statements can be literally compared with reality or with facts is to fall prey to an obfuscating metaphysics.

Now Neurath's remarks on this theme may appear, at first blush, to be simply a dogmatic denial of the obvious. His underlying thought may perhaps be interpreted more plausibly as a rejection of the philosophical tendency to read linguistic features into reality. The structure of language is not, after all, to be taken naively as a clue to the structure of reality. The correspondence suggested, for example, between atomic statements and atomic facts, and between molecular statements and molecular facts, is supported by nothing more than an anterior, and quite gratuitous hypostatization of objects to which certain elements of a language may be said to be directed. Facts, in general, understood as peculiar extra-linguistic entities precisely parallel to true statements, belong, in Neurath's scheme, to the class of "meaningless duplications... to be rejected" (SP, 291).[6]

Not only are such duplications superfluous; they mislead us into supposing that, in locating them independently and finding them to share the same structure with certain statements, we have a genuine method of justifying the acceptance of these statements. But facts, as entities distinct from the true statements to which they are presumed to correspond, have no careers of their own capable of sustaining such a method. These ghostly copies of true statements cannot be independently specified, confronted, or analyzed; their reality is no easier to determine than the truth of their respective parent sentences. Faced with the problem whether to *judge* a given sentence *as* true, it therefore does us no good to be told simply to ascertain whether there exists a structurally corresponding fact. If I am undecided about the truth of the sentence "The car is in the garage," I am equally undecided as to whether or not it is a fact that the car is in the garage: there are not two issues here, but one. Nor do I see how to go about resolving the latter indecision in a way that differs from my attempt to resolve the former. Appeal to the facts, taken strictly, thus turns out question-begging as a general method for ascertaining truth. For it requires, in effect, that the truth be determined as a condition of its own ascertainment.

The import of this line of reasoning may be illustrated strikingly by a consideration of prediction. The prevalent view is that in science, at any rate, a set of beliefs is put to the test by deriving therefrom a prediction that can be checked observationally against actual experience. When such a prediction is borne out by experience, the set of beliefs in question has passed a critical test; when the prediction is violated by experience, the test has been failed and the set must thereupon be revised so as to eliminate the prediction in question. The question that needs to be faced, however, is the question of how we can tell whether or not a prediction has been borne out or violated by experience. It must be stressed to begin with that the relations of logical consistency and contradiction hold only between certain statements and others, and *not* between statements and experiences. It may well be granted, therefore, that if a system S yields a prediction P and if we independently require S to be logically consistent with *Not-P*, we shall have to revise S. So far the issue concerns only the consistency relations among statements. But what, it may be asked, is our initial basis for setting logical consistency with *Not-P* as a constraint upon S? What can lead us to adopt *Not-P* in the first place?

To appeal to the logical consistency of *Not-P* with experience is nonsense. To say that we accept *Not-P* if it in turn yields predictions that are borne out by experience is to take the fatal first step in an infinite regress. To suggest that *Not-P* be judged true if and only if the corresponding fact represented by *Not-P* is real begs the question, as we have seen. To suppose, finally, that we have an infallible intuition of the truth of *Not-P* as a description of reality, that somehow its truth is immediately and indubitably evident to us upon intellectual inspection, is to revert to the myth of certainty. The conclusion to which we thus appear driven is that the whole idea of checking beliefs against experience is misguided. We do not go outside the realm of statements at all. What figures in the control of our system of beliefs is not experience, but purported statements of experience; not observation, but observation reports. Such is Neurath's conclusion, as we have already seen – a conclusion that, however well motivated, must surely be judged unacceptable as an account of science.

Convinced of the unacceptability of Neurath's account, Schlick insists that there must be an "unshakeable point of contact between knowledge and reality" (p. 226). To give up "the good old expression 'agreement with reality'" (p. 215), and to espouse instead a coherence theory such as that propounded by Neurath yields intolerable consequences:

> If one is to take coherence seriously as a general criterion of truth, then one must consider arbitrary fairy stories to be as true as a historical report, or as statements in a textbook of chemistry, provided the story is constructed in such a way that no contradiction ever arises. I can depict by help of fantasy a grotesque world full of bizarre adventures: the coherence philosopher must believe in the truth of my account provided only I take care of the mutual compatibility of my statements, and also take the precaution of avoiding any collision with the usual description of the world, by placing the scene of my story on a distant star, where no observation is possible. Indeed, strictly speaking, I don't even require this precaution; I can just as well demand that the others have to adapt themselves to my description; and not the other way round. They cannot then object that, say, this happening runs counter to the observations, for according to the coherence theory there is no question of observations, but only of the compatibility of statements.
>
> Since no one dreams of holding the statements of a story book true and those of a text of physics false, the coherence view fails utterly. Something more, that is, must be added to coherence, namely, a principle in terms of which the compatibility is to be established, and this would alone then be the actual criterion [pp. 215-216].

Since, in the case of conflict within a given set of statements, the coherence theory allows us to eliminate such conflict in various ways, "on one occasion selecting certain statements and abandoning or altering them and on another occasion doing the same with the other statements that contradict the first," the theory fails to provide an unambiguous criterion, yielding "any number of consistent systems of statements which are incompatible with one another" (p. 216). Schlick concludes that "The only way to avoid this absurdity is not to allow any statements whatever to be abandoned or altered, but rather to specify those that are to be maintained, to which the remainder have to be accommodated" (p. 216).

One might suppose, on the basis of such a conclusion, that Schlick would proceed to a defense of the certainty of protocol statements. Not so, however. He grants that such statements, as exemplified by familiar recorded accounts of scientific observation, and associated with "empirical facts upon which the edifice of science is subsequently built," are indeed subject to error and revision. "They are anything but incontrovertible, and one can use them in the construction of the system of science only so long as they are supported by, or at least not contradicted by, other hypotheses" (pp. 212-213). Even our own previously enunciated protocol statements may be withdrawn. "We grant," writes Schlick,

> that our mind at the moment the judgment was made may have been wholly confused, and that an experience which we now say we had two minutes ago may upon later examination be found to have been an hallucination, or even one that never took place at all.
>
> Thus it is clear that on this view of protocol statements they do not provide one who is in search of a firm basis of knowledge with anything of the sort. On the contrary, the actual result is that one ends by abandoning the original distinction between protocol and other statements as meaningless [p. 213].

Schlick thus agrees with Neurath in denying a privileged role to protocol statements. Like Neurath, he insists that they "have in principle exactly the same character as all the other statements of science: they are hypotheses, nothing but hypotheses" (p. 212). Where, then, is the fixed point of contact between knowledge and reality? Schlick's view is that it is to be located in a special class of statements that are not themselves within science but are nevertheless essential to its function and, in particular, to its confirmation. His special term for these statements is *Konstatierungen,* though he sometimes calls them "observation statements"; I shall here refer to them uniformly as "confirmation statements."[7]

A confirmation statement is a momentary description of what is simultaneously perceived or experienced. It provides an *occasion* for the production of a protocol statement proper, which is preserved in writing or in memory; it must, however, be sharply distinguished from the protocol statement to which it may give rise. For this protocol statement can no longer describe what is simultaneous with itself; the critical experience has lapsed during the time taken to fix it in writing or memory. The protocol statement, moreover, unlike the confirmation statement, does not die as soon as it is born; its own life extends far beyond the initial point nearest the experience in question. Though it has, to be sure, the advantage of providing an enduring account, the protocol statement is, thus, never more than a hypothesis, subject to interpretation and revision. "For, when we have such a statement before us, it is a mere assumption that it is true, that it agrees with the observation statements [i.e., the confirmation statements] that give rise to it" (pp. 220-221).

Confirmation statements may serve to stimulate the development of genuine scientific hypotheses, but they are too elusive to be construed as the ultimate and certain *basis* of knowledge. Their contribution consists rather in providing an absolute and indubitable culmination to the process of testing hypotheses. When a predicted experience occurs, and we simultaneously pronounce it to have occurred, we derive "thereby a feeling of *fulfilment,* a quite characteristic satisfaction: we are *satisfied*" (p. 222). Confirmation statements perform their characteristic function when we obtain such satisfaction.

> And it is obtained in the very moment in which the confirmation takes place, in which the observation statement [i.e., confirmation statement] is made. This is of the utmost importance. For thus the function of the statements about the immediately experienced itself lies in the immediate present. Indeed we saw that they have so to speak no duration, that the moment they are gone one has at one's disposal in their place inscriptions, or memory traces, that can play only the role of hypotheses and thereby lack ultimate certainty. One cannot build any logically tenable structure upon the confirmations, for they are gone the moment one begins to construct. If they stand at the beginning of the process of cognition they are logically of no use. Quite otherwise however if they stand at the end; they bring verification (or also falsification) to completion, and in the moment of their occurrence they have already fulfilled their duty. Logically nothing more depends on them, no conclusions are drawn from them. They constitute an absolute end [p. 222].

In bringing a cycle of testing to an absolute close, a confirmation statement helps to steer the further course of scientific investigation: a falsified hypothesis is rejected and the search for an adequate replacement ensues; a verified hypothesis is upheld and "the formulation of more general hypotheses is sought, the guessing and search for universal laws goes on" (p. 222). The cognitive culmination represented by confirmation statements had, originally, according to Schlick, a purely practical import: it indicated the reliability of underlying hypotheses as to the nature of man's environment, and thus aided man's adjustment to this environment. In science, the joy of confirmation is no longer tied to the "purposes of life" (p. 222), but is pursued for its own sake:

> And it is this that the observation statements [confirmation statements] bring about. In them science as it were achieves its goal: it is for their sake that it exists.... That a new task begins with the pleasure in which they culminate, and with the hypotheses that they leave behind

Epistemology of Objectivity · 243

does not concern them. Science does not rest upon them but leads to them, and they indicate that it has led correctly. They are really the absolute fixed points; it gives us joy to reach them, even if we cannot stand upon them [p. 223].

What is it, however, that enables confirmation statements to constitute "absolute fixed points"? In what does their special claim to certainty consist? Schlick conceives these statements as always containing demonstrative terms. His examples are, "Here yellow borders on blue," "Here two black points coincide," "Here now pain." The constituent demonstratives function as gestures. "In order therefore to understand the meaning of such an observation statement [confirmation statement] one must simultaneously execute the gesture, one must somehow point to reality" (p. 225). Thus he argues, one can understand a confirmation statement "only by, and when, comparing it with the facts, thus carrying out that process which is necessary for the verification of all synthetic statements" (p. 225). For to comprehend its meaning is simultaneously to apprehend the reality indicated by its demonstrative terms.

> While in the case of all other synthetic statements determining the meaning is separate from, distinguishable from, determining the truth, in the case of observation statements [confirmation statements] they coincide.... the occasion of understanding them is at the same time that of verifying them: I grasp their meaning at the same time as I grasp their truth. In the case of a confirmation it makes as little sense to ask whether I might be deceived regarding its truth as in the case of a tautology. Both are absolutely valid. However, while the analytic, tautological, statement is empty of content, the observation statement [confirmation statement] supplies us with the satisfaction of genuine knowledge of reality [p. 225].

The distinctiveness of confirmation statements lies, then, in their immediacy, that is, their capacity to point to a simultaneous experience, in the manner of a gesture. To such immediacy they "owe their value and disvalue; the value of absolute validity, and the disvalue of uselessness as an abiding foundation" (p. 225). It is of the first importance, for Schlick's view, to recognize the distinctiveness of confirmation statements and, in particular, to separate them from protocol statements, for this separation is the key to the problem as he sees it. "Here now blue" is thus not to be confused with the protocol statement of Neurath's type: "M.S. perceived blue on the nth of April 1934 at such and such a time and such and such a place." The latter is an uncertain hypothesis, but it is distinct from the former: it must mention a perception and identify an observer. On the other hand, one cannot write down a confirmation statement without altering the meaning of its demonstratives, nor can one formulate an equivalent without demonstratives, for one then "unavoidably substitutes...a protocol statement which as such has a wholly different nature" (p. 226).

In sum, if we consider simply the body of scientific statements, they are all hypotheses, all uncertain. To take into account also the relation of this body of statements to reality requires, however, that we acknowledge the special role of confirmation statements as well. An understanding of these statements enables us to see science as "that which it really is, namely, a means of finding one's way among the facts; of arriving at the joy of confirmation, the feeling of finality" (p. 226). These statements do not "lie at the base of science; but like a flame, cognition, as it were, licks out to them, reaching each but for a moment and then at once consuming it. And newly fed and strengthened, it flames onward to the next" (p. 227).

We have already expressed our own agreement with the critical side of Schlick's doctrine, namely, his rejection of a coherence theory such as Neurath's. We, too, have stressed the importance of acknowledging constraints upon our belief beyond those imposed by consistency alone. We can therefore sympathize with Schlick's effort to propose an account of scientific systems that relates them to the reality to which they purport to refer. And it must indeed be admitted that the spirit of his general view of science is in closer accord than Neurath's with our familiar conceptions as well as with the understandings of scientists themselves as to the purport of their own activities.

Yet Schlick's positive theory suffers from a variety of fundamental difficulties that render it altogether unacceptable. Let us consider, first of all, whether his doctrine of confirmation statements is capable of meeting the problem as he has diagnosed it. He seeks, after all, a principle beyond coherence "in terms of which the compatibility is to be established," insisting that the only way to avoid the difficulties of the coherence theory is to avoid allowing "any statements whatever to be abandoned or altered, but rather to specify those that are to be maintained, to which the remainder have to be accommodated" (p. 216).

But if this is indeed the only way to avoid the difficulties of the coherence view, then it must be doubted that Schlick's positive doctrine in fact succeeds in avoiding them. For the coherence view purports to be a theory of science – of "science as a system of statements," as Neurath puts it. Any attempt to restrict the arbitrariness of coherence along the lines of Schlick's diagnosis must specify fixed points to which the statements of science are to be adjusted. It must, that is, specify a fixity to which *science* is responsive, by which *scientific* spontaneity is contained. In particular, it must not permit every scientific statement whatever to be subject to revision but must, on the contrary, place definite limits upon statement revision within science.

It is just here that Schlick's doctrine fails. For he identifies as "absolute fixed points" only *confirmation statements,* which fall outside science, and he insists, moreover, that these statements provide no barrier whatever to the revision of scientific statements proper. In particular, Schlick stresses that protocol statements, which are the closest counterparts of confirmation statements within science, "have in principle exactly the same character as all the other statements of science: they are hypotheses, nothing but hypotheses... one can use them in the construction of the system of science only so long as they are supported by, or at least not contradicted by, other hypotheses" (pp. 212-213). We have here, it seems, a clear admission that, within the realm of science, coherence continues to rule, despite the certainty attributed to confirmation statements. The latter have in effect been so sharply sundered from the body of science that they can yield it no advantage derived from their own presumed fixity. If reality alone provides no fixed control over scientific systems, the postulation of intermediate confirmation statements thus accomplishes nothing in the way of achieving such control.

Nor is it easy to make Schlick's general account of the scientific role of such statements intelligible. They are described as having an essential role in scientific functioning – in particular, in the testing and verification of hypotheses. They are not to be thought of as constituting a logical basis or origin of science. "If they stand at the beginning of the process of cognition they are logically of no use" (p. 222). Rather, "they bring verification (or also falsification) to completion.... Logically nothing more depends on them, no conclusions are drawn from them. They constitute an absolute end" (p. 222). In marking the fulfillment of scientific predictions, confirmation statements are, however, said not only to yield a characteristic satisfaction, but to influence the course of subsequent inquiry: "the hypotheses whose verification ends in them are considered to be upheld, and the formulation of more general hypotheses is sought, the guessing and search for universal laws goes on" (p. 222). The problem is whether these various features ascribed to confirmation statements can be reconciled with one another.

For, on the one hand, these statements constitute an absolute end, having no logical function when standing at the beginning of further cognitive processes, since "the moment they are gone one has at one's disposal in their place inscriptions, or memory traces, that can play only the role of hypotheses and thereby lack ultimate certainty" (p. 222). On the other hand, they enable us to uphold the hypotheses they serve to verify and to reject those they falsify, in either case leading us to conduct subsequent inquiry in a significantly different manner. If, however, a confirmation statement truly constitutes an absolute end, how can it serve thus to qualify our further treatment of relevant hypotheses? Why, indeed, should a hypothesis, supposedly verified by a confirmation statement a moment before the last, be considered *now* to have been clearly upheld, leading us to search for broader hypotheses rather than simply to continue testing the original one? We now have, after all, only the fallible record of an alleged earlier verification and, as Schlick remarks, "it is a mere assumption that it is true" (p. 220), that it agrees with its parent confirmation statement. Similarly, why should the present protocol statement recording an alleged past falsification be taken as the trace of an absolute falsification by its parent confirmation statement, since it is itself no more than a hypothesis, subject only to the weak demands of coherence? In short, if the door closed by a given confirmation statement is indeed immediately reopened, this statement can constitute no absolute end; if, on the other hand, the door remains shut, the statement clearly has a logical bearing, in fact, an unwarranted logical bearing, upon subsequent investigation. Confirmation statements, it seems, cannot bring testing processes to absolute completion without qualifying further inquiry in a manner precluded by their momentary duration. However, unless they do bring such processes to absolute completion, they have, on Schlick's account, no function at all in the economy of science. The conclusion that Schlick's account of these statements is self-contradictory seems inescapable.

The notion that confirmation statements can have no logical function for subsequent cognitive processes rests on their radical immediacy, that is, on the idea that their function "lies in the immediate present" (p. 222). Schlick thus emphasizes their differentiation from the protocol statements to which they may give rise, statements which are "always characterized by uncertainty" (p. 226). To write down a confirmation statement or even to preserve it in memory is, strictly speaking, impossible, for the meaning of critical demonstratives is altered by preservation; replacement of these demonstratives "by an indication of time and place" moreover inevitably results in the creation of "a protocol statement which as such has a wholly different nature" (p. 226). But immediacy, one may feel, should cut both ways: if it eliminates logical bearing on subsequent processes, it must equally eliminate such bearing on earlier ones. Yet Schlick holds, as we have seen, that confirmation statements bring testing processes to an absolute completion:

> Have our predictions actually come true? In every single case of verification or falsification a "confirmation" [confirmation statement] answers unambiguously with a yes or a no, with joy of fulfilment or disappointment. The confirmations are final [p. 223].

Epistemology of Objectivity

How can this be? The prediction is, after all, a scientific hypothesis with "a wholly different nature" from that of the confirmation statement in question. How can it derive any benefit from the latter's certainty any more than a later protocol statement can?

Schlick gives, as an example of a prediction: "If at such and such a time you look through a telescope adjusted in such and such a manner you will see a point of light (a star) in coincidence with a black mark (cross wires)" (p. 221). Suppose we now have the confirmation statement, "Here now a point of light in coincidence with a black mark." For the sake of argument, let us grant that the latter statement is, at the critical moment, certain. Does it follow that it constitutes an unambiguous and final answer to the question of whether the prediction has in fact come true? Not at all. For the prediction stipulates, in its antecedent clause, certain conditions relating to physical apparatus, time, and the activity of an observer. Unless the experience reported by the confirmation statement is assumed to have occurred in accordance with the conditions thus stipulated, it cannot even be judged relevant to the prediction, much less to fulfill it with finality. On the other hand, if the assumption is made that these conditions have been satisfied in fact, this critical assumption itself shares in the uncertainty of the prediction, being itself clearly no more than a physical hypothesis. The question of whether a prediction has in fact come true is, then, just the question of whether a corrigible scientific statement, rather than a confirmation statement, is true. Such a question can always be reopened. The conclusion must be that the alleged certainty of confirmation statements no more enables them to provide absolutely certain completions for earlier scientific processes than it equips them to constitute absolute origins.

The alleged certainty of these statements must, finally, be called into question: What does such certainty amount to? According to Schlick, I cannot be deceived regarding the truth of my own confirmation statements, even though, as he writes, "the possibilities of error are innumerable" (p. 212). He admits, of course, that protocol statements are subject to error. Indeed, in one passage he concedes that a protocol statement saying "that N.N. used such and such an instrument to make such and such an observation" may be mistaken, because N.N. may "inadvertently... have described something that does not accurately represent the observed fact" (p. 212). Does this not imply that, in such a case, N.N. was himself deceived regarding the truth of his own confirmation statement? Such an implication is certainly unwelcome to Schlick, and contrary to his fundamental view. The question remains as to how it might reasonably be avoided. What is it, indeed, that precludes errors due to inadvertence and poor judgment in the case of confirmation statements alone, allowing them full scope in all other cases?

Schlick certainly does not suppose that one's own statements are generally immune to error. "Even in the case of statements which we ourselves have put forward," he writes, "we do not in principle exclude the possibility of error" (p. 213). However, he continues immediately to illustrate his point with specific and exclusive reference to protocol statements: "We grant that our mind at the moment the judgment was made may have been wholly confused, and that an experience which we now say we had two minutes ago may upon later examination be found to have been a hallucination, or even one that never took place at all" (p. 213). A statement by which we affirm the occurrence of our experience of two minutes earlier can, of course, in his scheme, not be a confirmation statement at all but only a protocol statement. The question persists, however: Is mental confusion possible only in the making of protocol statements? What protects us from confusion in the judgment of simultaneous experience? What necessity ensures us against error?

Concerning protocol statements occasioned by confirmation statements, Schlick writes that they are to be classed as hypotheses. "For, when we have such a statement before us, it is a mere assumption that it is true, that it agrees with the observation statements [confirmation statements] that give rise to it" (p. 220). The supposition is evident that if a protocol statement disagrees with the relevant confirmation statement, it must be the former and not the latter which is false. But why this asymmetry? What prevents the protocol statement from giving a truer account of a man's experience at a given moment than his own description at that moment? In one passage, Schlick suggests the curious doctrine that the protocol statement reports not a perception or an experience but rather the occurrence of a confirmation statement. The protocol statement, "M.S. perceived blue on the nth of April 1934 at such and such a time and such and such a place," he declares equivalent to "M.S. made ... (here time and place are to be given) the confirmation 'here now blue'" (p. 226). Under such a doctrine, to be sure, if M.S.'s actual confirmation statement at the relevant time and place was "Here now yellow," the protocol statement must be false. It does not follow, however, that the confirmation statement is true. And the proposed equivalence must itself be rejected, for to perceive blue and to say "Here now blue" are surely different. If it is false that M.S. *said* "Here now blue," it is not therefore false that he *perceived* blue. Despite the fact that he actually said "Here now yellow," he may have perceived blue: his confirmation statement itself may have been false. What eliminates such a possibility?

The fundamental answer, in Schlick's theory, is given by his reference to the demonstrative terms included in confirmation statements. As he puts it, " 'This here' has meaning only in connection with a gesture" (p. 225). To comprehend the meaning of a confirmation statement, "one must somehow point to reality" (p. 225). It follows, in his view, that I cannot understand a confirmation statement without thereby determining it to be true. Here is the fundamental source of the certainty of confirmation statements: their understanding presupposes their verification. Now it may perhaps be argued that the certainty thus yielded is weak because it is dependent upon the notion of "understanding." (I may, indeed, not be at all sure I understand a given confirmation statement when I make it, and may later decide that in fact I did not.) But the trouble, in any case, runs much deeper than this. Schlick's basic conception of the matter rests, I believe, upon a confusion.

For suppose it be granted that the meaning of demonstrative terms derives from their function as gestures, by which, as Schlick remarks, "the attention is directed upon something observed" (p. 225). Suppose it be admitted that "in order therefore to understand the meaning" of a confirmation statement, "one must simultaneously execute the gesture, one must somehow point to reality" (p. 225). What can be inferred from such admissions? They imply only that the comprehension of a confirmation statement requires attention to those observed elements indicated by its constituent demonstrative terms. In this and this sense only can comprehension of the statement be said to involve a "pointing to reality." By no means is it implied that we must point to reality in the wholly different sense of verifying the attribution represented by the statement as a whole. To attend to an indicated thing cannot be equated with determining the truth of any affirmation concerning it. It is therefore simply fallacious to infer, from the presumed necessity of attending to the things indicated by embedded demonstratives, the necessity of verifying the assertion made by a statement with the help of such demonstratives, in order to grasp its meaning.

The fallacy is concealed by the equivocal phrase "pointing to reality." For, in explaining the latter phrase, initially introduced to denote attention to the reference of a demonstrative term functioning as a gesture, Schlick writes: "In other words: I can understand the meaning of a 'confirmation' [confirmation statement] only by, and when, comparing it with the facts, thus carrying out that process which is necessary for the verification of all synthetic statements" (p. 225). Here is the easy, but illegitimate shift from the reference of a term to the reference of a statement, from things indicated to facts expressed, from attending to the object of a demonstrative to verifying the truth of a statement. Once this shift is exposed, Schlick's fundamental argument for the certainty of confirmation statements falls to the ground: the understanding of such statements does not, after all, presuppose their verification. I may understand a confirmation statement and be undecided as to its truth; what is more, I can understand, and even affirm, a confirmation statement that is false. Schlick's positive theory, no less than Neurath's, thus proves untenable.

The failure of both these theories may engender despair. For they seem, between them, to exhaust the possibilities for dealing with our basic dilemma between coherence and certainty. Either some of our beliefs must be transparently true of reality and beyond the scope of error and revision, or else we are free to choose any consistent set of beliefs whatever as our own, and to define "correctness" or "truth" accordingly. Either we suppose our beliefs to reflect the facts, in which case we beg the very question of truth and project our language gratuitously upon the world, or else we abandon altogether the intent to describe reality, in which case our scientific efforts reduce to nothing more than a word game. We can, in sum, neither relate our beliefs to a reality beyond them, nor fail so to relate them.

Despite this grim appraisal, I believe that despair is avoidable, and that the general approach of the previous lectures can be sustained. My view is that while rejecting certainty, it is yet possible to uphold the referential import of science; that to impose effective constraints upon coherence need beg no relevant questions nor people the world with ghostly duplicates of our language. In the remainder of the present lecture, I shall attempt to spell out the reasons for these views.

Let me turn first to the fundamental opposition between coherence and certainty. We have seen how central this opposition is in the thought of both Neurath and Schlick, who take contrary positions. Neurath insists that every scientific statement is subject to change, urging that "There is no way of taking conclusively established pure protocol sentences as the starting point of the sciences" (PS, 201). Rejecting certainty, he rejects, however, also the very idea of checking beliefs against experience or reality, taking coherence to be the only defensible alternative. Schlick, on the other hand, reacting against the absurdities of coherence, is driven to seek a class of "absolutely certain" (p. 223) statements capable of providing an "unshakeable point of contact between knowledge and reality" (p. 226). Affirming the need to relate science to fact, he assumes that only a doctrine of certainty can make such a relationship intelligible. Schlick and Neurath are thus agreed in binding extra-linguistic reference firmly to certainty and they join, therefore, in reducing the effective alternatives to two: (1) a rejection of certainty, as well as of appeals to extra-linguistic reference, yielding a coherence view, and (2) a rejection of the coherence view in favor of an appeal to extra-linguistic reference, yielding a commitment to certainty. Given such a reduction, we are driven to defend coherence if we find certainty repugnant, and impelled to defend certainty if appalled by the doctrine of coherence. The reduction makes it possible for these equally unpleasant alternatives thus to feed upon each other. But this reduction itself must be rejected. There is, in fact, no need to assume that the alternatives are exhausted by coherence and certainty; a third way lies open.

For what is required is simply a steady "referential" limitation upon unbridled coherence; certainty supplies much more than is required. In particular, it imports the notion of a fixity, a freedom from error and consequent revision, which cannot be defended for it is nowhere to be found. The supposition that certainty is required to restrain coherence is perhaps in part due to an ambiguity in the notion of "fixity": in one sense, a fixed point is simply one that is selectively designated, providing us with a frame of reference or a standard, for purposes of a given sort; in another sense, a fixed point is one that, in particular, does not undergo change over time. A point may be fixed in the first sense without being fixed in the second. There may be no temporally constant or *permanently* fixed points in a given context, and yet there may be fixed points that are relevant and effective in that context, at each moment. The scope of unrestrained coherence needs, indeed, to be supplemented by the introduction of relevantly "fixed" points at all times. However, the further supposition that there must be *some* points that are permanently fixed, that is, held forever immune to subsequent revision, is simply gratuitous.

We have earlier cited Schlick's remark that the coherence theory can be avoided only if we specify statements that "are to be maintained, to which the remainder have to be accommodated" (p. 216). That certain statements need to be *maintained* over time is an unwarrantedly strong demand. We need only recognize that statements have referential values for us, independent of their consistency relationships to other statements, and that these values, though subject to variation over time, provide us, at each moment, with sufficient "fixity" to constitute a frame of reference for choice of hypotheses.

How are these values of statements, compatible with lack of certainty, to be conceived? They may be thought of as representing our varied inclinations to affirm given statements as true or assert them as scientifically acceptable; equivalently, they may be construed as indicating the initial claims we recognize statements to make upon us, at any given time, for inclusion within our cognitive systems. A notion of this general sort has been put forward by Bertrand Russell in *Human Knowledge,* under the label "intrinsic credibility,"[8] and Nelson Goodman has spoken, analogously, of "initial credibility,"[9] the adjective serving in each case to differentiate the idea in question from the purely relative concept of "probability with respect to certain other statements." Goodman explains his conception as follows:

> Internal coherence is obviously a necessary but not a sufficient condition for the truth of a system; for we need also some means of choosing between equally tight systems that are incompatible with each other. There must be a tie to fact through, it is contended, some immediately

certain statements. Otherwise compatibility with a system is not even a probable indication of the truth of any statement.

Now clearly we cannot suppose that statements derive their credibility from other statements without ever bringing this string of statements to earth. Credibility may be transmitted from one statement to another through deductive or probability connections; but credibility does not spring from these connections by spontaneous generation. Somewhere along the line some statements, whether atomic sense reports or the entire system or something in between, must have initial credibility. So far the argument is sound.... Yet all that is indicated is credibility to some degree, not certainty. To say that some statements must be initially credible if any statement is ever to be credible at all is not to say that any statement is immune to withdrawal. For indeed, ... no matter how strong its initial claim to preservation may be, a statement will be dropped if its retention – along with consequent adjustments in the interest of coherence – results in a system that does not satisfy as well as possible the totality of claims presented by all relevant statements. In the "search for truth" we deal with the clamoring demands of conflicting statements by trying, so to speak, to realize the greatest happiness of the greatest number of them. These demands constitute a different factor from coherence, the wanted means of choosing between different systems, the missing link with fact; yet none is so strong that it may not be denied. That we have probable knowledge, then, implies no certainty but only initial credibility.[10]

There is much in the above account that is metaphorical, but the fundamental point for present purposes seems to me quite simple and persuasive: while certainty is untenable, it is also excessive as a restraint upon coherence. Such restraint does not require that any of the sentences we affirm be guaranteed to be forever immune to revision; it is enough that we find ourselves now impelled, in varying degrees, to affirm and retain them, seeking to satisfy as best we can the current demands of all. That these current demands vary for different, though equally consistent statements, and that we can distinguish, even roughly, the credibility-preserving properties of alternative coherent systems, suffices to introduce a significant limitation upon coherence. For it means that not all coherent systems are equally acceptable; we are not free to make an arbitrary choice among them. Nor, where what has hitherto been a satisfying system conflicts with a strong new candidate for inclusion, are we free to decide the matter arbitrarily, discarding one or the other at will. There is a price to be paid in either case in terms of overall credibility-preservation and, though it may be difficult to determine in given circumstances, it is a clearly relevant consideration.

In any event, it is the claims of sentences at a given time which set the problem of systematic adjudication at that time, and so restrain the arbitrariness of coherence. That these claims may vary in the future does not alter the present task. That a sentence may be given up at a later time does not mean that its present claim upon us may be blithely disregarded. The idea that once a statement is acknowledged as theoretically revisable, it can carry no cognitive weight at all, is no more plausible than the suggestion that a man loses his vote as soon as it is seen that the rules make it possible for him to be outvoted.

There is, to be sure, a certain awkwardness in expressing a statement's theoretical revisability in the same breath with asserting the statement itself, and the difficulty this presents may be one source of the deep-rooted feeling that certainty, rather than credibility, is required. If I say "There's a horse but it may not be", I may appear to be at odds with myself in a peculiar way; to remove the conflict I must presumably assert "There's a horse" unqualifiedly and defend the immunity of this statement to possible withdrawal at any future time. Such a cure would, however, be worse than the disease, for by this argument all statements would be certain. The problem is more easily handled by finding a suitably circumspect form of interpretation, e.g., "There's a horse; and I recognize the possibility that the statement just made might be withdrawn under other circumstances than those now prevailing". The recognition of the latter possibility clearly does not conflict with the initial assertion; such possibility surely does not need to be eliminated in order to make the way clear for assertion generally.

It follows that the basic dilemma with which we started, between coherence and certainty, collapses. That none of the statements we assert can be freed of the possibility of withdrawal does not imply that no statement exercises any referential constraint at any time. That none can be *guaranteed* to be an absolutely reliable link to reality does not mean that we are free to assert any statements at will, provided only that they cohere. That the statement "There's a horse" cannot be rendered theoretically certain does not permit me to call anything a horse if only I do not thereby contradict any other statement of mine. On the contrary, if I have learned the term "horse", I have acquired distinctive habits of individuation and classification associated with it; I have learned what Quine refers to as its "built-in mode . . . of dividing [its] reference."[11] These habits do not guarantee that I will never be mistaken in applying the term, but it by no means therefore follows that they do not represent selective constraints upon my mode of employing the term. On the contrary, such constraints generate credibility claims which enter my reckoning critically as I survey my system of beliefs. I seek not consistency alone, but am bound to consider also the relative inclusiveness with which a system honors initial credibilities.

It follows, therefore, that the emphasis of earlier lectures on the *independence* of observation statements as the primary locus of their control does not, after all, have the consequence of committing us to the coherence theory. Such statements, as we urged, are not isolated certainties, but must be accommodated with other beliefs in a process during which they may themselves be overridden. It is, however, enough for the purposes of control that they may clash with these other beliefs in such a way as to force an unsettling systematic review of the situation. Such a review is not (contrary to the hypothetical fears earlier expressed) motivated simply by the wish to restore consistency. The need is, of course, to maintain consistency, but also to sacrifice as little as possible of overall credibility.

An observational expectation induced in us by our heretofore satisfying system may, for example, be challenged by an experimental observation which drastically increases the credibility of a statement incompatible with this expectation while radically reducing the credibility of the expectation itself. The problem in such a situation is to determine which consistent alternative strikes a more inclusive balance of relevant credibility claims. To drop the initial expectation in favor of the more credible incompatible statement demands internal systematic revision in the interests of consistency. To exclude the incompatible statement and maintain the system intact lowers the overall credibility value of the latter, for the credibility loss of its constituent expectation reverberates inward. Every clash-resolution, in short, has its price. In some such situations, the choice may be relatively easy; in others it may be exceedingly delicate; and it may even, in some circumstances, defy resolution.

Nevertheless, it is clear that we are in no case free simply to choose at will among all coherent systems whatever. And, further, it is clear that the control exercised by observation statements does not hinge on certainty. It requires only that the credibility they acquire at particular times be capable of challenging, in the manner above described, the expectations flowing from other sources. Such control is, surely, not absolute, since any observation statement may itself be outvoted in the end; in pressing their independent claims, however, such statements nevertheless contribute, along with other statements, to the restraint of arbitrariness in the choice of beliefs. Control is, moreover, released from distinctive ties to any special sort of statement, and diffused throughout the realm of statements as a whole.

What shall now be said concerning the difficult notions of truth and reality? Eschewing certainty, some philosophers, Neurath included, have rejected all talk of reality and truth. Having pointed out that appeal to an immediate comparison with the facts as a method of ascertaining truth is question-begging, and that facts, construed literally as entities, are mere ghostly doubles of true sentences, they have proceeded to cast doubt upon all thought of external reference, as embodied in philosophically innocent talk of reality and fact, and in innocent as well as serious talk of truth. Such skepticism leads, however, to insuperable difficulties, for without external reference, science has no point. If we stay within the circle of statements altogether, we are trapped in a game of words, with which even Neurath (as indicated by his reference to science as an instrument for life) cannot be wholly satisfied. Taken in its extreme form, Neurath's doctrine understandably evokes the sort of criticism which Russell offers:

> Neurath's doctrine, if taken seriously, deprives empirical propositions of all meaning. When I say 'the sun is shining', I do not mean that this is one of a number of sentences among which there is no contradiction; I mean something which is not verbal, and for the sake of which such words as 'sun' and 'shining' were invented. The purpose of words, though philosophers seem to forget this simple fact, is to deal with matters other than words.... The verbalist theories of some modern philosophers forget the homely practical purposes of everyday words, and lose themselves in a neo-neo-Platonic mysticism. I seem to hear them saying 'in the beginning was the Word,' not 'in the beginning was what the word means.' It is remarkable that this reversion to ancient metaphysics should have occurred in the attempt to be ultra-empirical.[12]

One source of the trouble is a persistent confusion between truth and estimation of the truth, between the import of our statements and the processes by which we choose among them. If, for example, appeal to reality or direct comparison with the facts is defective as a method of *ascertaining* truth, this does not show that the *purport* of a true statement cannot properly be described, in ordinary language, as "to describe reality" or "to state the facts." We may have no certain intuition of the truth, but this does not mean that our statements do not purport to be true. Now, if the sentence "Snow is white" is true, then snow is (really, or in fact) white, and vice versa, as Tarski insists.[13] As Quine has remarked, "Attribution of truth in particular to 'Snow is white', for example, is every bit as clear to us as attribution of whiteness to snow."[14] We may be unclear as to how to decide whether the sentence "Snow is white" is true, but the sentence in any case *refers* to snow and claims it to be white, and if we decide to hold the sentence to be true, we must be ready to hold snow to be (really, or in fact) white. The question of how we go about deciding what system of descriptive statements to accept in science is the question of how to estimate what statements are true. Whatever method we employ, our statements will refer to things quite generally and will purport to attribute to them what, *in reality, or in fact,* is attributable to them.

There is thus no way of staying wholly within the circle of statements, for in the very process of deciding which of these to affirm as true, we are deciding how to refer to, and describe things, quite generally. The *import* of our statements is inexorably referential. It is, however, quite another matter to suppose that, because this is so, the *methodology* by which we accept statements may be philosophically described in terms of an appeal to such supposed entities as facts, with which candidate statements are to be directly compared. For facts, postulated as special entities corresponding to truths, are generally suspect, and the determination of their existence is question-begging if proposed literally as a method of ascertaining the truth.

We thus separate the question of the *import* of scientific systems from the question of the *methods* by which we choose such systems. Can such methods be described without dependence upon the notion of direct comparison with the facts? Both Neurath and Schlick assume that any conception of science as referential must display such dependence. This supposition, however, is simply false. The conception of credibility above sketched represents a notion of choice among systems of statements, yet makes no use of any idea such as that of comparison with the facts. It is, nevertheless, perfectly compatible with the recognition that any system selected is referential in its *import*. Moreover, credibility considerations rest on the referential values which statements have for us at a given time, that is, on the inclinations we have, at that time, to affirm these statements as true.

Such inclinations as to statements are, surely, tempered by habits of individuation and classification acquired through the social process of learning our particular vocabulary of *terms*. In learning the term "horse", for example, I have incorporated selective habits of applying and withholding the term; these habits, operating upon what is before me, incline me to a greater or lesser degree to affirm the statement "There's a horse". If I have learned the term "white" as well as "horse," I may, further, be strongly inclined, on a given occasion, to affirm "That horse is white", and my inclination, on such an occasion, will be understandable in part, as a product of my applying both "horse" and "white" to what I see. I need, however, surely not recognize any such additional entity as *the fact that that horse is white,* nor need I have an applicable term for such a supposed entity in my vocabulary. Though it hinges in various ways on referential habits associated with a given vocabulary, the notion of initial credibility thus requires no reference to *facts,* in particular.

To be sure, whether I am to accept a statement depends not only on my initial inclination to accept it, but rather on its fitting coherently within a system of beliefs that is sufficiently preserving of relevant credibilities. Here again, however, there is no reference to any such entities as facts, with which statements are to be compared. In accepting a system I nevertheless take its *import* to be referential: I hold its statements to be true, and genuinely accept whatever attributions it makes to the entities mentioned in these statements. In a philosophically harmless sense, I may then say that I take the system as expressive of the facts. I have, at no time, any guarantees that my system will stand the test of the future, but the continual task of present evaluation is the only task it is possible for me to undertake. Science, generally, prospers not through seeking impossible guarantees, but through striving to systematize credibly a continuously expanding experience.

NOTES

1. Otto Neurath, "Sociology and Physicalism," tr. Morton Magnus and Ralph Raico. Reprinted with permission of The Free Press from *Logical Positivism* by A. J. Ayer, ed., p. 285. Copyright © 1959 by The Free Press. A Corporation. Originally appeared as "Soziologie im Physikalismus," *Erkenntnis* 2 (1931-32). Page references to this article in the text will be preceded by "SP."

2. Moritz Schlick, "The Foundation of Knowledge." tr. David Rynin. Reprinted with permission of The Free Press from *Logical Positivism* by A. J. Ayer, ed., p. 226. Copyright © 1959 by The Free Press, A Corporation. Originally appeared as "Über das Fundament der Erkenntnis," *Erkenntnis* 4 (1934). Page references to Schlick in the text refer to this article.

3. Otto Neurath, "Protocol Sentences," tr. Frederic Schick. Reprinted with permission of The Free Press from *Logical Positivism* by A. J. Ayer, ed., p. 201. Copyright © 1959 by The Free Press, A Corporation. Originally appeared as "Protokollsätze," *Erkenntnis* 3 (1932-33). Page references to this article in the text will be preceded by "PS."

4. Otto Neurath, *Foundations of the Social Sciences* (Chicago: University of Chicago Press, 1944), 5.

5. *Foundations of the Social Sciences*, 13.

6. On the parallelism of language and reality, see Ludwig Wittgenstein, *Tractatus Logico-Philosophicus* (London: Routledge & Kegan Paul, 1922). For further discussion, see Edna Daitz, "The Picture Theory of Meaning," *Mind* (1953), reprinted in A. G. N. Flew, ed., *Essays in Conceptual Analysis* (London: Macmillan & Co., 1960); John Passmore, *A Hundred Years of Philosophy* (London: Gerald Duckworth & Co. 1957); and Nelson Goodman, "The Way the World Is," *Review of Metaphysics* 14 (1960): 48-56.

7. There are problems in choosing a suitable translation; see David Rynin's note on these problems in Ayer, ed., *Logical Positivism*, 221. I choose "confirmation statements" to emphasize that statements are thus denoted, in preference to Rynin's "confirmations," though I believe the latter choice follows Schlick's own usage more closely.

8. Bertrand Russell, *Human Knowledge* (New York: Simon and Schuster, 1948), Part 2, chap. 11, and Part 5, chaps. 6 and 7.

9. Nelson Goodman, "Sense and Certainty," *Philosophical Review* 61 (1952): 160-67.

10. Ibid., 162-63.

11. Willard Van Orman Quine, *Word and Object* (New York and London: The Technology Press of the Massachusetts Institute of Technology, and John Wiley & Sons, 1960), 91.

12. Bertrand Russell, *An Inquiry into Meaning and Truth* (London: Allen and Unwin, 1940), 148-49; Penguin ed. (Harmondsworth: Penguin Books, 1962), 140-41.

13. Alfred Tarski, "The Semantic Conception of Truth," *Philosophy and Phenomenological Research* (1944); reprinted in Herbert Feigl and Wilfrid Sellars, eds., *Readings in Philosophical Analysis* (New York: Appleton-Century-Crofts, 1949), 52-84.

14. Willard Van Orman Quine, *From A Logical Point of View* (Cambridge: Harvard University Press, 1953), 138.

13. VISION AND REVOLUTION: A POSTSCRIPT ON KUHN

IN CHAPTER 4 of *Science and Subjectivity*, I offered several arguments critical of Professor Thomas Kuhn's views as expressed in his influential book *The Structure of Scientific Revolutions*.[1] His recent replies to these criticisms[2] seem to me so inadequate as to suggest that he, and therefore others as well, may have failed to grasp their full import. Accordingly, I shall, in the first part of this paper, briefly recapitulate my earlier arguments and offer a short rejoinder to Professor Kuhn's replies.[3] The second part of the paper will expand upon my earlier discussion to consider the basic metaphors of *vision* and *revolution*, offered by Kuhn to replace the traditional notion of deliberation. My argument here will be that these new metaphors are incongruous in critical respects, and my discussion will conclude by considering their relations to the contrast between understanding and accepting a theory.

1. RECAPITULATION AND REJOINDER

In criticizing Professor Kuhn's view, my discussion in *Science and Subjectivity*[4] offered the following arguments:

1. *Self-refutation:* If paradigm debates are characterized by an "incompleteness of logical contact" (p. 109) between proponents of rival paradigms, and the transition to a new paradigm does not occur "by deliberation and interpretation" (p. 121), it is self-defeating to justify this view itself by deliberation appealing to factual evidence from the history of science. If it is true that "paradigm changes do cause scientists to see the world of their research-engagement differently ... that after a revolution scientists are responding to a different world" (p. 110), there can be no appeal to ostensibly paradigm-neutral factual evidence from history in support of Kuhn's own new paradigm. Conversely, if *historians* can transcend particular paradigms and evalu-

---

This paper appeared in *Philosophy of Science* 39 (1972): 366-74.

ate them by appeal to neutral evidence, so can *scientists,* i.e., they can engage in rational paradigm debates which are perfectly intelligible. [See *SS,* 21-22, 53, 74]

2. *Observational difference:* It does not follow, from the fact that different paradigms organize their observations differently, that they are directed to different objects – that "after a revolution scientists are responding to a different world" (p. 110). From the fact that certain items are seen in varying ways under diverse categorizations, it cannot be inferred that they are not identical. There is a contrast between seeing *x* and seeing *x as* something or other. [See *SS,* 41]

3. *Meaning variation:* Paradigm change does not inevitably alter constituent meanings through altering the language or definitions of basic terms, with the result that "communication across the revolutionary divide is inevitably partial" (p. 148). The supposition of *inevitable* meaning change confuses the notion of a language as consisting simply of a vocabulary and grammar, with that of a language construed as a system of assertions. Or else it overlooks the fact that variation of definition or sense is consistent with referential stability and that only the latter is required for the stability of deduction. [See *SS,* 58-62, 64]

4. *Gestalt switch:* Assuming that adoption of a paradigm is accompanied by a "relatively sudden" and "intuitive" alteration of perception such as the "gestalt switch," it does not follow that there are no public procedures of assessment by which such a paradigm is evaluated after it is originated. If, as Kuhn says, "no ordinary sense of the term 'interpretation' fits these flashes of intuition through which a new paradigm is born" (p. 122), it may not be inferred that the term 'interpretation' does not apply to the processes by which the scientific community debates the merits of the new paradigm. [See *SS,* 78-79]

5. *Perception vs. debate:* If intuitive processes of perception characterize the psychology of the paradigm originator, he himself does not defend his paradigm by simple appeal to such processes. He engages in debate and proposes arguments. If he is deluded about the force of these arguments, the delusion cannot be demonstrated simply by arguing from the psychology of perception, but requires appeal to the content of the relevant debates. Nor are *particular examples* of debates at cross-purposes sufficient. For common criteria may have borderline regions of indeterminacy, and even where determinate, there may well be differences of judgment in application, as well as misunderstanding. [See *SS,* 79-80]

Vision and Revolution: A Postscript on Kuhn · 261

6. *Inadequacy of normal science:* Paradigm choice cannot be resolved by normal science, according to Kuhn, for if differing scientific schools disagree about the character of problems and solutions, "they will inevitably talk through each other when debating the relative merits of their respective paradigms" (p. 108). Thus, deliberation gives way to persuasion and conversion. But to assume that deliberation and interpretation are *restricted* to normal science begs the very question at issue. "If paradigms are not corrigible by normal science, it does not follow that they are not corrigible at all. If scientific schools inevitably talk through each other when arguing from within their respective paradigms, it is not further inevitable that they do always argue from within their respective paradigms." [*SS*, 80-81]

7. *Incommensurability of paradigms:* Competing paradigms, for Kuhn, are addressed to different problems, embody different standards and different definitions of science (p. 147); they are based on different meanings and operate in different worlds (pp. 148-149) [*SS*, 81]. Therefore, he argues, "The proponents of competing paradigms are always at least slightly at cross-purposes" (p. 147). Before they can communicate, "one group or the other must experience the conversion that we have been calling a paradigm shift. Just because it is a transition between incommensurables, the transition ... cannot be made a step at a time, forced by logic and neutral experience" (p. 149). But if the paradigms are indeed so different how can they be in competition? If they are indeed rivals, they must be accessible to some shared perspective within which they can be compared. Incommensurability does not imply incomparability. [See *SS*, 82-83]

8. *Second-order incommensurability:* Proponents of different paradigms, says Kuhn, acquire different criteria for determining relevant problems and solutions. The debate between them is circular or inconclusive because "each paradigm will be shown to satisfy more or less the criteria that it dictates for itself and to fall short of a few of those dictated by its opponent" (pp. 108-109). "Paradigm differences are thus inevitably reflected upward, in criterial differences at the second level" [*SS*, 84]. This argument, however, confuses *internal* criteria, by which paradigms determine problems and solutions, with *external* criteria by which they are themselves judged. The latter are independent of the former, and, hence, the argument that paradigms must inevitably be self-justifying collapses. [See *SS*, 84-86]

9. *Resistance to falsification:* Paradigms are not rejected when counterinstances occur, until an alternative is available, argues Kuhn,

while acceptance of a new and untried paradigm depends on faith (pp. 77, 157). Thus, he concludes "the competition between paradigms is not the sort of battle that can be resolved by proofs" (p. 147). But proof is in any case irrelevant, and faith in a new hypothesis is compatible with acknowledgement of shared procedures of evaluation by which the hypothesis is to be assessed. Nor is loss of faith in the truth of a hypothesis inconsistent with its continued use as a practical tool of research or application, or with belief that some modification of the hypothesis is true. Kuhn himself seems to admit the critical point by allowing that, although scientists do not renounce the paradigm provoking crisis, they "may begin to lose faith and then to consider alternatives" (p. 77). The global use of terms such as 'acceptance', 'rejection' and 'paradigm' itself, prevents his full appreciation of this point. [See *SS*, 86–88]

10. *Cumulativeness:* In denying the cumulative character of scientific change, Kuhn says "the successful new theory must somewhere permit predictions that are different from those derived from its predecessor. That difference could not occur if the two were logically compatible" (p. 96). But if the two are incompatible logically, they must be, at least in part, commensurable, and *a fortiori,* comparable. It follows also that accumulation, while not necessary, is at least logically possible. Further, despite his criticism of the notion of cumulative science, he says that "new paradigms . . . usually preserve a great deal of the most concrete parts of past achievement and they always permit additional concrete problem-solutions besides" (p. 168). [See *SS,* 89, 65]

11. *Reintroduction of criticized notions:* Notions criticized by Kuhn reemerge often under new labels, in his theory. Thus, as just noted, commensurability is implied by his emphasis on logical incompatibility. Falsification returns under the guise of anomaly, crisis, and loss of faith. Cumulativeness is acknowledged in the preservation and extension of past achievement, as noted. Interpretation and deliberation are acknowledged in all the above ways as well as his emphasis on the promise of a new paradigm "to resolve some outstanding and generally recognized problem that can be met in no other way" (p. 168). The critical distinction between theory-genesis and theory-justification is thus, in effect, reinstated. (If we take Professor Kuhn's denials alone, they form a radical and interesting departure from older views, but they seem clearly untenable. If we take just his affirmations of older concepts under new labels, we have a plausible but no longer novel view. If we take both the denials and the affirmations, we find an

inconsistent account. The strong and exciting part of this account can be defended by judicious retreat to the weaker and unexciting part, but in no case do we have both a radical and a tenable view of science.) [See *SS*, 88-89]

In his recent replies, Professor Kuhn responds to my criticisms and those of others by insisting that his book does contain "a preliminary codification of good reasons for theory choice. These are ... reasons of exactly the kind standard in philosophy of science: accuracy, scope, simplicity, fruitfulness, and the like."[5] They are, however, he insists, "values" rather than "rules of choice," and *may be applied* differently by different scientists.[6] Nowhere, however, have I argued that there must be rules *applied uniformly* by scientists. If these reasons are not, in being conceived as *values,* to be construed as utterly free of all constraints or themselves paradigm-dependent, then they allow the comparison of rival paradigms and make paradigm debates intelligible. Such intelligibility is, however, at odds with Kuhn's statement that the proponents of competing paradigms "will inevitably talk through each other" (p. 108).

In his *Postscript,* he refers to my discussion along with others' as follows: "Because I insist that what scientists share is not sufficient to command uniform assent about such matters as the choice between competing theories or the distinction between an ordinary anomaly and a crisis-provoking one, I am occasionally accused of glorifying subjectivity and even irrationality."[7] Now I cannot be certain what others have said but it is clear that my arguments above enumerated do not include a demand for uniform assent. Indeed, I explicitly stated [*SS*, 80], that "the existence of common evaluative criteria is compatible with borderline regions in which these criteria can yield no clear decisions. And ... even the objective availability of clear decisions is consistent with honest differences of judgment, not to mention plain misunderstandings." The issue is not *uniformity* but *objectivity,* and objectivity requires simply the possibility of intelligible debate over the comparative merits of rival paradigms.[8]

2. VISION AND REVOLUTION

I turn now to a consideration of certain broader aspects of Professor Kuhn's treatment, which I have not previously discussed in any detail. My earlier critique does note the "striking way in which Kuhn's account applies psychological, political, and religious categories to the description of scientific change. The older references to logical system, observational evidence, theoretical simplicity, and experimental test have given way, in his account, to mention of the gestalt switch, conversion, faith, decision, and death" [*SS*,

78]. We have a replacement of the categories of interpretation and deliberation by new categories or metaphors in treating of science. The two fundamental and controlling metaphors are those of *vision* and *revolution*. What I shall now argue is that these are incongruous with one another in a philosophically significant way. Neither, taken alone, supports the view of paradigm change offered by Kuhn to replace the traditional conception. What supports Kuhn's view is rather a hybrid that resembles neither vision nor revolution, but is violated by both. This question seems to me of general interest, independently of the particulars of arguments earlier reviewed, for what is at stake is the fundamental aspect under which we attempt to make science intelligible to ourselves, the controlling analogies by which we tend to think of it.

Take first the topic of *vision*. Kuhn assimilates theoretical change to a gestalt reorganization of vision. "What were ducks in the scientists' world before the revolution are rabbits afterwards. The man who first saw the exterior of the box from above later sees its interior from below" (p. 110). The patterned alteration of thought and experience concomitant with theoretical change is plausibly compared with the intuitive and spontaneous shift in perception of a reversible figure, and resembles it better than it does the piecemeal articulations associated with deliberation. Anomalies and crises, says Kuhn, "are terminated, not by deliberation and interpretation, but by a relatively sudden and unstructured event like the gestalt switch. Scientists then often speak of the 'scales falling from the eyes' or of the 'lightning flash' that 'inundates' a previously obscure puzzle, enabling its components to be seen in a new way that for the first time permits its solution.... No ordinary sense of the term 'interpretation' fits these flashes of intuition through which a new paradigm is born" (pp. 121-122).

The metaphor of *revolution* is employed in *conjunction* with that of the gestalt switch. "What were ducks... before the revolution are rabbits afterwards" (p. 110). But the notion of revolution is ramified in further ways. Competition is visualized as combat, with victory the prize. The conflict is a matter of "techniques of persuasion, or about argument and counterargument in a situation in which there can be no proof" (p. 151). Progress always accompanies victory because the winning camp is in a position to rewrite the textbooks and the implicit history of the subject. And Kuhn cites Planck's statement that "a new scientific truth does not triumph by convincing its opponents and making them see the light, but rather because its opponents eventually die, and a new generation grows up that is familiar with it" (p. 150).

Now a closer look at these two controlling metaphors reveals a critical incongruity between them. While there is, both in the case of the reversible figure and the case of revolutionary conflict a certain mutual exclusiveness of elements, the notion of an opposition of *claims* applies only to the latter. Consider the two views of the reversible cube: These two views are, as a matter of natural fact, exclusive at any given moment. We can flip from one to the other and may even come to shuttle back and forth between them with a certain amount of familiarity. However, they cannot both be seen simultaneously. Such exclusivity may be superficially assimilated to that of the revolutionary situation in which our allegiance must, at any time, be given to one side or the other – where we may shift loyalties but where it is impossible to be loyal to both sides at once or to join segments of each in some form of compromise. However – and this is the critical point – a revolution is a matter of opposed *loyalties and allegiances,* of conflicting *judgments and claims,* whereas there are no analogous questions of loyalty or allegiance or of conflicting claims in the case of alternative views of a reversible figure.

Accordingly, having appreciated the reversibility of the duck-rabbit, there is no question of *arguing* over the relative merits of the duck or the rabbit as the *proper and exclusive* view of the duck-rabbit figure, nor is there any clash of opposed loyalties in the interpretation of the reversible cube. No one *advocates* exclusive acceptance of the exterior-superior view of the cube as better than the interior-inferior view, nor is there any question of deliberate assent or commitment to one, rather than the other. As there are no arguments over the exclusive merits of the duck or the rabbit, there is no duck-party seeking victory over the rabbit-party or aiming to consign it to the dustbin of history. Nothing in this situation indeed corresponds to paradigm debates in the case of theoretical controversies. Nor are there anything like "good reasons" offered in such controversies.

The case of revolution is quite different. Each side seeks victory, demands exclusive allegiance, claims superiority, expresses commitment, propounds arguments, engages in interpretation and persuasion, formulates its rationales, rebuts the arguments of the opposition. Nor is each party totally enclosed in its own conceptual and rhetorical box. It expresses its own view, to be sure, but it attacks the views of its opponents, claiming to understand them well enough to refute them. It addresses its arguments not only to its convinced adherents and to susceptible opponents but to as-yet neutral bystanders whose allegiances are sought, and to whose thought-world the revolutionary line needs to be made to relate. Revolutionary argument is, to be sure, not necessarily scientific in spirit: it may be dogmatic, metaphysical, deceptive or otherwise defective from a scientific point of view. However, it claims commitment, it expresses advocacy, it offers arguments, it gives reasons, it propounds interpretations, it demands acceptance. To reduce the combat of revolutionary parties to a gestalt switch is to *leave out* the critical aspect of *advocacy and opposed loyalties;*

it is to omit the notion of a *claim* and that of a *rationale*. Whereas, conversely, to offer the gestalt switch as a case of revolutionary transition is to *import* the inapplicable concepts of advocacy, commitment, and party combat to merely phenomenally alternative perceptual configurations.

The latter point is particularly important. For if we imagine that the relatively spontaneous intuitive process by which one such configuration replaces another, and to which deliberation seems irrelevant, carries with it a *commitment that is exclusive,* we have formed a hybrid notion that is true *neither* to perception *nor* to revolution: Visual apprehension is perhaps plausibly described as intuitive, but it neither demands nor presupposes exclusive commitment. Revolution is plausibly described as demanding such exclusive commitment but it emphatically is not a spontaneous nor intuitive process to which deliberation is, or can be assumed to be, irrelevant.

If we apply such a hybrid notion to the case of scientific theory change, we are led to develop a view that emphasizes the *intuitive and spontaneous* shift of thought and leaves no room for deliberation or interpretation, but we also emphasize that this shift requires exclusive commitment, that it represents a victory for one side in a conflict of loyalties. Such a conception of paradigm change, as *determinative of commitment but itself immune to deliberation,* approximates Kuhn's view of paradigms: they are not altered through interpretation. As he says, "Rather than being an interpreter, the scientist who embraces a new paradigm is like the man wearing inverted lenses" (p. 121). Anomalies and crises "are terminated, not by deliberation and interpretation, but by a relatively sudden and unstructured event like the gestalt switch" (p. 121). Moreover, adopting a new paradigm is not *just* like the vision of the duck for the first time in the duck-rabbit figure: the paradigmatic vision demands commitment, acceptance; it is jealous of other paradigms and seeks to triumph over them. Indeed, its victory enables it to celebrate its triumph by stamping itself into the records, textbooks, and chronicles under its control.

Such a view, insofar as it derives any plausibility from the analogies with vision and with revolution, does not deserve such plausibility. For these analogies, *taken in combination,* form a hybrid which is true to neither. Nor is this simply a matter of the pragmatic *incongruity* of models, a gauche mixing of philosophical metaphors. The mixture perpetuates an old philosophical mistake that consists in deriving values from visions; Plato thus supposed a vision of the form of the good to yield wisdom in life and imagined the role of philosopher-king as that of social authority resting upon superior insight.[9] But values cannot be seen; nor are visions judgments or the premises of judgments. Value judgments, on the other hand, are certainly *judgments;* they make claims and demand acceptance and commitment. Insofar as this is true, they invite discussion, interpretation, and

Vision and Revolution: A Postscript on Kuhn · 267

deliberation: they allow of a weighing of evidence and a scrutiny of reasons *pro* and *con*. This process of *deliberation over a claim* cannot be short-circuited by a spontaneous vision or insight. Philosophers have tried hard, to be sure, in the interests of one or another dogmatic and authoritarian conception, to end the deliberative discussion of *pros* and *cons* by authoritative appeals to a favored superior vision. But the attempt can hardly be sustained upon reflection.

Nor can the use of the further analogy with conversion improve the situation. For if converts rest only with the occurrence of mystical experiences, they cannot hope thereby to propound *claims* and *judgments* even though their perspective on life may be fundamentally altered. If, on the other hand, they appeal to evidence and stated considerations, their claims invite debate logically, no matter how resistant they may be to such debate. The matter is further confused by the fact that conversion itself is sometimes defended on the analogy with vision.

How then shall we construe the process of theory-change? If the metaphors of vision and revolution cannot be applied simultaneously, can one or the other be chosen in itself to represent the process, or can both be employed, but not in combination? Let us consider the situation once again and recall also the special features of the analogies in question.

The duck and the rabbit do not make rival claims upon our acceptance. Once you have seen both the duck and the rabbit, you may perhaps be puzzled by various aspects of the reversible figure, but the question as to which is to be exclusively accepted does not arise. There is no choice to be made, no common problem or standard by reference to which such a choice could be made intelligible. You accept both "readings" of the puzzle figure as facts and make no decisions of exclusive acceptance. The "readings" themselves make no claims upon your commitment. They may be mutually exclusive as natural occurrences but they are not logically incompatible nor do they generate rival claims and loyalties. Having seen them both, you can rise above them and contain them without logical or moral conflict.

Where theories in science are in opposition, the same cannot be said; the conflict is not simply a matter of inability to hold both theories in view simultaneously, so to speak, but a conflict of claims demanding resolution. Here there is a choice to be made, a resolution to be sought which will determine acceptances and commitments. You may not be able to *reach* a decision comfortably, or at all, but you cannot tolerantly rise above the situation and adopt a sophisticated nonchalance. In these respects the vision metaphor is clearly inadequate.

Further, the process of resolution is neither relatively sudden nor intuitive nor individual. Kuhn speaks of the "flashes of intuition through which a new paradigm is born" (p. 122). But the birth of a paradigm in the mind does not exhaust the period of its ascendancy and triumph. If "interpretation" does not properly describe the process of *birth,* it certainly describes the processes by which the paradigm is submitted to public scrutiny and runs the gauntlet of debate and criticism which precede its victory. Crises are not, contrary to Kuhn, terminated by gestalt switches (p. 121). The gestalt switch is only the beginning. The new paradigm idea has to be formulated, published, argued, defended, tested, and submitted to examination by colleagues with different preconceptions and with access to various sorts of evidence. In this respect too, the metaphor of the gestalt switch is inadequate, and that of revolution, with its notion of prolonged struggle and ultimate decision, seems superior.

Yet, once we recognize that the *adoption* of a new theory or paradigm is not an instantaneous or individual affair, we may find a certain point to the vision metaphor after all. It is a myth to suppose that paradigms are *units,* that they are simply *accepted,* or simply *rejected* as such, and in a momentary spontaneous process at that. This picture needs complication in many directions. At least one complication involves separation of the *birth* of a paradigm from its *testing* in the public arena.

Now we must not suppose that the *originator* gives it over to others to test; that he himself is, of necessity, fully convinced through the origination itself. On the contrary, the originator's own conviction is, in principle, subject to the vicissitudes of the public process of scientific discussion and evaluation. He originates the idea and there is no way of telling how he does it, nor any way of reducing creativity to rule, but his conviction in his idea depends, at least in part, upon how it fares. When others contribute to testing the originator's idea they must first appreciate it: they seek to determine its content and its logical bearing upon the available evidence.

The process of *grasping or understanding* a theory (whether by the originator or his colleagues) may perhaps plausibly be compared to vision. Indeed *seeing* is a standard way of describing *comprehension.* Intuition is a matter of *seeing the point.* It may indeed be relatively sudden, and like creativity itself, impossible to reduce to routine or mechanical rule. Understanding does not, however, in itself, imply advocacy or commitment, nor is it excluded by rejection of the theory in question.

This is indeed a central point of the scientific attitude. Understanding a theory, our acceptance or denial of it is not thereby prejudged. Our advocacy or rejection itself depends upon the outcome of tests and argumentation; it is not predetermined simply by our comprehension. Conversely, rejecting a theory does not *imply* that we do *not* understand it.

In sum, vision may perhaps appropriately serve as a metaphor for *comprehension* of a paradigm or theory, though not for its testing and acceptance or rejection. The latter involve advocacy and claims to commitment; they involve debates and counterdebates, a period of testing and ultimate ascendancy or decline. In these latter respects, the revolutionary metaphor is appropriate, as earlier argued. But as also remarked earlier, revolutionary debate is not necessarily scientific debate. Advocacy *in itself* is not necessarily fair, logical, or responsive to competent argument or relevant evidence. In these respects, science can only be compared in very limited ways to revolution. The *quality* of scientific deliberations makes for a special and rare form of argumentation.

It demands responsibility to the evidence, openness to argument, commitment to publication, loyalty to logic, and an admission, in principle, that one may turn out to be wrong. These special features of deliberation go far beyond the gestalt switch analogy and they outstrip the revolution metaphor as well. An understanding of science requires appreciation of these special features, a recognition that science itself marks a revolution in the quality of human thought.

NOTES

1. Thomas S. Kuhn, *The Structure of Scientific Revolutions* (Chicago and London: University of Chicago Press, 1962). Because of the numerous citations in the present paper, page references to passages in *The Structure of Scientific Revolutions* will hereafter be given directly following quoted portions in the text, enclosed within parentheses. Note that all such references are to the first edition of 1962, rather than to the second edition of 1970.

2. In the second edition of *The Structure of Scientific Revolutions,* 1970, "Postscript-1969," pp. 174-210; and in "Reflections on my Critics" in I. Lakatos and A. Musgrave, eds., *Criticism and the Growth of Knowledge* (Cambridge: Cambridge University Press, 1970), 231-78).

3. I do not offer a comprehensive examination of Kuhn's replies to his critics, however. Dudley Shapere, in "The Paradigm Concept," *Science* 172 (1971): 706 ff., reviews the general effect of Kuhn's recent replies in a discussion I find persuasive.

4. Hereafter, page references to this book will be given directly in the text, prefixed by the letters *"SS,"* and enclosed in parentheses. (Note that *parenthesized* page references *without* the prefix *"SS" always* refer to Kuhn's first edition, as explained above in note 1.)

5. I. Lakatos and A. Musgrave, eds., *Criticism and the Growth of Knowledge,* 261.

6. Ibid., 262.

7. *The Structure of Scientific Revolutions,* 2d ed., 186.

8. For a discussion of Kuhn's replies relating also to other issues, see Shapere's review "The Paradigm Concept" (note 3 above).

9. Compare K. R. Popper, *The Open Society and Its Enemies* (London: Routledge and Kegan Paul, 1945; 3d. ed. revised, 1957), chaps. 7 and 8.

14. THE WONDERFUL WORLDS OF GOODMAN

1. WHAT ARE WORLDS?

"WORLDMAKING", Goodman tells us, "begins with one version and ends with another."[1] Is worldmaking, then, simply the making of versions, that is, descriptions, depictions or other representations, and are worlds to be construed just as versions? The answer does not lie on the surface. The term 'world' is nowhere defined in the book and an examination of the passages in which the term appears yields two conflicting interpretations: On the first, or *versional,* interpretation, a world is a true (or right) world-version and the pluralism defended simply reflects, and extends to versions generally, the *Structure of Appearance* doctrine that conflicting systematizations can be found for any prephilosophical subject matter. On the second, or *objectual* interpretation, a world is a realm of things (versions or non-versions) referred to or described by (119) a right world-version. Pluralistic talk of worlds is here not simply talk of conflicting versions; "multiple actual worlds" is Goodman's watchword and he cautions us that it should not "be passed over as purely rhetorical." (110)[2]

2. WORLDS AS VERSIONS

Each of these two interpretations of "worlds" can call upon implicit as well as explicit statements in support. Take first the versional interpretation. After suggesting that sometimes a cluster of versions rather than a single version may constitute a world, (itself a nonobjectual view) Goodman says, "but for many purposes, right world-descriptions and world-depictions and world-perceptions, the ways-the-world-is, or just versions, can be treated as our worlds." (4) "In what non-trivial sense", he goes on to ask, "are there... many worlds?" And he answers, "Just this, I think: that many different world-versions are of independent interest and importance, without any requirement or presumption of reducibility to a single base." (4) Worlds are here right world-versions, and the multiplicity of worlds is the multiplicity of such world-versions. Clinching this interpretation, Goodman then intro-

This paper appeared in *Synthèse* 45 (1980): 201-209.

duces his treatment of "ways of worldmaking" as follows: "With false hope of a firm foundation gone, *with the world displaced by worlds that are but versions*... we face the questions how *worlds* are made, tested, and known." (7, my italics) The basic discussion that follows of "processes that go into worldmaking" (7) is, then, to be understood as concerned with *versions* rather than with *things, objects, or realms* described by them, and the testing and making of *worlds* is to be construed as the testing and making of *versions*.

That it is versions that are at stake is implicit throughout this discussion, where the individuation of worlds is said at times to hinge on the concepts and distinctions available to relevant groups of persons (9), on emphasis and accent (11), on relevant kinds (11), on ordering (12), and on modes of organization "*built into a world*" (14, italics in the original). "Worlds not differing in entities...may differ in ordering" (12), says Goodman, thus distinguishing worlds where there is no difference whatever in the things denoted. He allows indeed that "...a green emerald and a grue one, *even if the same emerald*...belong to worlds organized into different kinds." (11, my italics, see also 101) Now since Goodman explicitly upholds the nominalistic principle "no difference without a difference of individuals" (95), when he here differentiates worlds simply by the order or emphasis of versions or the kinds indicated by them, he must be referring neither to the realms of individuals described, nor, surely, to various abstract entities associated with them, but rather to the versions themselves.

The point is strikingly illustrated by two contrasting discussions of the question of variant histories – one in Goodman's early paper "A World of Individuals",[3] and the other in the present book. In the first of these discussions, he writes, "We do not take the varied histories of the Battle of Bull Run as recounting different occurrences. In daily life a multiplicity of descriptions is no evidence for a corresponding multiplicity of things described."[4] On the other hand, in *Ways of Worldmaking*, he says of "two histories of the Renaissance: one that, without excluding the battles, stresses the arts; and another that, without excluding the arts, stresses the battles" that "This difference in style is a difference in weighting that gives us *two different Renaissance worlds.*" (101-102, my italics) Consistency with the nominalist principle demands that the worlds mentioned in this last quotation not be construed as comprising the described occurrences, but that they be taken rather as versional.

3. WORLDS AS OBJECTS

Let us now turn to the objectual interpretation of worlds. Goodman speaks of "the many stuffs – matter, energy, waves, phenomena – that worlds are made of" (6), and the presumption of the passage is that he is not simply referring to the inscriptions constituting versions. He uses the adjective "actual" to modify "worlds", characterizing the "multiple worlds" he countenances as "just the actual worlds... *answering to* true or right versions," (94, my italics) the natural reading of "answering to" being: "denoted by", "referred to", "compliant with", or "described by". Goodman indeed expressly distinguishes between "versions that do and those that do not refer", and he insists that we want "to talk about the things and *worlds*, if any, *referred to* ... " (96, my italics). Furthermore, he introduces the notion of truth "in a given actual world", holding that a statement is true in such a world if "true insofar as that world alone is taken into consideration." (110) Here, "world" presumably cannot be intended as "world-version", as is further implicit in the following consideration: he remarks that conflicting statements cannot be taken as "true in the same world without admitting all statements whatsoever ... as true in the same world, and that world itself as impossible." (110) Were "world" to be taken in this passage as "world-version", there would here be no impossibility whatever – only inconsistency.

Goodman explains both truth and rightness in terms of *fitting a world:* "...a statement is true, and a description or representation right, for a world it fits" (132), he declares. Like the notion of "answering to", that of "fit" appears also to be a semantic idea, and the related use of "world" clearly objectual rather than versional.

The objectual interpretation is necessitated, finally, by those passages in which Goodman explicitly treats worlds as comprised of *ranges of application* of predicates, or as consisting of the *realms* of different versions. In this vein, he writes,

> the statements that the Parthenon is intact and that it is ruined are both true – for different temporal parts of the building; and the statement that the apple is white and that it is red are both true – for different spatial parts of the apple.... In each of these cases, the two *ranges of application* combine readily into a recognized kind or object; and the two statements are true in different parts or subclasses of the same *world.* (111, my italics)

Clearly, the reference here is not to different parts or subclasses of the same version.

This example concerned ranges of application; consider now the reference to worlds as *realms*. Discussing two geometrical systems with rival accounts of points, Goodman asserts that if they are both true they are so in different realms – the first "in our sample space taken as consisting solely of lines", and the second "in that space taken as consisting solely of points." For more comprehensive versions that conflict similarly, he says that "their *realms* are thus less aptly regarded as within one world than as *two different worlds*..." (116, my italics). The reference of "worlds" in this passage is not to versions but to things to which versions apply; the interpretation here, in short, is objectual.

4. ARE WORLDS MADE?

Now the versional and the objectual interpretations of worlds do not mix; they are in conflict. As we have seen, the idea of *different Renaissance worlds* emerging from variant histories cannot be objectual, since their realms of application are assumed identical. Conversely, the versional interpretation is precluded by the notion of actual worlds *referred to* by true versions, since such versions in fact refer to all sorts of things, non-versions as well as versions.

Goodman seems to hold, indeed, that these conflicting interpretations of "worlds" reflect the vacillations of antecedent theoretical practice. The line drawn by such practice between "versionizing" and "objectifying" is, he believes, not a hard but a variable line, motivated by convenience and convention. "In practice," he writes, "we draw the line wherever we like, and change it as often as suits our purposes. On the level of theory, we flit back and forth between extremes as blithely as a physicist between particle and field theories. When the verbiage view threatens to dissolve everything into nothing, we insist that all true versions describe worlds. When the right-to-life sentiment threatens an overpopulation of worlds, we call it all talk." (119) Yet, the availability of these two interpretations – however the line may be drawn – makes it important to examine closely Goodman's thesis that worlds are made. I can accept this thesis with "worlds" taken versionally, but I find it impossible to accept otherwise.

5. WORLDMAKING: VERSIONAL YES, OBJECTUAL NO

That Goodman himself intends worldmaking to be taken both ways is shown in a variety of passages. In a summary statement toward the end of the book, he says,

Briefly, then, truth of statements and rightness of descriptions, representations, exemplifications... is primarily a matter of fit: fit to what is *referred to* in one way or another, or *to other renderings,* or to modes and manners of organization. The differences between *fitting a version to a world, a world to a version, and a version together or to other versions* fade when the role of versions in *making the worlds they fit* is recognized. (138, my italics)

Moreover, Goodman specifically speaks of worlds, taken objectually, as made. In a crucial passage, he writes, "...we make worlds by making versions.... The multiple worlds I countenance are just the actual worlds made by and answering to true or right versions." (94) That this passage requires the objectual interpretation is shown by the mention of worlds as *answering to* true versions. Thus, in saying we make worlds by making versions, Goodman is not uttering the triviality that we make versions by making them. Can he then be asserting rather that in making right versions we make what they refer to – that is, in making true descriptions we make what they describe, in making applicable words we make what they denote?

Apparently, the answer is yes. "Of course," he writes, "we want to distinguish between versions that do and those that do not refer, and to talk about the things and worlds, if any, referred to: but these things and worlds and even the stuff they are made of – matter, anti-matter, mind, energy, or what not – are fashioned along with the versions themselves." (96) Here he clearly says that we make not only versions but also the things they refer to and even the material of which these things are made.

Now the claim that it is we who made the stars by making the word "star" I consider absurd, taking this claim in its plain and literal sense. It mistakes a feature of discourse for a feature of the subject of discourse – a mistake Goodman himself has warned against in an earlier paper,[5] and it seems to conflict with his own insistence on the difference between a version and what it refers to. Goodman himself emphasizes (94) that his "willingness to accept countless alternative true or right world-versions does not mean that everything goes... that truths are no longer distinguished from falsehoods..." Since, as I believe, the claim that we made the stars is false if anything is, his version of versions is itself *false* if it implies this claim. Nor is it helpful to say that we made the stars *as* stars – that before the word 'star' existed, stars did not exist *qua* stars. For, in the first place, that stars did not exist *qua* stars does not imply that they did not exist, or that we made them. And, in the second place, the existence of stars *qua* stars is just their existence plus their being called 'stars'. No one disputes that before we had the word 'stars', stars weren't called 'stars', but that doesn't mean they didn't exist. It would be altogether misleading on this basis alone to say we *made* them.[6]

But a deeper philosophical motivation underlies Goodman's notion of worldmaking. A pervasive theme in his work is the rejection both of the given and the notion of a "ready-made world" that "lies waiting to be described." (132) He urges again and again that the organization of our concepts and categories is not unique, that such "modes of organization ... are not 'found in the world' but *built into a world*". (12-14, esp. 14, italics in original) The supposition is perhaps that unless we take our star-versions to have made the stars, we will be driven to accept either a neutral given without concepts altogether or else the pre-existence of our conceptual scheme to the exclusion of all others. While agreeing with the underlying philosophical motivation, I cannot, however, see that the latter supposition is sound. That stars existed before people implies nothing about concepts, their uniqueness or pre-existence. Star-concepts did not, but stars did, antedate the emergence of living creatures. Star-concepts were surely not ready-made, waiting to be used; they were indeed made by us. It doesn't follow the stars were therefore made by us rather than in fact (but in a metaphorical sense) waiting to be described. To reject the given and to allow a multiplicity of conceptual schemes does not require objectual worldmaking.

The objectual version of worldmaking may, however, perhaps have another philosophical source in Goodman's view of *facts* – more particularly his recognition of how vocabularly constrains and shapes our factual descriptions. The topic arises in his discussion of the phenomenon of apparent motion, that is, the seeing of a moving light where there are, physically, just two distinct flashes, the one following the other a short distance away. Discussing the case of certain subjects who report not seeing the apparent motion, Goodman asks whether they are not perhaps indeed aware of it, but taking it as a *sign* of the physical sequence of light flashes – that is, *looking through* the phenomenal to the physical state "as we take the oval appearance of the table top as a sign that it is round." (92) Can this possibility be tested? Can such subjects be brought to report directly on their actual perceptual experience? To ask them "to avoid all conceptualization" would be useless, since it would leave them "speechless". Rather, as Goodman suggests, "the best we can do is to specify the sort of terms, the vocabulary" to be used, instructing the subjects to describe what they see "in perceptual or phenomenal rather than physical terms." And this, says Goodman,

> casts an entirely different light on what is happening. That the instruments to be used in fashioning the facts must be specified makes

pointless any identification of the physical with the real and of the perceptual with the merely apparent. (92)

He concludes, further, that we must not say "both are versions of the same facts" in any sense that implies "there are independent facts of which both are versions." (93)

There are then, for Goodman, no independent facts, construed as entities discrete from versions and their objects. What then does his talk of "fashioning the facts" (92) come to? Presumably this: that the true reports of observations giving descriptions of such objects are constrained by the vocabularies employed; these vocabularies are thus instruments for creating factual descriptions. Since all our knowledge of objects is, moreover, embodied in such descriptions, our knowledge is, itself, in the same way, shaped by our vocabularies. But what are objects themselves? We have no access to objects aside from our knowledge of them; they are therefore themselves shaped by our vocabularies. It is thus we can say that in making our versions we make their objects. Possibly some such line of reasoning motivates Goodman's objectual worldmaking.

Whether it does or not, I do not myself find it convincing. Even were it true that we have no access to objects aside from our knowledge of them, it would not follow that objects are made by our knowledge. Moreover, to say we have no access to, or contact with objects aside from our knowledge of them is true only if by "access" we intend such things as understanding and awareness, that is, "cognitive access". Thus the statement is trivial; it assures us that we can have knowledge of objects only in having knowledge of them. And to say that our knowledge of objects is shaped by our vocabularies boils down to saying that the descriptions we compose are made up of the words we have. From this triviality it clearly does not follow that we create or shape the things to which our words refer, or determine that our descriptions shall be true. In making the true statement that there were stars before men, we do not also make the stars that were there then.

Now Goodman himself insists on the separation of truth from falsehood; as we have seen, he denies "that everything goes". (94) There are, he asserts, false as well as true versions; he rules out the idea that any version can be made true at will. And his discussion of fiction indeed offers concrete examples of such constraints. "Some depictions and descriptions," he writes, do not literally denote anything. Painted or written portrayals of Don Quixote, for example, do not denote Don Quixote – who is *simply not there* to be denoted." (103, my italics) The creation of a Don Quixote-version evidently does not automatically create an object for it. The mere making of the word does not guarantee it will be non-null. Whether there is or is not an object satisfying a version of our making is thus not, in general, up to us.

Whether a world answers to a version is, in general, independent of what we may wish or will. How then can Goodman describe his "actual worlds" as both "made by" and "answering to true or right versions"? How can he say "we make worlds by making versions"? (94) I conclude that he cannot and that, despite his disclaimer (110), objectual talk of worldmaking had *better* be taken as "purely rhetorical".

NOTES

This paper was presented at a symposium sponsored by the American Philosophical Association on Goodman's *Ways of Worldmaking*, on December 28, 1979. (Note 6 is added here for the first time.) I am grateful to Samuel Scheffler for discussion and criticism.

1. Nelson Goodman, *Ways of Worldmaking* (Indianapolis: Hackett Publishing Company, 1978), 97. From here on, all page references to this book will be given in parentheses following their respective citations in the text.

2. For the *versional* but not the *objectual* interpretation, 'world' is always, strictly speaking, short for 'world-version', a compound in which the constituent 'world' is syncategorematic and non-referential, its position inaccessible to variables of quantification.

3. N. Goodman, "A World of Individuals", *The Problem of Universals* (Notre Dame, Indiana: University of Notre Dame Press, 1956), 13-31, now reprinted in Goodman, *Problems and Projects* (Indianapolis: Bobbs-Merrill, 1972), 155-72.

4. *Problems and Projects*, 164.

5. "Philosophers sometimes mistake features of discourse for features of the subject of discourse. We seldom conclude that the world consists of words just because a true description of it does, but we sometimes suppose that the structure of the world is the same as the structure of the description." *Problems and Projects*, 24.

6. In his *Languages of Art* (Indianapolis: Hackett Publishing Company, 1968, 1976), 88, Goodman defends himself against the charge that he makes what a picture expresses depend upon what is said about it, thus "crediting the expression achieved not to the artist but to the commentator." He writes: " 'Sad' may apply to a picture even though no one ever happens to use the term in describing the picture; and calling a picture sad by no means *makes* it so." (my italics) Exactly. 'Star' may apply to something even though no one ever happens to use the term in describing it; and calling something a star by no means makes it one.

PART THREE
Learning & Acting

THE PAPERS in this part address various themes relating to action – action as guided by knowledge, embodied in skill, justified through commitment, expressing perception, organized by emotion, shaped by community, and acquired by learning. Evident throughout Part 3 is an effort to bridge the inherited divisions between knowledge and conduct; cognition and feeling; learning, understanding, and doing. Equally evident is an effort to reconcile rational control with change, to avoid both untrammeled coherence and blind certainty, whether in construing science, conduct, or society. Both these efforts reflect familiar emphases of pragmatic philosophy, which is accorded explicit consideration at the close of this part.

The first three papers belong to ethics. Each of them, in its own way, argues for a naturalistic view of the subject, that is to say, for a view that relates ethics to, rather than sundering it from, science. The first of these papers attacks four antinaturalist restrictions on the analysis of ethical terms; the third rebuts the claim of absurdity brought against Dewey's particular interpretation of desirability as a form of dispositional desire. The second paper, however, offers a general view of justification as a key to the understanding of normative ethics. Set forth in such a way as to encompass not only the realm of action but also that of cognition, this view indeed incorporates the latter within the former. The resulting analysis construes the process of justification as bridging scientific and practical thought – in neither case, moreover, resting upon certainty or evaporating into coherence.

All the other papers, with the exception of the last, fall within the philosophy of education, broadly speaking, but they ramify into epistemology as well. Thus, the fourth essay distinguishes three models of teaching largely on the basis of their contrasting conceptions of mind and knowledge; the fifth examines Ryle's account of knowing, charging it with inconsistency;

the seventh discusses understanding, method, and skill in mathematics education, arguing that to distinguish between deductive and strategic reasons is important for such education; the eighth develops a concept of "cognitive emotions," of which surprise is a crucial example; and the last includes a critique of Peirce's theory of belief, doubt, and inquiry, as well as a discussion of his comparison of methods. Even the sixth and ninth selections, concerned as they are primarily with moral and social aspects of education, relate such aspects to critical thought generally, in the sphere of science and elsewhere.

It is evident from this account that I understand the philosophy of education not as a thing apart, a separate academic enterprise unified by method or doctrine. Rather, directed as it properly is toward problems of learning, teaching, and schooling, it makes inevitable connection with epistemology, ethics, and the philosophy of mind, with the philosophies of the various teaching subjects (such as philosophy of language, of science, of art, of history, and so forth), and with yet other branches of philosophical inquiry. Bridges are indeed required, not only between the substantive areas of life earlier mentioned (knowledge and conduct, cognition and feeling) but also between the various studies bearing on such areas.

I want now to comment on the general notion of bridges. Bridges relate, but relation does not require reduction. The effort to connect realms, which I have praised in pragmatism, by no means requires a blinking of their independence. Yet the upshot of this point is not some general antireductionism in principle. Reduction, wherever possible, yes, by all means. Whether or not it succeeds in a given instance, however, let us still prize whatever relations of significance there are to be found.

These somewhat abstract comments may be brought to earth by a closer look at the Ryle and Dewey selections included here. There is, as has occasionally been remarked, a certain philosophical affinity between Ryle and Dewey, which overrides their fundamental disparities in style, culture, temperament, and method. It is, I suppose, some sense of that affinity which drew me to Ryle after my earlier studies of Dewey and the other pragmatists. Both thinkers reject Cartesian views of the mind, both adopt a dispositional interpretation, finding the mental in the patterning of overt activity, both understand intelligence not as intuitive but rather as procedural, and both are concerned with educational matters – with learning and knowing rather than just with the learned and the known.

The work of each can be seen as an effort to bridge the Cartesian chasm, to relate body and mind, intelligence and skill, theory and practice, learning and doing. Both indeed identify significant relations between the members of each pair, illuminating many dark corners of the philosophy of mind and of knowledge. But, as I argue in the two selections concerning

Introduction

them, they do not succeed in effecting the reductions they are occasionally tempted to claim.

Ryle's attempted reduction of propositional knowing to capacity thus seems to me to fail. Nonetheless, such knowing is widely linked with capacities and other dispositions in description and explanation. The comparison I would make is to theoretical terms in science which, although not reducible to observational terms, are yet widely related to them in description and explanation.

Dewey, in accord with the other pragmatists, lays great stress upon activity in the interpretation of knowledge. Knowing is not simply the work of a disembodied mind; it grows out of the mind's bodily interactions with observable nature. The body's sensory and motor apparatus are instruments of, rather than obstacles to inquiry. It is through the controlled activity of experiment, yielding observable effects, that scientific theory advances, and such advance serves to reorient consequent practice in turn.

The relations Dewey thus establishes between theory and practice, learning and doing, are indeed, as I believe, of the first importance. Taken, among other things, as a way of marking the existence of such relations, the pragmatic emphasis on continuity is unexceptionable. Yet this emphasis seems at times to turn reductionistic – to construe theory as wholly absorbed into the sphere of action and observation, its role solely to mediate between practical problem and practical resolution. Analogously, the school's autonomy is underplayed, its primary function conceived as the solving of social problems.

Thus, Dewey's remark that ideas in science are "generated within" scientific procedures as well as tested by scientific operations, seems to me, as I argued in my *Four Pragmatists,* too strong by far.[1] The generation of theoretical ideas, as distinct from their testing, is not reducible to scientific procedures. Creative processes in science are independent of such procedures and hardly therefore "integral with the course of experience itself." Nor does it seem to me correct to say, as Dewey does, that scientific "problems are solved when changes are interconnected with one another." Rather, to make sense of what Dewey calls "the happening of experienced things," and their changes, at the level of observation or practice, science develops theoretical structures that go beyond this level and are incapable of being reduced to it. To conceive experimental knowledge as, in Dewey's words, "a mode of doing" overstates the case. Experimental knowledge, I submit, is born both of doing and of theorizing, and theorizing is itself independent of the constraints of activity or observation at the level of practice.

As for the school, I argue that, integral as it may be with its own society, its role cannot properly be reduced to that of an agency for social improvement. It is also an agency for the defence and advancement of intellectual concerns and critical standards that have their own worth, and in the light of which social problems may themselves take on altered perspectivies. If the school may indeed be viewed as an intermediary agency helping to improve society in the long run, society, I believe, "may equally be viewed as an intermediary agency to be judged by its dedication to the autonomous values of intelligence, criticism, knowledge, and art, of which the school is the guardian."[2]

NOTES

1. I. Scheffler, *Four Pragmatists* (London: Routledge & Kegan Paul, 1974).

2. Ibid., 254.

1. ANTINATURALIST RESTRICTIONS IN ETHICS

THE PROTECTION of whole domains from investigation by the use of general restrictive arguments is a familiar story in modern philosophy. The literature bristles with arguments purporting to show why this or that field is, in principle, inaccessible to ordinary rational inquiry. In most cases, the strategy is identical: some characteristic of the domain-to-be-protected is asserted to render it distinctive, and, moreover, so distinctive as to exempt it from study by typical methods.

Ethical language has enjoyed the dubious benefits of such wholesale protection. And since it is the ethical naturalists who have most stressed the continuity of ethics with other domains, it is not surprising that restrictive arguments based on an assertion of ethical uniqueness have generally taken the political form of antinaturalism. Indeed, since the usual philosophical, as contrasted with the political, meaning of "antinaturalism" is far from clear, it might be suggestive to construe it just in terms of wholesale methodological restrictiveness. An ethical antinaturalist, then, would be one who demanded some wholesale restriction in the kinds of method appropriate to ethical analysis, based on some allegedly distinctive feature of ethical language.

Four antinaturalist arguments have been repeatedly offered, with numerous variations, in recent philosophy, all demanding severe restrictions on analytic procedures in ethics. I shall try to show why I think they are all fallacious or inconclusive. Since each recurs in various contexts with change of detail, I shall treat each schematically, rather than make a close textual examination of some particular formulation. Nevertheless, I am not aware of any such formulation to which my critical remarks are inapplicable.

A. The argument from descriptiveness.

The first argument we consider purports to show that ethical terms are not descriptive, in some vital sense, and can thus not be analyzed in descriptive terms. For example, the predicate "is blue," it is argued, applies to things in virtue of some specific sensory quality which they exhibit, and

---

This paper appeared in the *Journal of Philosophy* 50 (16 July 1953): 457-66.

its correct use is thus descriptive. If we introspect carefully and try to isolate some such sensory content in virtue of which "is good" is ascribed to things, we fail. We may classify all the qualia of our sense experience into their several clusters and exhaust them all without coming upon a quality of goodness. Having no constant relations to the sensory characters of things, ethical terms are not descriptive of their denotata in the sense in which other terms are, and whether or not they be cognitive in some other sense, they cannot be analyzed into ordinary, empirically descriptive terms. Ethical naturalism is not merely inadequate in detail; it is fallacious in principle.

The intent of this argument is, ostensibly, to restrict ethical analysis while allowing analysis free play elsewhere. Yet, a closer examination of its use of "descriptive" reveals not only that it has strangely awkward consequences but that it fails to separate ethical from non-ethical language, as ordinarily conceived.

We might notice, first, that the argument seems cast in phenomenal terms. "Is blue," as a dispositional predicate of objects, however we construe dispositional terms, is surely not applicable in virtue of a shared sensory quality. It is predicable in virtue of any of a variety of sensory outcomes of different standard tests which, moreover, do not exhaust its significance. When we reflect that science is largely made up of dispositional terms, the denial of descriptiveness to all but phenomenal predicates seems absurd indeed.

Even limiting ourselves to the phenomenal framework, however, we shall find difficulty in understanding how the argument's use of "descriptive" is to do the work demanded of it. Presumably, no phenomenal predicate is to be taken as descriptive unless its denotata share a specific sensory quality. Now, short of metaphysics, a criterion formulated in terms of sensory simples relativizes the notion of descriptiveness to certain systems. For some phenomenal systems speak of no sensory simples at all, but construe what are presystematically taken to be such simples as classes of other elements. Further, even among phenomenal systems with sensory simples, choice of a particular set of simples varies. Finally, the criterion is inapplicable to many-placed predicates altogether. This use of "descriptiveness" is not simply arbitrary and untrue to ordinary usage. Its relativity renders the argument indeterminate, for a term not descriptive in one system may be descriptive in another, while so-called non-ethical terms are subject to the same type of variation.

However, even given any particular system, the criterion fails to distinguish ethical from non-ethical while resulting in generally awkward consequences. Assume, for example, that we consider a system in which the colors blue, yellow, green, and red are taken as simple. I suppose we should consider it curious that, though "is blue" is descriptive by the proposed criterion, "is not-blue" is non-descriptive. And while we might not be too troubled (because of its universal applicability) to find that "is blue or not-blue" is non-descriptive, we might find it paradoxical that "is blue and not-blue" *is* descriptive, since all its denotata share a simple. Further, not all non-descriptive predicates are universally applicable; consider the predicate "is bi-colored", which must be denied descriptiveness because a blue-and-green presentation shares no simple with a red-and-yellow one though they both satisfy the predicate, or "is tri-colored," which is non-descriptive because, though every pair of its denotata shares a simple, the same simple is not shared by every two pairs.[1] Consider, finally, predicates of sensory qualities themselves, e.g., "is a simple sensory quality." Need we say there is a subtle second-order sensory quality shared, for instance, by lavender, bitterness, velvetyness, pungency, and shrillness, in order that "is a sensory quality" may be said to be descriptive? Or do we have to include the latter predicate among those not descriptively analyzable?

In trying to understand the argument itself, aside from considerations relating to the use of "descriptive," we come upon difficulties of a different order. Take, for example, the predicate "is colored," which must be non-descriptive by our criterion, yet analyzable, for our sample system, as "is blue or red or green or yellow." Now though the components of the analysans are here singly descriptive, the analysis cannot be considered as one *in descriptive terms,* since this would violate the claim that no non-descriptive term is analyzable in descriptive terms, and ruin the case against naturalist ethics. Yet if, in order to save this claim, we insist that the analysans is here non-descriptive (since a difference in descriptiveness of analysandum and analysans means a difference in application and hence a false analysis), what is now the force of the restriction being urged against ethical naturalism? If a naturalist analysis is incorrect by independent criteria, e.g., non-coextensiveness, it may be sinning against the argument's dictum. As soon as it is shown correct by the same independent criteria, its analysans becomes, tautologously, non-descriptive. So that the argument, while virtuously disdainful of incorrect analyses, excludes no analysis as incorrect except the ones we have already decided are incorrect by independent criteria, while welcoming all we are ready to admit independently. Masking as a restriction, it excludes no analytic procedure and no actual analysis. It merely defines "analysis in descriptive terms" so as not to apply to any analysans of a non-descriptive predicate, where the latter term and its counterpart "descriptive" are construed in a strangely arbitrary sense.

It may be said that the argument means that, by independent criteria, naturalist analyses are incorrect. This is, of course, a reversal; we argue now not from unequal descriptiveness to fallaciousness, but from fallaciousness proper. This, however, begs the question, for it is precisely the issue of fallaciousness which is in debate. That this or that analysis is incorrect may be admitted by all. That no naturalist analysis is possible cannot be shown by assuming it.

Suppose, finally, that, considering the last set of difficulties, we decide to kill two birds with one stone, by admitting "is colored" as descriptive, and relaxing our criterion of descriptiveness to avoid some of its awkwardness. We shall say, now, that a predicate is descriptive if and only if it is *coextensive with some complex predicate,* each predicative component of which singly has denotata all sharing some simple. We may, indeed, now admit both "is colored" and its analysans as descriptive, but we run into worse trouble. We are arguing, it turns out, that no ethical term is descriptively analyzable since none is equivalent to some logically complex predicate whose components denote in virtue of sensory simples. But this is precisely one of the points at issue. To assume it as a reason for itself is to collapse the argument into a *petitio principii.* I conclude that ethical language has not been shown incapable of analysis in standard fashion by this argument.[2]

B. The argument from simplicity.

The second argument we consider claims that, since the quality referred to by "is good" is simple, this predicate must be altogether indefinable. Analysis of terms is here conceived as parallel to analytic decomposition of perceptual units. When dealing with a complex perceptual unit, we can decompose it as well as define its name. When, however, we have reached the atomic indivisibles of the perceptual domain, analysis in both the concrete and the linguistic senses must cease.

Insofar as this argument assumes a unique structure of the perceptual field, in abstraction from a system, it is on unsafe ground. There is no metaphysical necessity which attaches to one particular way of dividing up the experiential field, and there are no absolute criteria of perceptual simplicity. Are qualitative universals simpler than concrete events? Which of these, if any, is simpler than cross-sectional time-slices of experience? The argument is equally inadequate in using "is indefinable" as a one-place predicate. Terms are neither definable nor indefinable in the abstract. They become so in relation to a particular apparatus and a particular systematic basis.

Suppose, then, that we relativize the argument in both respects; we speak now of the simplicity of entities relative to a system, i.e., the atoms or ground-elements which are indivisible within it, and we refer to indefinability with respect to this system. Let the argument now be emended to read, "No name[3] of an entity which is atomic relative to a system is definable within that system." This considerably weakens the original claim, for it allows that the name of an entity atomic relative to one system may be definable within another system where it is non-atomic. But even the weakened claim is false, for names of ground-elements may be, and generally are, defined in terms of the primitive relations of the system. Thus, in Carnap's *Der logische Aufbau der Welt,* his ground-elements are *erlebs* and his primitive is a two-place predicate *"Er."* "Something x is an *erleb"* is now defined as "There is some y which either bears *Er* to x or to which x bears *Er.*" In Goodman's system, in which qualia are atomic, and "W" is primitive, "X is a *quale*" is defined as "There is some y to which x bears W, and there is no W-related z which forms a proper part of x."[4] Thus, even this weakened claim turns out to be false.

In an effort to save some of this claim, we may wish to construe it not as an argument from the simplicity of ground-elements but as an argument from the simplicity of the names themselves, i.e., "no simple name can be defined." When we relativize this claim so that it refers to simplicity within a system and definability therein, it is hard to see what "simple" can mean but "primitive." If so, the claim is true but hopelessly trivial, i.e., "no name primitive within a system can be defined within that system." The only conceivable relevance this might have as an antinaturalist claim would be on the assumption that "is good" must necessarily be primitive to every ethical system. But this, of course, begs the question, besides invoking a mythical necessity. In sum, I think the arguments for indefinability of ethical terms on grounds of simplicity are no serious obstacles to the application of analytic methods in ethics.

Perhaps a few words should be said, in this connection, about the alleged irreducibility of ethics to science, where this thesis is defended on the grounds of relative indefinability rather than the atomicity of ethical qualities. White[5] has called Moore's antinaturalism obscurely infinitistic, since natural predicates are not enumerated nor adequately characterized. The same obscurity surrounds the notion of science and befogs this thesis of irreducibility. As soon as a particular scientific vocabulary is adequately specified, either by enumeration or by some pragmatic description, e.g., "the vocabulary used by American scientists in 1950," the bogus universality of the claim is exposed, for who would venture to predict from such a vocabulary the character of the scientific vocabulary at any other time? Analogous remarks are often relevant to claims made for the irreducibility of one scientific domain to another.

There is, however, an additional and special confusion in the issue of ethical irreducibility which does not arise in questions of the reduction of one limited scientific domain to another. This involves the ambiguity of the notion of irreducibility and the notion of being scientific, and may best be illustrated by a contrasting example: Where the reducibility of biology to physics is discussed for instance (assuming that here also some specification of vocabularies is made), indefinability of biological terms in physical terms implies irreducibility in a quite precise sense. Since both biology and physics are acknowledged scientific disciplines in the sense of sharing common logical and inductive canons, however, irreducibility in this sense does not imply that biology is less scientific in methods and procedures, but only that the body of science requires biological primitives in addition to the physical. Even those who make the most obscurely infinitistic claims for biological irreducibility thus intend, for the most part an emancipation from physical models and vocabulary, as they conceive them, in the interests of biology *as a science;* irreducibility in the sense of indefinability clearly has nothing to do with scientific status.

When the question concerns the reducibility of ethics to science, however, a subtle confusion enters. Those who claim irreducibility on the grounds of indefinability do not now simply expand the domain of science to include ethical primitives. They assume that the alleged irreducibility of ethics to scientific vocabulary shows that ethics is unscientific in the totally different sense of being incapable of treatment by scientific canons of logic and method, and becoming absorbed into the body of science. Irreducibility in the sense of indefinability has become illicitly transformed into irreducibility in the sense of inaccessibility to scientific method. Thus, even if a satisfactory way were known of characterizing all non-ethical terms in all future scientific vocabularies, and of proving the thesis of indefinability, it would still be completely fallacious to conclude that ethics is not capable of study by scientific methods, or that ethics, with its own indefinables, could not be absorbed into the body of science, just as the eternal irreducibility of biology, if proved, would not imply the non-scientific nature of its procedures. Of course, indefinability does not imply scientific status either; this depends on other considerations, among them the clarity of application of the indefinable terms in question. In any event, relative indefinability is irrelevant to the question of scientific status, and those who fallaciously assume the opposite and go on to urge methods of intuition or dogmatism in ethics generally fail to realize that indefinability does not absolve them of the responsibility to give some informal explanation of their indefinable terms and to meet standards of clarity and consistency of application, i.e., to present some sort of analysis anyway. To sum it all up, to claim indefinability of ethical predicates on the ground of simplicity is either confused, false, or question-begging; to argue that ethics is above scientific criticism because of relative indefinability is a *non-sequitur* even if the assertion of such indefinability could be made clear and proved.

C. The argument from synonymy.

The third argument claims that a definition of ethical predicates in terms of non-ethical ones must always be unsuccessful because it must always fail to express synonymy. Definiendum and definiens may be coestensive, it is admitted, but can never have the same meaning. That this is so, it is argued, may be seen from the fact that a definition of an ethical predicate, e.g., "is good," in non-ethical terms, e.g., "is desired," is never accepted without question as a stipulation; always we may meaningfully ask whether all objects designated by the definiens are truly designated by the definiendum, i.e., "are all desired things truly good?" This constantly recurring question, the so-called open question, indicates that the definiendum has a meaning prior to the definition, for otherwise this query would be meaningless. In addition, the fact that the question recurs *as a factual option* indicates that synonymy is not achieved in the definition.

It is important to note the lack of either an enumeration or a general characterization of non-ethical predicates, recently pointed out by Professor White. We are asked by the antinaturalist, as White puts it, "to 'see' the absolute difference between any given natural predicate and any given ethical predicate as well as to see what it is that all natural predicates have in common."[6] All this on the basis of examining a small number of proposed definitions.

Sometimes this argument is presented as if the very fact that the definiendum has an independent meaning shows that no definition is possible. Of course, such a view would make all constructive definitions equally impossible. And, since philosophy is largely a clarification of terms in ordinary use, a clarification subject to the control of pre-analytic usage, this consequence is rather drastic.

Perhaps more usually, however, the stress is laid on the intuited fact that in every case the recurring question is a factual one, i.e., one requiring a synthetic answer. This is an indication that no definiens is synonymous with an ethical definiendum, and hence that no definition of ethical terms is adequate. Now, even if we overlook the infinitistic obscurity of the argument, it contains two dubious semantic assumptions, which are independently vulnerable. The first is the suitability of the notions of factuality, synonymy, and analyticity for natural languages. I need not repeat the points made in this connection by Professors Quine,[7] White,[8] and Goodman;[9] I merely record my opinion that they have collectively made this assumption at least debatable. But this is not all. For there is a second assumption, not implied by the first, which is equally shaky, i.e., the assumption that synonymy or intensional identity is a *sine qua non* of adequate definition. Again, I shall not cover well-worn ground. I merely call attention to Nelson Goodman's analysis in *The Structure of Appearance* and his conclusion that even extensional identity is too strict as a criterion of definitional adequacy. In sum, it seems to me that no form of this argument escapes devastating criticism, and hence that no form of this argument need occasion ethical naturalists any worry.

D. The argument from emotiveness.

This argument claims that, since distinctive uses of ethical terms are expressive or incitive, rather than cognitive, standard types of analysis are misapplied in this domain. Those who offer this argument vary in the extent to which they also rely on the previous ones, and hence in the scope which they envisage for this emotive thesis. Those who repeat the argument from synonymy, for instance, hold that naturalistic or cognitive analyses are always misapplied for so-called typically ethical usage, while those who rely on none of the previous arguments generally offer a more moderate emotivism. If our comments on the previous arguments were well taken, we need reckon only with this moderate form, according to which there is *some* sense of ethical terms in which they function emotively entirely, and for which, hence, cognitive analysis is mistaken.

This moderate thesis is, of course, quite compatible with naturalistic analyses, since it claims no universality for itself. This compatibility does not, however, guarantee its truth, which must be shown on independent grounds. Now it seems to me that both the interpretation of this weak thesis and the evidence thus far advanced for it are obscure on several counts.

First, the claim that ethical terms function entirely emotively in some contexts is apparently not intended to deny that they may have the grammatical forms of predicates even in those contexts. Nor is it denied that even in their solely emotive use, they are applied to some things and withheld from others, e.g., "This is good." Now, it seems to me that this isolation of some group of things from an environing group by a term grammatically functioning as a predicate is precisely a cognitive function, which may be equivalently performed by other predicates, not ordinarily called "ethical." Of course, recourse may be had to some special sensory descriptiveness which makes some predicates more truly predicates than others, though both are applicable to things. As already pointed out, however, such a notion has its own flaws. Now if these comments are well taken, even the claim of *solely* emotive meaning for some contexts is too strong and should be replaced by a claim of *primarily* emotive meaning, with the corresponding admission that cognitive analysis cannot be altogether ruled out for any context of ethical terms, ordinarily conceived. Emotive meaning will be considered now never as an alternative to designative function, but as an additional factor of psychological charge.

It is just in this area of connection with psychology that we encounter the second set of obscurities. For the state of psychological theories of cognition, belief, emotion, and attitude is presently too imprecise to permit of sharp statement of the hypotheses of emotivism and of clear experimental test. Such test there must be eventually, and perhaps the most important aspect of emotivism is its own incitive function in stimulating research in this no man's land of science. But in advance of such research, emotive theses remain hypotheses, fruitful though they may turn out to be.

It is often said, as Stevenson himself remarks, that emotive theses (even in advance of detailed scientific warrant) derive a certain clarity from ordinary usage and common-sense observations. It certainly appears plausible, for instance, that terms vary in emotivity, and that some may function primarily emotively. However, it is equally plausible that emotivity is much wider than ethicality of terms, that there is a continuum of emotivity rather than two sharply separated groups, and that some non-ethical terms function primarily emotively too. So that if the defining characteristic of ethical terms is taken to be their emotivity or even some minimum degree of emotivity, we will probably have a different and much wider class than the one usually spoken of as ethical and, moreover, a class not sharply separated except by stipulation. To believe, in advance of inquiry, that emotivity, or some degree of it, will neatly slice off just those terms called ethical by philosophers and that the degree of relevance of cognitive analysis will conform tidily to this scheme seems to me premature at best.

In sum, only a combination of emotivism with some preceding argument is incompatible with a cognitive analysis. More moderate forms, compatible in any event, seem programmatic at best and require both theoretical refinement and empirical test.

NOTES

1. See Goodman's discussion of what he calls "the difficulty of imperfect community" in *The Structure of Appearance* (Cambridge: Harvard University Press, 1951).

2. I wish to thank Mr. A. N. Chomsky, with whom I discussed some points related to this section.

3. Or predicate.

4. Goodman, 174.

5. M. G. White, "A Finitistic Approach to Philosophical Theses," *Philosophical Review* 60 (1951): 299-316.

6. White, 309.

7. W. V. Quine, "Two Dogmas of Empiricism," *Philosophical Review* 60 (1951): 20-43.

8. M. G. White, "The Analytic and Synthetic: An Untenable Dualism," in *John Dewey: Philosopher of Science and Freedom,* ed. S. Hook (New York: Dial Press, 1950).

9. Nelson Goodman, "On Likeness of Meaning," *Analysis* 10 (Oct. 1949): 1-7.

2. ON JUSTIFICATION AND COMMITMENT

THE IDEA of justification is one of the keys to normative ethics. To say of some act that it is right, warranted, or valid, is to say that it is justified. To predicate goodness of something is to hold its approval justified. To describe an act as obligatory is to say not only that it is justified but also that there is no feasible alternative equally justified. Understanding justification, then, we are in a position to unlock some of the rustiest and most heavily-bolted doors in ethical theory. I want, in what follows, to propose an interpretation of this key idea.

What we ordinarily may be said to justify, strictly, are actions, deliberate moves, controllable act-patterns, items of our behavior for which we are responsible. Indeed, justification does not apply to anything for which no one is responsible, while to be responsible for something is just to be subject to the demand for its justification. Since what we are, strictly, responsible for is our controllable behavior, what we are called upon to justify is such behavior. The apparent exception as regards the justification of cognitive *statements* is a real exception only if we split cognition from action, and deny that belief, affirmation, and assertion are kinds of deliberate behavior, for which we are responsible. As a matter of fact, a study of the justification of cognitive belief, affirmation, or sentence-acceptance turns out to be illuminating for all cases of justification.

How then do we justify the acceptance of some sentence, A? Coherence of A with some system of sentences, in the sense of derivability or inclusion within the system, is not sufficient, for the negation of A is also coherent with *some* system, though the systems with which A and $-A$, respectively, are coherent are mutually incompatible, if consistent. Clearly we need some way of choosing, among internally coherent but mutually incompatible systems, one which is to serve as a standard system. The correspondence of A with fact is equally unsatisfactory as a description of what is essential to cognitive justification. It is unclear what it means for a statement to correspond with fact; moreover, no matter how we agree to understand it, it is not likely to be sufficient either, for the most conscientiously corresponding statement has systematic import, i.e., is subject to withdrawal under pres-

---

This paper appeared in the *Journal of Philosophy* 51 (1954): 180-90.

sure from incompatible sister statements which we happen to be interested in saving. Somehow, it seems, we need to supplement coherence with correspondence in such a way that correspondence will select a standard system which will, in turn, represent just those sentences with which *A* must come to terms, or cohere.

A proposal for doing just this was recently suggested by Professor Goodman in another connection, during the course of a discussion of empirical certainty.[1] Agreeing with Professor Lewis that probability with respect to certain premises is never sufficient to render any sentence credible, Goodman argues that no sentence need nevertheless be certain; it need only have some degree of underived or initial credibility. The correspondence factor, that is, may be minimally conceived as *some* degree of initial credibility attaching to sentences. Yet no sentence is ever immune from withdrawal, or, what amounts to the same thing, is a necessarily fixed description of any particular fact of experience. Thus, the sentence "There is now a sheet of paper before me" is highly credible but may be replaced, if need arises, by any of a number of other sentences with lower initial credibility, *provided there is a total gain.* Choice among internally coherent, mutually incompatible systems is accomplished, that is, by noting which of them maximizes initial credibility. The justification for accepting *A* at a given time may now be made not on the grounds of its own initial credibility, nor of some unspecified coherence, but on the basis of its coherence with the system which maximizes initial credibility at that time, while, together with its sister sentences, *A* indirectly controls the choice of this standard system. Circularity is avoided because, whereas it is each single sentence which is judged by coherence, it is the totality of sentences which exercises control by correspondence, or initial credibility.

An example or two may help to clarify this conception. Suppose we have two incompatible sentences differing in initial credibility, both independent of our heretofore standard system, both seeking entry to the system. Our choice will go to the sentence with higher initial credibility, not by consideration of these two sentences alone, but by anticipation of systematic effect, i.e., because the opposite choice would mean a lower total credibility value for our standard system.

Of course, since no sentence is a necessary description of a given experience, neither are we ever limited to just two alternatives. We may have an indefinite number of applicants vying for entry to the system. We may, if we like, consider every sentence as a candidate for inclusion, since every sentence may trivially be said to have some degree, perhaps zero, or a negative value, of initial credibility. As initial credibility rises, the temptation to add to our total credibility by a simple inclusion of the sentence increases. But no matter how urgent it becomes, it is counterbalanced by the demand that the total system must not be lowered in credibility by the repercussions of such inclusion. Whether, then, we consider two or an indefinite number of candidates for inclusion, systematic consequences for total credibility are dominant in our inclusion-policy.

Not only are all sentences short of certainty, but the degree of their initial credibility is subject to change over time. Let us take, as a simplified illustration of such change, a case of systematic overhauling precipitated by a disconfirming crucial experiment. At time t_1, before the experiment, one of the sentences, Z, deductively entailed by a given theory and hence part of the system including it, has a given degree of initial credibility. At time t_2, following the experiment, $-Z$ jumps high in credibility, while Z drops drastically. The resulting choice is often a complicated one. Including the negate and revising the system internally may increase the latter's credibility in this area, but has repercussions in other areas where accompanying theoretical revision, in the interests of coherence, demands a reshuffling of sentences. In addition, pragmatic factors of inertia, convenience, and simplicity must be taken into account. It is clear, nevertheless, that if maintaining the system intact jeopardizes its total credibility, a drastic overhauling is indicated. Now, a systematic overhauling, though heightening the total credibility, may oust some sentences with higher initial credibilities in favor of counterparts with lower. The acceptance of the latter is justified clearly not on the basis of their own isolated merits but rather on the basis of their systematic connections.

Summing up the discussion so far, then, we justify the acceptance of A at time t by showing that A belongs to the maximally credible system of sentences at time t; we justify accepting a particular system at time t by showing that its total credibility value at time t is not less than that of any of its contemporary rivals. Since A exercises partial control over choice of the standard system, moreover, its coherence with the latter must not be construed as a passive meshing with any systematic status quo. A may fail to cohere with the accepted system at a given time, and help to force a drastic overhauling and a change in system-acceptance.

So far I have spoken of the justification of sentence-acceptances and of system-acceptances, and have used the notion of coherence rather uncritically. It may be said, however, that the coherence-rules of a system, though not themselves systematic sentences but rather extra-systematic devices, need justification for their acceptance as well, since they exercise some control over admission of sentences. As Professor Quine points out, though they may be altered reluctantly or perhaps never, they are theoretically always subject to modification. Our examples show just how changes in the requirements of coherence might be useful. We spoke of choice among incompatible sentences or systems. Now a weakening of the rules of incompatibility might always be made in such a way as to eliminate the need for choice, and enable us to dispose of all such problems trivially, were it not for the fact that we have compunctions about changing rules. Now the fact that these compunctions operate to prevent our solving *all* problems by weakening the rules of the game, though we may solve *some* of our problems in this way, shows that rule-acceptance requires justification too, that rules are rationally controlled though credibility is inapplicable to them.

I think that pragmatic considerations are as relevant here as elsewhere. Some rules are more habitual, seem more natural, more economical of effort than others. However, I think that for rules as well as systems, initial sentence-credibility exercises considerable control. If one set of rules, roughly speaking, admits fewer low-credibility sentences than another, it is preferable, other things being equal. Abandoning a given coherence requirement, for example, to accommodate two erstwhile incompatible sentences we should like to save, we may lower the total credibility value of our system by admitting a host of undesirable sentences otherwise excluded. Thus, rules may be said to be justified to the degree in which they maximize credibility in the systems in which they are applied.

It will be recalled that I began by limiting the notion of justifiedness to deliberate, responsible behavior. Now I should not like to be misunderstood as denying that, in ordinary discourse, we often do speak of justifying *sentences,* or, for that matter, of being responsible for some object, situation, or other non-behavioral entity. I am suggesting, however, that this mode of speech is an extension, that we are, strictly speaking, not responsible for the situation, but for our behavior in bringing it about or preserving it; that we, correspondingly, do not justify sentences, but rather sentence-acceptances. Independent evidence for this suggestion is, perhaps, afforded by the fact that, even in ordinary discourse, we do not speak of responsibility for anything to which our action is considered to lack the appropriate relation of bringing about, contributing toward, or helping to preserve.

As a matter of fact, this suggestion accords well with familiar ethical tradition. Together with responsibility for our controllable, deliberate behavior go sanctions directed toward a selective nurturing of some kinds of action as over against others; responsibility of the agent derives its meaning from the significance of certain of his actions. If I am right in supposing that we may not properly be said to justify anything which imposes no responsibility, then only our deliberate behavior is properly justifiable, while apparently contrary locutions may be construed as extensions. The parallelism is strengthened by our speaking, ordinarily, not only of the justifiedness of some entity simply, but also of an agent's justifiedness in performing some act, just as we speak of an agent's responsibility for some act.

Justification is, then, I should say, in every case applied to behavior, because we are vitally interested in the control of behavior and the classification which facilitates control. Interested in controlling the future history of sentence-acceptance, we justify sentence-acceptances, not sentences. Understandably, however, we extend the notion of justification to the latter because of their close relations to sentence-acceptances. Some reflections on these relations will perhaps serve to clarify the present conception.

According to our previous discussion, the acceptance of A at time t is justified if A belongs to the maximally credible system of sentences at time t. Now it is obviously not necessary that, for every sentence, S, which belongs to this system, there should in actual historical fact be an S-acceptance, though every sentence-acceptance involves a sentence. Thus, we can refer to any actual sentence-acceptance by reference to a corresponding sentence, although not conversely, even for sentences included in the maximally credible set. Furthermore, if inclusion in this set be termed "groundedness," then justifying an acceptance will always involve reference to a corresponding, grounded sentence, while showing that a sentence is grounded will not always imply that there is, in point of fact, a corresponding, justified acceptance. Justification, then, should be distinguished from groundedness, and reserved for application to behavioral entities. Since each act of justified sentence-acceptance is, however, connected to a grounded sentence, we may refer to each act by way of its associated sentence, while the utility of such reference, as contrasted with an independent psychological description of the act, is a precision, a specificity, a stability otherwise unattainable. The transfer of the notion of justification from affirmations to sentences is, then, reasonable, convenient, and harmless for the most part.

I have, however, dwelt at some length upon this apparently unimportant point because the close connection between sentence-grounding and affirmation-justification has led to understandable confusion of the two, and has caused considerable trouble in the study of ethical and legal justification. On the one hand, those anxious to avoid subjectivism in ethics and to treat ethics cognitively seek a general notion of justification in terms of sentences. Just as science justifies sentences, so, it is thought, ethics must justify sentences if it is to be objective. On the other hand, opponents of this position, in an effort to account for the tie-up of ethics with action and to deny *this* parallelism with science, feel that they must deny ethical objectivity altogether, as well as the cognitive nature of ethical judgment. The first group says, in effect, "If ethics is to be objective, its goal must be a system of justified sentences." The second group retorts, in effect, "Since ethics is concerned with action, and hence not with a listing of justified sentences, it can itself *be* no more than action, a form of non-cognitive stimulation."

Actually, it seems to me that both arguments are *non sequiturs*. Even if we grant, for the moment, that justifiedness is attributable to sentences, we do not thereby deny that it may also hold of acts. If so, the goal of ethics may be a listing of justified acts, just as that of science is a listing of justified sentences; ethics can be cognitive and objective though concerned with action. On the further assumption of this paper that justifiedness is never properly applicable to sentences at all, and that it is to be distinguished from groundability, this confused dilemma between sentential non-activism and active non-cognitivism never arises. Furthermore, a new, pervasive parallel unites the realms of cognition and action. Cognition is a kind of action, and justifiability applies to the one as it does to the other.

If I am not mistaken in thinking that justification is of behavior in the interests of control, that in cognitive justification we are interested in controlling acceptance of grounded sentences, and that we refer to such acts by way of their linguistic counterparts to achieve stability and descriptive precision, a number of features in the landscape of our problem begin to take on more definite form. It becomes clear that while justification may have the same general features and motivation when what are justified are not acceptance-acts, nevertheless, because these other acts have no relatively stable and precise linguistic counterparts, we lack what appears to be the best way presently available for characterizing them. We are reduced to characterizing them in both motivational and socio-cultural terms, and both these ways are presently more primitive, more complex, and more evanescent than description by way of linguistic relations. Thus, though the justificational pattern is constant, we may expect at the outset a tremendously greater amount of vagueness and fogginess in ethico-legal justification simply because we have no well-developed referential scheme. Our difficulties in denoting and classifying non-affirmational acts mean that we are unclear even in specifying what it is we are interested in controlling. It is not that we are perfectly clear about our interests but lack a language. It is rather that, having no precise language, we cannot possibly be clear about our interests. And this is so because interests themselves are not clearly specifiable at present in abstraction from their objects. To the extent that we can't specify these objects, we simply don't know what our interests are. Thus, the transfer from affirmational to non-affirmational justification is bound to involve an increase in vagueness, though the pattern remain constant.

Furthermore, we noted that for acceptance-justification, the necessary and sufficient condition is some maximal pattern of associated sentence-credibility. Where we are dealing with non-affirmational acts, credibility is clearly irrelevant, since attributable to sentences only. Yet the notion of some degree of initial credibility attaching to sentences, which it is our purpose to save the most of in acceptance, is suggestive. For to rank sentences in the order of their initial credibility is to rank them at the same time in the order of our initial commitment to their acceptance. And to save the maximum of initial credibility is to save the maximum of our initial commitment to acceptance. To justify acceptances, we might say, is to follow a sort of psychological inertia, or law of least action. We start with an indefinite number of initial acceptance-commitments of various intensities, and we try to conserve as much of the total as we can. Forced to change our acceptance-pattern, we change in such a way as to continue to preserve a maximum of initial commitment at any given time. The reference to credibility, then, I suggest, is not something ultimate in itself for this context, but serves as an indication of initial acceptance-commitment, just as the reference to sentences serves to characterize acceptance behavior. Finally, the notion of the groundedness of sentences and systems in terms of maximum credibility gives us a way of harmonizing and continually reequilibrating the totality of our acceptance-commitments in such a way as to conserve the maximum and achieve the greatest stability.

Now if this is true of acceptance-commitments, it may also be true, I suggest, of all sorts of commitment. At any given time, we have all kinds and degrees of committedness to actions and action-patterns at various levels of generality. Our purpose is to harmonize these commitments; to single out, for each moment, a set of acts with maximum initial commitment at that time, which may serve as a standard of justification. I expect to be reminded at this point that there is no law of excluded middle or rule of incompatibility for acts as for sentences, and hence no reason for singling out one standard set of acts, exclusion from which will preclude justification. Without the notion of incompatibility, that is, how can anything be excluded? This, I think, puts the cart before the horse. It is because the degree of credibility of sentence-incompatibles corresponds to the degree of initial commitment to their joint acceptance that we may use the former as a relevant measure of the latter in justifying acceptance. Sentence incompatibility, then, is, in an important sense, a reflection of act-incompatibility. For the sake of clarity, let us reserve the notion of incompatibility for sentences, and speak of acts as being *incongruous,* to the extent that the initial commitment to their sum is lower than the sum of initial commitments to each singly. The crucial incongruity occurs, of course, when the initial commitment to their sum is lower than each single initial commitment.

We are all familiar with the phenomenon of personal consistency, which makes two acts jointly valueless for us though each singly holds promise. "The philosopher and the lady-killer cannot both keep house in the same tenement of clay," said William James, and I doubt that he meant here to deny the attractiveness of either of these careers, taken singly and pursued consistently. The point here is that, for acts as well as sentences, we cannot simply justify everything if we want to maximize initial commitment, for the incongruity of acts means that a smaller set may have more initial commitment than a larger. While rules of coherence or congruity are justifiable here, as before, to the extent to which they maximize initial commitment, sets of acts are justified when maximal in this respect, and individual acts are justified when they belong to the maximal set. Circularity is avoided here, as before, since, while each act singly is justified by inclusion in the standard act-set, it is the totality of acts which exerts control over the choice of a standard act-set. This description, in terms of commitment, coincides for acceptances with our previous explanation in terms of credibility, as noted, but is generalizable to legal and ethical contexts where credibility is inapplicable.

An important point brought out by this analysis is the systematic import of acts, which forms the basis of ethical and legal justification. Justification is never a question of an isolated act just as it is never a question of an isolated sentence-acceptance, nor is it a non-rational stimulation, as the extreme emotivists would have it. It is the systematic rechanneling of initial commitments in such a way that each act is judged in terms of all others. We do not start from scratch, but always with initial commitments of some degree; but neither do we rest content with the latter. We modify and transform them into derived commitments of various sorts by systematic pressure which is channeled through principles of congruence. These derived commitments to acts, action-patterns, and rules are always changing, yet always subject to control. Whoever looks at ethics through law, or whoever recognizes the complex interplay of initial attraction, derived commitment, and the drive for personal consistency in individual moral choice, will acknowledge the rational and systematic structuring of justification.

The principle of least action operates also in non-affirmational justification. We seek the maximum preservation of initial commitments and make the smallest changes consonant with the continual preservation of this maximum. Legal or ethical reform may involve, however, a kind of systematic overhauling when small changes no longer suffice for maximizing commitment. New social conditions, corresponding in a way to crucial experiments, may radically alter the initial commitments to acts of various kinds, or bring new acts into being which demand taking into account. Radical change in congruence-rules may occur, with a transitional period of unsettlement before a new stability forms. Rapid personal growth may in this way also involve a basic reorientation and new congruence rules.

I may be asked how I justify the maximization of commitment. My answer can only be that I am trying to describe what I take to be the meaning of rational justification. Only a thorough testing of the present proposal will reveal whether or not it is accurate. I cannot further justify the maximization of commitment within my analysis, for on my account such justification is meaningless.

Perhaps a word should be said at this point about the much-debated issue of subjectivity in ethics. On the present account, disagreements over justification are rationally soluble only if initial commitments are constant. Now it is clearly reasonable to assume that degrees of such commitment are assigned differently by different people, and especially by members of different cultures. But this seems to me highly realistic. Legal and ethical outlooks in different cultures may develop justificational schemes which follow the same rational pattern, and yet, since they start from different initial positions, may conflict beyond the possibility of rational adjudication. The same is obviously true of persons within the same cultural environment, or of the same person at different stages of growth. Such subjectivity is not, however, tantamount to the irrationality of the domain of ethics. According to the prevalent stereotype, the rational realm is the realm in which all must eventually come to agree, and the model of such a realm is science. I fail to see, however, what Providence guarantees universal agreement in any domain. Certainly, if the present analysis is correct, subjectivity reigns in the same sense, though perhaps to a lesser degree, in the cognitive or scientific domain, since all justification rests upon initial commitments, which may vary from time to time and from person to person. How to bring a hallucinatory schizophrenic, by rational means, to agree to the truth of physics is, I think, a hopeless problem; one which cannot be decided by defining physics as a rational domain. Rationality, in any event, does not create commitments, but only sets up communication among them, so that we may be guided by a controlled totality, rather than by any single one gone wild. Though disagreements, then, over initial commitments are not rationally soluble, this subjectivity is inevitable in all domains and hence cannot entail a distinguishing irrationality for ethics.

Furthermore, there is a practical factor which offsets the theoretical subjectivity in question. We cannot determine with finality at any given time, regarding any given disagreement, that we have exhausted rational means of adjudication and gotten down to the rock bottom of all relevant initial commitments. Theoretically, we may always continue to expand our attention, originally focused on the circumscribed area of conflict, so that it takes in more and more of the totality of our acts. We may hope to encounter some area of shared commitment, of systematic centrality, such that the original disagreement will be overshadowed. Thus, subjectivity, in the sense indicated, is compatible with a constant practical relevance of shared search for areas of agreement.

NOTES

Read before the Harvard Philosophical Club on February 19, 1953. I wish to thank Professor N. Goodman, Mr. A. N. Chomsky, and Mr. S. Morgenbesser for helpful comments.

1. N. Goodman, "Sense and Certainty," *Philosophical Review* 61 (April 1952): 160-67.

3. IS THE DEWEY-LIKE NOTION OF DESIRABILITY ABSURD?

FOLLOWING an incisive discussion of Dewey's contrast between the desired and the desirable,[1] Professor Morton White concludes that Dewey has failed to construe "is desirable" in the sense of "ought to be desired." He argues that if, with Dewey, we take the relation between "is desirable" and "is desired" to be identical with that between "is objectively red" and "appears red," we must take "is desirable" as "is desired under normal conditions." If we now construe the latter predicate as an expansion of "ought to be desired," argues Professor White, how can we escape the absurdity of construing "appears red under normal conditions" as "ought to appear red"?

In this paper, I want to suggest a way of escape. I do not claim that this way out is in accord with Dewey's original intentions, nor am I personally committed to the Deweyan treatment as a generally adequate analysis of "is desirable." For one thing, the unanalyzed use of the disposition-notion (shared with so much of current philosophizing) seems to me to be a defect, while considerations of ontology and conformity with usage require further examination. Yet, I do wish to maintain that my proposal for escape preserves Dewey's construction of "is desirable" as dispositional in contrast with "is desired," and is, in this sense, Dewey-like, while at the same time avoiding the peculiar absurdity pointed out by Professor White.

This absurdity arises, I think, only if we interpret the equation of "is desired under normal conditions" with "ought to be desired" as a specific application of a *general analysis* of " – ought...." Our difficulty, that is, is not with the application but with the inferred general equation of " – ought... " with " – ... under normal conditions" (where the " – " is in both cases to be replaced by the same name or description, while the first "... " takes the infinitive and the second "... " the appropriate indicative of a given form). If the specific equation of "is desired under normal conditions" with "ought to be desired" were indeed to commit us to this general

---

This paper appeared in the *Journal of Philosophy* 51 (1954): 577-82.

schema, then we could, in fact, not avoid taking "*a* ought to appear red" as "*a* appears red under normal conditions," and vice versa.

However, the specific equation need not commit us to this general schema at all. Instead of taking the "ought" in "ought to be desired" as a separate unit to be analyzed, let us take "ought to be desired" as a single, indivisible predicate (hereafter referred to as "o-t-b-d"), from which "ought" cannot be detached. Instead of sanctioning the general schema of the previous paragraph, in which "ought" is analyzed contextually, we now sanction only the explicit analysis:

(I) " – o-t-b-d" *for* " – is desired under normal conditions."

It is immediately apparent that this analysis gives us no right to do anything with " – ought . . . " in general; specifically, we no longer have official license absurdly to equate "*a* ought to appear red" with "*a* appears red under normal conditions." At the same time, both analysandum and analysans of (I) are validly contrasted, as dispositional, with "is desired." Hence, this suggestion enables us to avoid absurdity while remaining Dewey-like.

It seems, however, impossible to rest here. For, if *this* analysis gives us no right to do anything with " – ought . . . " in general, we need some supplementary analysis, if our ethics is to be adequate. What *shall* we do with "*a* ought to appear red" after all, short of treating each combination of "ought" with some infinitive as a new whole predicate requiring separate analysis?

I think that reflection on the nature of the absurdity we wish to avoid is helpful here. It seems fantastic to analyze "*a* ought to appear red" as "*a* appears red under normal conditions" because the so-called normativeness or psychological magnetism of the "ought" is unprovided for. Now Dewey's way (and the way of other naturalists) is, I take it, to account for this magnetism by reference to desire; in (I) we have explicitly connected the magnetism of "ought to be desired" with a certain pattern of actual desire. In "*a* appears red under normal conditions," however, no reference to a desire-pattern is made at all, and the magnetism of the analysandum is left dangling, as it were, even by naturalistic standards. Whether or not reference to desire is *ultimately* satisfactory in accounting for normativeness, that is, we haven't even given it a try in "*a* appears red under normal conditions." That is why the absurdity is a challenge specifically to naturalists; those who reject the whole program of accounting for normativeness by desire are unhappy even with the first analysans, "*a* is *desired* under normal conditions."

What we need, then, is to analyze " – ought . . . " generally so as to account for its normativeness at least as this is done in "*a* is desired under normal conditions," i.e., by some reference to desire. We can do this by

Is the Dewey-like Notion of Desirability Absurd?

saying that something ought to be such-and-such not if it actually is such-and-such under normal conditions, but rather if its being such-and-such is desired under normal conditions. Specifically, we will analyze "*a* ought to appear red" as "that *a* appear red is desired under normal conditions," which, by (I), becomes "that *a* appear red ought to be desired." But aren't we now going in a complete circle? We have analyzed one "ought" by another. The appearance of circularity is altogether specious, however, if we remember to interpret "ought to be desired" as our single predicate, "o-t-b-d," introduced before. Starting with "o-t-b-d," we analyze all " – ought . . . " 's in its terms, thus injecting the desire-reference into all of them, and explaining their normativeness in naturalistic fashion.

One modification is, however, necessary. In (I), we thought of "o-t-b-d" as a predicate of individuals; if we want now to apply it to *a*'s appearing red in a sentence like "that *a* appear red o-t-b-d," we need ostensibly to construe it as a predicate of states-of-affairs, and enlarge our ontology accordingly. This may only be an apparent need, however, for the expansion of the above sentence according to (I), i.e., "that *a* appear red is desired under normal conditions," may itself eventually be interpreted as "there is a red-appearing-*a*-desire under normal conditions" rather than as asserting something about some state-of-affairs. Thus, in calling "o-t-b-d" a predicate of states-of-affairs, we shall not mean to imply any ultimately necessary ontology, but shall refer merely to grammatical form. We shall, that is, want, in any event, to be able to form sentences like "That *a* appear red o-t-b-d." Given the right to do this, on whatever ontological grounds, we now propose, instead of (I), the following two-step analysis:

(A) "That – . . . o-t-b-d" *for* "That – . . . . is desired under normal conditions."
(B) " – ought to . . . " *for* "That – . . . o-t-b-d."

According to this analysis, all cases of " – ought to . . . " are analyzed by reference to "o-t-b-d," while the latter itself reduces to some sort of dispositional desire. Thus all normativeness is hitched up to desire, and we have a way of taking care of "*a* ought to appear red" without utter absurdity. At the same time, both "o-t-b-d" and " – ought . . . " are validly distinguished from nondispositional predicates, preserving the Dewey-like pattern. If anyone is dissatisfied with the general reduction of normativeness to desire, he will now need to argue on other grounds than internal absurdity.

It is, of course, not necessary to insist on "o-t-b-d" as basic in the analysis of " – ought. . . . " The fact that our discussion refers to Dewey's notion of desire should not, that is, obscure the possibility of constructing an alternative analysis taking some other psychological-attraction notion as basic, for example, that of prizing, or approving. So long as the analysis ties all " – ought . . . " 's up to some such attraction in the manner indicated, we may achieve the requisite generality without absurdity, while remaining ethical naturalists.

Following the choice of a particular basic term, there will, no doubt, need to be a decision as to how to treat the rejected candidates, if any. Shall we, for instance, assuming our previous analysis (A) and (B), sanction the asymmetry of taking "That people be kind o-t-b-d" as "That people be kind is desired under normal conditions," while construing "That people be kind ought to be approved" as "That people's being kind be approved o-t-b-d"? Or shall we introduce an "o-t-b-a" as equivalent to "o-t-b-d," and so for every psychological-attraction notion we hold equally powerful for grounding normativeness? Such a decision is, in any case, an independent matter, but it is perhaps worth remarking that if we take the latter alternative, we do not thereby trivialize our procedure, for we demand such special ought-compounds only for a small number of sufficiently powerful psychological-attraction predicates, while leaving the "ought" free in all other cases, e.g., "ought to appear red." Even if, on the other hand, we sanction the former alternative of asymmetry, this does not violate the Dewey-like pattern, while accord with Dewey's intentions is, in any case, not our concern here. That Professor White, interpreting Dewey, speaks of *the* relation between "is desirable" and "is desired" as identical with *that* between "is objectively red" and "appears red" should not obscure the fact that in neither case is a *unique* relation involved, but rather many relations, and that the particular one in question is that expressed by " – is dispositional in contrast to. . . . " Analogously, "ought to be desired" need not parallel "ought to be approved" *in every respect;* it is sufficient for our purposes that both be dispositional. In any event, our decision in this matter is an independent question to be settled on independent grounds.

An important consequence of our two-step analysis needs yet to be indicated. We have required of "o-t-b-d" a grammatical subject of the form, "That – . . . ," and explained that the usual ontological interpretation in terms of states-of-affairs need not be ultimately involved. With this proviso in the back of our minds, let us be bold to use the ordinary terminology now, so as to condense exposition. An explicit result of our treatment is that "is desirable" (construed as "ought to be desired") now has two expansions, depending on whether it is used as a predicate of states-of-affairs, or as a predicate of individuals.

(1) "o-t-b-d" is a predicate of states-of-affairs. States-of-affairs are, by (A), desirable when and only when they are desired under normal conditions, since it is in just such cases that states-of-affairs ought-to-be-desired.

(2) "ought to be desired," however, is a predicate of individuals, requiring analysis by (B) just as "ought to appear red" does. Individuals are, by (B), desirable when and only when their being desired o-t-b-d, i.e., is desired under normal conditions, since it is in just such cases that individuals ought to be desired.

The distinction between knowing individuals and knowing propositions (between "is known" as a predicate of individuals and "is known" as a predicate of propositions, or between "knowing..." and "knowing that – ") is widely recognized. An analogous distinction for many other words seems equally apparent, e.g., between "seeing..." and "seeing that – ," "imagining..." and "imagining that – ," "recognizing..." and "recognizing that – ," "regretting..." and "regretting that – ," "desiring..." and "desiring that – ." The explicit preservation of a similar distinction in the case of "is desirable," and its attempted clarification, are thus virtues of our two-step analysis.[2]

For both individuals and states-of-affairs, however, it is not enough to be desired in order to be desirable. Whereas, however, for states-of-affairs it *is* enough to be desired under normal conditions, this requirement is still insufficient for the desirability of individuals. For individuals we require that their being desired is desired under normal conditions, whether or not they themselves are desired under normal conditions. Thus, not only does "*a* is desired now" not entail "*a* is desirable," but even "*a* is desired under normal conditions" does not entail "*a* is desirable," removing us two degrees from Mill's error. Since also, conversely, "*a* is desirable" does not entail "*a* is desired under normal conditions," we make the connection of normativeness with desire more tortuous and subtle still, thus helping to explain the feeling of some sort of mysterious magnetism while not relinquishing the firm anchorage of desire.

The question to be asked of this analysis now is: Is it generally adequate by standard criteria, such as conformity to usage, ontological commitments, fruitfulness, etc.? The present paper implies no answer to *this* question. Its claim is merely that, even if every Dewey-like analysis is inadequate, there are some that are at least not absurd.

NOTES

1. In "Valuation and Obligation in Dewey and Lewis," by Morton G. White, *Philosophical Review* 58 (No. 4, 1949): 321-29. See also Professor White's discussion in ch. 13 of his *Social Thought in America* (New York: Viking Press, 1949).

For critical comment, see Professor Hook's paper, "The Desirable and

Emotive in Dewey's Ethics," included in *John Dewey: Philosopher of Science and Freedom,* ed. Sidney Hook (New York: Dial Press, 1950), and " 'Desirability' and 'Normativeness' in White's Article on Dewey," by John Ladd, *Philosophical Review* 60 (No. 1, 1951): 91-99.

2. I am indebted to Professor Morton White for suggesting this parallelism, and for a discussion of certain related points.

4. PHILOSOPHICAL MODELS OF TEACHING

Teaching may be characterized as an activity aimed at the achievement of learning, and practiced in such manner as to respect the student's intellectual integrity and capacity for independent judgment. Such a characterization is important for at least two reasons: first, it brings out the intentional nature of teaching, the fact that teaching is a distinctive goal-oriented activity, rather than a distinctively patterned sequence of behavioral steps executed by the teacher. Second, it differentiates the activity of teaching from such other activities as propaganda, conditioning, suggestion, and indoctrination, which are aimed at modifying the person but strive at all costs to avoid a genuine engagement of his judgment on underlying issues.

This characterization of teaching, which I believe to be correct, fails, nevertheless, to answer certain critical questions of the teacher: What sort of learning shall I aim to achieve? In what does such learning consist? How shall I strive to achieve it? Such questions are, respectively, normative, epistemological, and empirical in import, and the answers that are provided for them give point and substance to the educational enterprise. Rather than try to separate these questions, however, and deal with each abstractly and explicitly, I should like, on the present occasion, to approach them indirectly and as a group, through a consideration of three influential models of teaching, which provide, or at any rate suggest, certain relevant answers. These models do not so much aim to *describe* teaching as to *orient* it, by weaving a coherent picture out of epistemological, psychological, and normative elements. Like all models, they simplify, but such simplification is a legitimate way of highlighting what are thought to be important features of the subject. The primary issue, in each case, is whether these features are indeed critically important, whether we should allow our educational thinking to be guided by a model that fastens upon them, or rather whether we should reject or revise the model in question. Although I shall mention some historical affiliations of each model, I make no pretense to historical accuracy. My main purpose is systematic or dialectical, that is, to outline and

---

This paper appeared in *Harvard Educational Review* 35 (1965): 131-43.

examine the three models and to see what, if anything, each has to offer in our own quest for a satisfactory conception of teaching. I turn, then, first to what may be called the 'impression model.'

THE IMPRESSION MODEL

The impression model is perhaps the simplest and most widespread of the three, picturing the mind essentially as sifting and storing the external impressions to which it is receptive. The desired end result of teaching is an accumulation in the learner of basic elements fed in from without, organized and processed in standard ways, but, in any event, not generated by the learner himself. In the empiricist variant of this model generally associated with John Locke, learning involves the input by experience of simple ideas of sensation and reflection, which are clustered, related, generalized, and retained by the mind. Blank at birth, the mind is thus formed by its particular experiences, which it keeps available for its future use. In Locke's words:[1]

> Let us then suppose the mind to be, as we say, white paper, void of all characters, without any ideas: how comes it to be furnished? Whence comes it by that vast store which the busy and boundless fancy of man has painted on it with an almost endless variety? Whence has it all the materials of reason and knowledge? To this I answer, in one word, from experience; in that all our knowledge is founded, and from that it ultimately derives itself. Our observation, employed either about external sensible objects, or about the internal operations of our minds, perceived and reflected on by ourselves, is that which supplies our understandings with all the materials of thinking. These two are the fountains of knowledge, from whence all the ideas we have, or can naturally have, do spring.

Teaching, by implication, should concern itself with exercising the mental powers engaged in receiving and processing incoming ideas, more particularly powers of perception, discrimination, retention, combination, abstraction, and representation. But, more important, teaching needs to strive for the optimum selection and organization of this experiential input. For potentially, the teacher has enormous power; by controlling the input of sensory units, he can, to a large degree, shape the mind. As Dewey remarked,[2]

> Locke's statements ... seemed to do justice to both mind and matter ... One of the two supplied the matter of knowledge and the object upon which the mind should work. The other supplied definite mental powers, which were few in number and which might be trained by specific exercises.

Philosophical Models of Teaching · 311

The process of learning in the child was taken as paralleling the growth of knowledge generally, for all knowledge is constructed out of elementary units of experience, which are grouped, related, and generalized. The teacher's object should thus be to provide data not only useful in themselves, but collectively rich enough to support the progressive growth of adult knowledge in the learner's mind.

The impression model, as I have sketched it, has certain obvious strong points. It sets forth the appeal to experience as a general tool of criticism to be employed in the examination of all claims and doctrines, and it demands that they square with it. Surely such a demand is legitimate, for knowledge does rest upon experience in some way or other. Further, the mind is, in a clear sense, as the impression model suggests, a function of its particular experiences, and it is capable of increased growth with experience. The richness and variety of the child's experiences are thus important considerations in the process of educational planning.

The impression model nevertheless suffers from fatal difficulties. The notions of absolutely simple ideas and of abstract mental powers improvable through exercise have been often and rightly criticized as mythological:[3] simplicity is a relative, not an absolute, concept and reflects a particular way of analyzing experience; it is, in short, not given but made. And mental powers or faculties invariant with subject matter have, as everyone knows, been expunged from psychology on empirical as well as theoretical grounds. A more fundamental criticism, perhaps, is that the implicit conception of the growth of knowledge is false. Knowledge is not achieved through any standard set of operations for the processing of sensory particulars, however conceived. Knowledge is, first and foremost, embodied in language, and involves a conceptual apparatus not derivable from the sensory data but imposed upon them. Nor is such apparatus built into the human mind; it is, at least in good part a product of guesswork and invention, borne along by culture and by custom. Knowledge further involves *theory,* and theory is surely not simply a matter of generalizing the data, even assuming such data organized by a given conceptual apparatus. Theory is a creative and individualistic enterprise that goes beyond the data in distinctive ways, involving not only generalization, but postulation of entities, deployment of analogies, evaluation of relative simplicity, and, indeed, invention of new languages. Experience is relevant to knowledge through providing tests of our theories; it does not automatically generate these theories, even when processed by the human mind. That we have the theories we do is, therefore, a fact, not simply about the human mind, but about our history and our intellectual heritage.

In the process of learning, the child gets not only sense experiences but the language and theory of his heritage in complicated linkages with discriminable contexts. He is heir to the complex culture of belief built up out of innumerable creative acts of intellect of the past, and comprising a patterned view of the world. To give the child even the richest selection of sense data or particular facts alone would in no way guarantee his building up anything resembling what we think of as knowledge, much less his developing the ability to retrieve and apply such knowledge in new circumstances.

A *verbal* variant of the impression model of teaching naturally suggests itself, then, as having certain advantages over the *sensory* version we have just considered: what is to be impressed on the mind is not only sense experience but language and, moreover, accepted theory. We need to feed in not only sense data but the correlated verbal patterning of such data, that is, the *statements* about such data which we ourselves accept. The student's knowledge consists in his stored accumulation of these statements, which have application to new cases in the future. He is no longer, as before, assumed capable of generating our conceptual heritage by operating in certain standard ways on his sense data, for part of what *we* are required to feed into his mind is this very heritage itself.

This verbal variant, which has close affinities to contemporary behaviorism, does have certain advantages over its predecessor, but retains grave inadequacies still, as a model of teaching. To *store* all accepted theories is not the same as being able to *use* them properly in context. Nor, even if some practical correlation with sense data is achieved, does it imply an understanding of what is thus stored, nor an appreciation of the theoretical motivation and experimental evidence upon which it rests.

All versions of the impression model, finally have this defect: they fail to make adequate room for radical *innovation* by the learner. We do not, after all, feed into the learner's mind all that we hope he will have as an end result of our teaching. Nor can we construe the critical surplus as generated in standard ways out of materials we supply. We do not, indeed cannot, so construe insight, understanding, new applications of our theories, new theories, new achievements in scholarship, history, poetry, philosophy. There is a fundamental gap which teaching cannot bridge simply by expansion or reorganization of the curriculum input. This gap sets *theoretical* limits to the power and control of the teacher; moreover, it is where his control ends that his fondest hopes for education begin.

THE INSIGHT MODEL

The next model I shall consider, the 'insight model,' represents a radically different approach. Where the impression model supposes the teacher to be conveying ideas or bits of knowledge into the student's mental treasury, the insight model denies the very possibility of such conveyance. Knowledge, it insists, is a matter of vision, and vision cannot be dissected into elementary sensory or verbal units that can be conveyed from one person to another. It can, at most, be stimulated or prompted by what the teacher does, and if it indeed occurs, it goes beyond what is thus done. Vision defines and organizes particular experiences, and points up their significance. It is vision, or insight into meaning, which makes the crucial difference between simply storing and reproducing learned sentences, on the one hand, and understanding their basis and application, on the other.

The insight model is due to Plato, but I shall here consider the version of St Augustine, in his dialogue, 'The Teacher,'[4] for it bears precisely on the points we have dealt with. Augustine argues roughly as follows: the teacher is commonly thought to convey knowledge by his use of language. But knowledge, or rather *new* knowledge, is not conveyed simply by words sounding in the ear. Words are mere noises unless they signify realities present in some way to the mind. Hence a paradox: if the student already knows the realities to which the teacher's words refer, the teacher teaches him nothing new. Whereas, if the student does not know these realities, the teacher's words can have no meaning for him, and must be mere noises. Augustine concludes that language must have a function wholly distinct from that of the signification of realities; it is used to *prompt* people in certain ways. The teacher's words, in particular, prompt the student to search for realities not already known by him. Finding these realities, which are illuminated for him by internal vision, he acquires new knowledge for himself, though indirectly as a result of the teacher's prompting activity. To *believe* something simply on the basis of authority or hearsay is indeed possible, on Augustine's view; to *know* it is not. Mere beliefs may, in his opinion, of course, be useful; they are not therefore knowledge. For knowledge, in short, requires the individual himself to have a grasp of the realities lying behind the words.

The insight model is strong where the impression model is weakest. While the latter, in its concern with the conservation of knowledge, fails to do justice to innovation, the former addresses itself from the start to the problem of *new* knowledge resulting from teaching. Where the latter stresses atomic manipulable bits at the expense of understanding, the former stresses primarily the acquisition of insight. Where the latter gives inordinate place to the feeding in of materials from the outside, the former stresses the importance of firsthand inspection of realities by the student, the necessity for the student to earn his knowledge by his own efforts.

I should argue, nevertheless, that the case offered by Augustine for the prompting theory is not, as it stands, satisfactory. If the student does not know the realities behind the teacher's words, these words are, presumably, mere noises and can serve only to prompt the student to inquire for himself. Yet if they *are* mere noises, how can they even serve to prompt? If they are not understood in any way by the student, how can they lead him to search for the appropriate realities that underlie them? Augustine, furthermore, allows that a person may believe, though not know, what he accepts on mere authority, without having confronted the relevant realities. Such a person might, presumably, pass from the state of belief to that of knowledge, as a result of prompting, under certain conditions. But what, we may ask, could have been the content of his initial belief if the formulation of it had been literally unintelligible to him? The prompting theory, it seems, will not do as a way of escaping Augustine's original paradox.

There is, however, an easier escape. For the paradox itself rests on a confusion of the meaning of *words* with that of *sentences.* Let me explain. Augustine holds that words acquire intelligibility only through acquaintance with reality. Now it may perhaps be initially objected that understanding a word does not always require acquaintance with its signified reality, for words may also acquire intelligibility through definition, lacking such direct acquaintance. But let us waive this objection and grant, for the sake of argument, that understanding a word *always* does require such acquaintance; it still does not follow that understanding a true sentence similarly requires acquaintance with the state of affairs which it represents. We understand new sentences all the time, on the basis of an understanding of their constituent words and of the grammar by which they are concatenated. Thus, given a sentence signifying some fact, it is simply not true that, unless the student already knows this fact, the sentence must be mere noise to him. For he can understand its meaning indirectly, by a synthesis of its parts, and be led thereafter to inquire whether it is, in reality, true or false.

If my argument is correct, then Augustine's paradox of teaching can be simply rejected, on the ground that we *can* understand statements before becoming acquainted with their signified realities. It follows that the teacher can indeed *inform* the student of new facts by means of language. And it further seems to follow that the basis for Augustine's prompting theory of teaching wholly collapses. We are back to the impression model, with the teacher using language not to prompt the student to inner vision, but simply to inform him of new facts.

The latter conclusion seems to me, however, mistaken. For it does *not* follow that the student will *know* these new facts simply because he has been *informed;* on this point Augustine seems to me perfectly right. It is knowing, after all, that Augustine is interested in, and knowing requires something more than the receipt and acceptance of true information. It requires that the student earn the right to his assurance of the truth of the information in question. New *information,* in short, can be intelligibly conveyed by statements; new *knowledge* cannot. Augustine, I suggest, confuses the two cases, arguing in effect for the impossibility of conveying new knowledge by words, on the basis of an alleged similar impossibility for information. I have been urging the falsity of the latter premiss. But if Augustine's premiss is indeed false, his conclusion as regards knowledge seems to me perfectly true: to *know* the proposition expressed by a sentence is more than just to have been told it, to have grasped its meaning, and to have accepted it. It is to have earned the right, through one's own effort or position, to an assurance of its truth.

Augustine puts the matter in terms of an insightful searching of reality, an inquiry carried out by oneself, and resting in no way on authority. Indeed, he is perhaps too austerely individualistic in this regard, rejecting even legitimate arguments from authority as a basis for knowledge. But his main thesis seems to me correct: one cannot convey new knowledge by words alone. For knowledge is not simply a storage of information by the learner.

The teacher does, of course, employ *language,* according to the insight model, but its primary function is not to impress his statements on the student's mind for later reproduction. The teacher's statements are, rather, instrumental to the student's own search of reality and vision thereof; teaching is consummated in the student's own insight. The reference to such insight seems to explain, at least partially, how the student can be expected to apply his learning to new situations in the future. For, having acquired this learning not merely by external suggestion but through a personal engagement with reality, the student can appreciate the particular fit which his theories have with real circumstances, and, hence, the proper occasions for them to be brought into play.

There is, furthermore, no reason to construe adoption of the insight model as eliminating the impression model altogether. For the impression model, it may be admitted, does reflect something genuine and important, but mislocates it. It reflects the increase of the culture's written lore, the growth of knowledge as a public and recorded possession. Furthermore, it reflects the primary importance of conserving such knowledge, as a collective heritage. But knowledge in this public sense has nothing to do with the process of learning and the activity of teaching, that is, with the growth of knowledge in the individual learner. The public treasury of knowledge constitutes a basic source of materials for the teacher, but he cannot hope to transfer it bit by bit in growing accumulation within the student's mind. In conducting his teaching, he must rather give up the hope of such simple transfer, and strive instead to encourage individual insight into the meaning and use of public knowledge.

Despite the important emphases of the insight model which we have been considering, there are, however, two respects in which it falls short. One concerns the simplicity of its constituent notion of insight, or vision, as a condition of knowing; the other relates to its specifically cognitive bias, which it shares with the impression model earlier considered. First, the notion that what is crucial in knowledge is a vision of underlying realities, a consulting of what is found within the mind, is far too simple. Certainly, as we have seen, the knower must satisfy *some* condition beyond simply being informed, in order to have the right to his assurance on the matter in question. But to construe this condition in terms of an intellectual inspection of reality is not at all satisfactory. It is plausible only if we restrict ourselves to very simple cases of truths accessible to observation or introspection. As soon as we attempt to characterize the knowing of propositions normally encountered in practical affairs, in the sciences, in politics, history, or the law, we realize that the concept of a *vision of reality* is impossibly simple. Vision is just the wrong metaphor. What seems indubitably more appropriate in all these cases of knowing is an emphasis on the processes of deliberation, argument, judgment, appraisal of reasons *pro* and *con*, weighing of evidence, appeal to principles, and decision-making, none of which fits at all well with the insight model. This model, in short, does not make adequate room for principled deliberation in the characterization of knowing. It is in terms of such principled deliberation, or the potentiality for it, rather than in terms of simple vision, that the distinctiveness of knowing is primarily to be understood.

Second, the insight model is specifically cognitive in emphasis, and cannot readily be stretched so as to cover important aspects of teaching. We noted above, for example, that the application of truths to new situations is somewhat better off in the insight than in the impression model, since the appropriateness of a truth for new situations is better judged with awareness of underlying realities than without. But a judgment of appropriateness is not all there is to application; habits of proper execution are also required, and insight itself does not necessitate such habits. Insight also fails to cover the concept of character and the related notions of attitude and disposition. Character, it is clear, goes beyond insight as well as beyond the impression of information. For it involves general principles of conduct logically independent of both insight and the accumulation of information. Moreover, what has been said of character can be applied also to the various institutions of civilization, including those that channel cognition itself. Science, for example, is not just a collection of true insights; it is embodied in a living tradition composed of demanding principles of judgment and conduct. Beyond the cognitive insight, lies the fundamental commitment to principles by which insights are to be criticized and assessed, in the light of publicly available evidence or reasons. In sum, then, the shortcoming of the insight model may be said to lie in the fact that it provides no role for the concept of *principles,* and the associated concept of *reasons.* This omission is very serious indeed, for the concept of principles and the concept of reasons together underlie not only the notions of rational deliberation and critical judgment, but also the notions of rational and moral conduct.

THE RULE MODEL

The shortcoming of the insight model just discussed is remedied in the 'rule model,' which I associate with Kant. For Kant, the primary philosophical emphasis is on reason, and reason is always a matter of abiding by general rules or principles. Reason stands always in contrast with inconsistency and with expediency, in the judgment of particular issues. In the cognitive realm, reason is a kind of justice to the evidence, a fair treatment of the merits of the case, in the interests of truth. In the moral realm, reason is action on principle, action that therefore does not bend with the wind, nor lean to the side of advantage or power out of weakness or self-interest. Whether in the cognitive or the moral realm, reason is always a matter of treating equal reasons equally, and of judging the issues in the light of general principles to which one has bound oneself.

In thus binding myself to a set of principles, I act freely; this is my dignity as a being with the power of choice. But my own free commitment obligates me to obey the principles I have adopted, when they rule against me. This is what fairness or consistency in conduct means: if I could judge reasons differently when they bear on my interests, or disregard my principles when they conflict with my own advantage, I should have no principles at all. The concepts of *principles, reasons,* and *consistency* thus go together and they apply both in the cognitive judgment of beliefs and the moral assessment of conduct. In fact, they define a general concept of rationality. A rational man is one who is consistent in thought and in action, abiding by impartial and generalizable principles freely chosen as binding upon himself. Rationality is an essential aspect of human dignity and the rational goal of humanity is to construct a society in which such dignity shall flower, a society so ordered as to adjudicate rationally the affairs of free rational agents, an international and democratic republic. The job of education is to develop character in the broadest sense, that is, principled thought and action, in which the dignity of man is manifest.

In contrast to the insight model, the rule model clearly emphasizes the role of principles in the exercise of cognitive judgment. The strong point of the insight model can thus be preserved: the knower must indeed satisfy a further condition beyond the mere receiving and storing of a bit of information.

But this condition need not, as in the insight model, be taken to involve simply the vision of an underlying reality; rather, it generally involves the capacity for a principled assessment of reasons bearing on justification of the belief in question. The knower, in short, must typically earn the right to confidence in his belief by acquiring the capacity to make a reasonable case for the belief in question. Nor is it sufficient for this case to have been explicitly taught. What is generally expected of the knower is that his autonomy be evidenced in the ability to construct and evaluate fresh and alternative arguments, the power to innovate, rather than just the capacity to reproduce stale arguments earlier stored. The emphasis on innovation, which we found to be an advantage of the insight model, is thus capable of being preserved by the rule model as well.

Nor does the rule model in any way deny the psychological phenomenon of insight. It merely stresses that insight itself, wherever it is relevant to decision or judgment, is filtered through a network of background principles. It brings out thereby that insight is not an isolated, momentary, or personal matter, that the growth of knowledge is not to be construed as a personal interaction between teacher and student, but rather as mediated by general principles definitive of rationality.

Furthermore, while the previous models, as we have seen, are peculiarly and narrowly *cognitive* in relevance, the rule model embraces *conduct* as well as cognition, itself broadly conceived as including processes of judgment and deliberation. Teaching, it suggests, should be geared not simply to the transfer of information nor even to the development of insight, but to the inculcation of principled judgment and conduct, the building of autonomous and rational character which underlies the enterprises of science, morality and culture. Such inculcation should not, of course, be construed mechanically. Rational character and critical judgment grow only through increased participation in adult experience and criticism, through treatment that respects the dignity of learner as well as teacher. We have here, again, a radical gap which cannot be closed by the teacher's efforts alone. He must rely on the spirit of rational dialogue and critical reflection for the development of character, acknowledging that this implies the freedom to reject as well as to accept what is taught. Kant himself holds, however, that rational principles are somehow embedded in the structure of the human mind, so that education builds on a solid foundation. In any event, the stakes are high, for on such building by education depends the prospect of humanity as an ideal quality of life.

There is much of value in the rule model, as I have sketched it. Certainly, rationality is a fundamental cognitive and moral virtue and as such should, I believe, form a basic objective of teaching. Nor should the many historical connotations of the term 'rationality' here mislead us. There is no intent to suggest a faculty of reason, nor to oppose reason to experience or to the emotions. Nor is rationality being construed as the process of making logical deductions. What is in point here is simply the autonomy of the student's judgment, his right to seek reasons in support of claims upon his credibilities and loyalties, and his correlative obligation to deal with such reasons in a principled manner.

Moreover, adoption of the rule model does not necessarily exclude what is important in the other two models; in fact, it can be construed quite plausibly as supplementing their legitimate emphases. For, intermediate between the public treasury of accumulated lore mirrored by the impression model, and the personal and intuitive grasp of the student mirrored by the insight model, it places general principles of rational judgment capable of linking them.

Yet, there is something too formal and abstract in the rule model, as I have thus far presented it. For the operative principles of rational judgment at any given time are, after all, much more detailed and specific than a mere requirement of formal consistency. Such consistency is certainly fundamental, but the way in which its demands are concretely interpreted, elaborated, and supplemented in any field of inquiry or practice, varies with the field, the state of knowledge, and the advance of relevant methodological sophistication. The concrete rules governing inference and procedure in the special sciences, for example, are surely not all embedded in the human mind, even if the demands of formal consistency, as such, *are* universally compelling. These concrete rules and standards, techniques and methodological criteria evolve and grow with the advance of knowledge itself; they form a live tradition of rationality in the realm of science.

Indeed, the notion of tradition is a better guide here, it seems to me, than appeal to the innate structure of the human mind. Rationality in natural inquiry is embodied in the relatively young tradition of science, which defines and redefines those principles by means of which evidence is to be interpreted and meshed with theory. Rational judgment in the realm of science is, consequently, judgment that accords with such principles, as crystallized at the time in question. To teach rationality in science is to interiorize these principles in the student, and furthermore, to introduce him to the live and evolving *tradition* of natural science, which forms their significant context of development and purpose.

Scholarship in history is subject to an analogous interpretation, for beyond the formal demands of reason, in the sense of consistency, there is a concrete tradition of technique and methodology defining the historian's procedure and his assessment of reasons for or against particular historical accounts. To teach rationality in history is, in effect, here also to introduce the student to a live tradition of historical scholarship. Similar remarks might be made also with respect to other areas, e.g., law, philosophy and the politics of democratic society. The fundamental point is that rationality cannot be taken simply as an abstract and general ideal. It is embodied in *multiple evolving traditions,* in which the basic condition holds that issues are resolved by reference to *reasons,* themselves defined by *principles* purporting to be impartial and universal. These traditions should, I believe, provide an important focus for teaching.

CONCLUSION

I have intimated that I find something important in each of the models we have considered. The impression model reflects, as I have said, the cumulative growth of knowledge in its *public* sense. Our aim in teaching should surely be to preserve and extend this growth. But we cannot do this by storing it piecemeal within the learner. We preserve it, as the insight model stresses, only if we succeed in transmitting the live spark that keeps it growing, the insight which is a product of each learner's efforts to make sense of public knowledge in his own terms, and to confront it with reality. Finally, as the rule model suggests, such confrontation involves deliberation and judgment, and hence presupposes general and impartial principles governing the assessment of reasons bearing on the issues. Without such guiding principles, the very conception of rational deliberation collapses, and the concepts of rational and moral conduct, moreover, lose their meaning. Our teaching needs thus to introduce students to those principles we ourselves acknowledge as fundamental, general, and impartial, in the various departments of thought and action.

We need not pretend that these principles of ours are immutable or innate. It is enough that they are what we ourselves acknowledge, that they are the best we know, and that we are prepared to improve them should the need and occasion arise. Such improvement is possible, however, only if we succeed in passing on, too, the multiple live traditions in which they are embodied, and in which a sense of their history, spirit, and direction may be discerned. Teaching, from this point of view, is clearly not, as the behaviorists would have it, a matter of the teacher's shaping the student's behavior or of controlling his mind. It is a matter of passing on those traditions of principled thought and action that define the rational life for teacher as well as student.

As Professor Richard Peters has written,[5]

> The critical procedures by means of which established content is assessed, revised, and adapted to new discoveries have public criteria written into them that stand as impersonal standards to which both

teacher and learner must give their allegiance... To liken education to therapy, to conceive of it as imposing a pattern on another person or as fixing the environment so that he 'grows,' fails to do justice to the shared impersonality both of the content that is handed on and of the criteria by reference to which it is criticized and revised. The teacher is not a detached operator who is bringing about some kind of result in another person which is external to him. His task is to try to get others on the inside of a public form of life that he shares and considers to be worthwhile.

In teaching, we do not impose our wills on the student, but introduce him to the many mansions of the heritage in which we ourselves strive to live, and to the improvement of which we are ourselves dedicated.

NOTES

1. *Essay Concerning Human Understanding,* Book 2, Chap. 1, Sect. 2.

2. John Dewey, *Democracy and Education* (New York: Macmillan, 1916), 72.

3. Dewey, '... the supposed original faculties of observation, recollection, willing, thinking, etc., are purely mythological. There are no such ready-made powers waiting to be exercised and thereby trained.'

4. J. Quasten and J. C. Plumpe (eds), *Ancient Christian Writers,* No. 9, St Augustine, 'The Teacher,' translated and annotated by J. M. Colleran (Westminster, Maryland: Newman Press, 1950); relevant passages may also be found in Kingsley Price, *Education and Philosophical Thought* (Boston: Allyn & Bacon, 1962), 145-59.

5. *Education as Initiation* (London: Evans, 1964); an inaugural lecture delivered at the University of London Institute of Education, 9 December 1963.

5. ON RYLE'S THEORY OF PROPOSITIONAL KNOWLEDGE

CONCERNED to emancipate intelligent performance from dependence upon the apprehension of truths, *The Concept of Mind*[1] naturally places greater stress upon *knowing how* than upon *knowing that*. I want here to reconsider its treatment of the latter. I shall argue, first, that this treatment is inconsistent and, secondly, that certain remedies suggested by selected passages in the book are untenable, on independent grounds.

1. KNOWING AS AN ACHIEVEMENT

Ryle is concerned with the proper categorization of mental powers and operations. How do we determine, however, to what category a thing properly belongs? Ryle offers no general account, but provides a well-known series of illustrations of category mistakes (16). His guiding principle seems to be that things belonging to the same category should be subject to the same sorts of qualification, and accessible also to the same sorts of question. In particular, if something is an activity or performance, it should be capable of the same sorts of modification and accessible to the type of questioning normally applicable to activities or performances. Application of this principle leads to a denial that *knowing that* is an activity. To suppose it an activity is, indeed, a category mistake. How, alternatively, should *knowing that* be construed?

In certain important passages (130, 150/1) Ryle presents a general distinction between task verbs and achievement verbs. The failure to distinguish between them leads to dire epistemological consequences. "Special cognitive acts and operations have been postulated to answer to" achievement words "as if to describe a person as looking and seeing were like describing him as walking and humming instead of being like describing him as angling and catching, or searching and finding" (151). Furthermore, since knowing is not qualifiable by adverbs such as 'erroneously' or 'incorrectly', if it is further construed as an operation or performance, it

This paper appeared in the *Journal of Philosophy* 65 (No. 22, 1968): 725-32.

seems to follow that there is some cognitive performance immune to mistake. "It has long been realized," writes Ryle,

> that verbs like 'know', 'discover', 'solve', 'prove', 'perceive', 'see', and 'observe'... are in an important way incapable of being qualified by adverbs like 'erroneously' and 'incorrectly'. Automatically construing these and kindred verbs as standing for special kinds of operations or experiences, some epistemologists have felt themselves obliged to postulate that people possess certain special inquiry procedures in following which they are subject to no risk of error.... So men are sometimes infallible (152).

"The fact that doctors cannot cure unsuccessfully," he writes, "does not mean that they are infallible doctors; it only means that there is a contradiction in saying that a treatment which has succeeded has not succeeded" (238). The conclusion is, then, that knowing is not a mode of inquiry any more than curing is a mode of treatment. There is no more reason to suppose there is some infallible investigative procedure than there is to suppose that there is some medical strategy which never fails. Inquiry is a matter of our efforts to attain knowledge, whereas knowledge requires the satisfaction of independent conditions holding as a matter of fact. It is "the distinction between task verbs and achievement verbs" that here "frees us from... [a] theoretical nuisance" (152). The upshot of Ryle's discussion in these passages, then, is that *knowing that* is an achievement.[2] Let us now, however, turn to another theme of Ryle's general account.

2. KNOWING AS A CAPACITY

A fundamental contrast that runs through *The Concept of Mind* is that of dispositions and occurrences. Occurrences are concrete events, happenings, or episodes, whereas dispositions are latent properties, abstract sets, potentialities, habits, tendencies, or capacities (116). Now a major division among dispositions is that separating capacities from tendencies. Attribution of a tendency tells us that something will very likely be the case, whereas attribution of a capacity denies the likelihood that something will not be the case. As Ryle puts it, " 'Fido tends to howl when the moon shines' says more than 'it is not true that if the moon shines, Fido is silent'. It licenses the hearer not only not to rely on his silence, but positively to expect barking" (131).

In a very important later passage, Ryle declares that 'know' is a capacity verb, "signifying that the person described can bring things off, or get things right," whereas " 'believe'... is a tendency verb and one which does not connote that anything is brought off or got right" (133-134). He is clearly referring, in this passage, to *knowing that* rather than *knowing how* and affirms explicitly that *knowing that* and *believing that* operate in the same field and have propositional significance. Yet he declares that "to know is to be equipped to get something right and not to tend to act or react in certain manners." His reason for the distinction is that belief "can be qualified by such adjectives as 'obstinate', 'wavering', 'unswerving'," etc. whereas knowing cannot. But it is not at all clear that the distinction he offers in explanation can be coherently maintained.

He declares belief to consist in propensities "to make certain theoretical moves but also to make certain executive and imaginative moves, as well as to have certain feelings" (135), illustrating the point by elaborating the variety of responses to which someone is prone if he believes that the ice is thin; e.g., he tells others the ice is thin, he skates warily, he imagines possible disasters, and so forth. But what of the person who *knows* the ice is thin: does he not share the same propensities? Ryle admits that "A person who knows that the ice is thin, and also cares whether it is thin or thick, will, of course, be apt to act and react in these ways too" (135). He argues, however, that to say someone keeps to the edge because he *knows* the ice is thin is to employ a different sense of 'because' from that used in saying he keeps to the edge because he *believes* the ice is thin. To which one might reply that this latter consideration is quite beside the point. It is not the sense of 'because' that is in question but rather the interpretation of knowledge as capacity: if *knowing that* involves the same *tendencies or propensities* as believing, how can it be contrasted so sharply with believing, as belonging to the category of *capacities?* Moreover, it is generally held that to *know that* something is the case implies to *believe that* it is; on general grounds, one would then suppose *knowing that* to involve whatever attributions are accomplished by *believing that*.

These considerations show that there is at least a problem of interpretation for Ryle on this point. Possibly, there is some way of meeting it. It might, perhaps, be suggested that knowing, unlike believing, does not consist *solely* in propensities, but comprises also a surplus of capacities. It is questionable whether this suggestion is likely to provide an adequate solution, but I shall not argue the point here, since, in any event, a much more serious difficulty looms: Capacities and propensities are *both* dispositional, and it is inconsistent to offer any sort of *dispositional* account of knowing and also to consider knowing an *achievement*. To this fundamental difficulty I now turn.

3. ACHIEVEMENT VERSUS CAPACITY

Ryle, as we have seen, considers *knowing that* an achievement. He also considers it to be dispositional. The trouble is that these properties exclude each other, so the joint attribution is self-contradictory. For Ryle explicitly introduces his main discussion of achievement words (149-153) by stressing their episodic character, i.e., their character as signifying occurrences, while he stresses equally that dispositional statements "narrate no incidents" (125), holding that to classify a word as dispositional is to say at least "that it is not used for an episode" (116). Nor does he ever suggest that 'know' is a hybrid word which is both episodic and dispositional, of the sort he talks about under the label of "semi-hypothetical" or "mongrel categorical" statements. He is perfectly clear and explicit in classifying 'know' as a dispositional word, without qualification (116). He is also perfectly straightforward in calling achievement words "genuine episodic words" (149). It follows that 'know' cannot be both an achievement word and a dispositional word. Ryle's total account is thus literally inconsistent.

In *Conditions of Knowledge,* I erroneously suggested (26-28) that Ryle's notion of achievement led to a conception of knowing as an abstract *state.* In this I was misled by my own view that belief is, roughly speaking, a state which aims at the truth, while knowing succeeds in this aim (25), knowing constituting an achievement state relative to belief. Further, my primary motivation was to discuss Ryle's argument against infallibility, an argument which does not peculiarly depend upon an episodic conception of achievement. So I did not sufficiently clearly distinguish my own abstract notion of achievement from Ryle's episodic notion. Ryle does, however, imply clearly that knowing is episodic, that (barring lucky achievements) it consists in the performance of a bit of inquiry with a certain upshot. Unfortunately, as we have seen, he also says that knowing is dispositional, hence not episodic at all. This is the inconsistency I promised to show.

4. CAPACITY FOR ACHIEVEMENT

Are there any remedies for Ryle's predicament? A first proposal that suggests itself immediately is almost as immediately seen to be inadequate. Of achievements Ryle says in one of the places above noted, "They are not acts, exertions, operations or performances, but... the fact that certain acts, operations, exertions or performances have had certain results" (151). Perhaps, then, an achievement should thus be literally identified with the *fact that* an act has had some result, in which case it will no longer be episodic, since *facts* are not episodes or occurrences but abstract entities presumed to correspond with truths. But if achievements are thus abstract, there is no longer any incompatibility in holding knowing to be both achievement and disposition, inasmuch as both are properly conceived as nonepisodic. To this proposal it may be objected that, though facts and dispositions are both nonepisodic, it does not follow that knowing or anything else can be coherently construed as both a fact and a disposition. Dispositions are, if anything, properties: To ascribe a disposition to x is to attribute to x a property. *That x has a certain property* may be a fact, but it is not in turn another attributable property. Conversely, to say it is a fact is to hold that a sentence asserting that x has this property is true; but to say that smoking or knowing is a disposition makes no appeal to truth, for there is no corresponding sentence that is even a candidate for truth.

The passage we are here concerned with runs counter to the preponderance of Ryle's statements on the subject, in which he forthrightly declares the episodic nature of achievements; in fact he explicitly treats achievement words as a class of performance words. "Many of the performance-verbs with which we describe people," he writes, ". . . signify the occurrence not just of actions but of suitable or correct actions. They signify achievements" (130). I conclude that the first remedy rests on a loose statement of the episodic theory and does not, moreover, offer a viable alternative.

A second suggestion relates to the fact that the acquisition of a capacity is itself governed by criteria. A capacity may be learned gradually through performance, and will be judged to have been acquired only when some stipulated standard has been met. Thus, to *have* a learned capacity is to have achieved success in meeting a relevant standard of performance. Knowing, as a capacity, in this way represents also an achievement. This idea is, in itself, reasonable enough in the case of *knowing how,* but it is not at all evident in the case of *knowing that.* Standards are certainly relevant in the latter case as in the former, but it is not clear that they involve *performance* in both cases, nor, *a fortiori,* a gradual refinement of performance. Practice is not directly relevant to propositional knowledge as distinct from skills; a boy may practice swimming, but not knowing that 3 and 5 make 8.[3] Nor, if he knows he has a headache, has he had to master some ingredient evidential procedure for determining that he has a headache. To be sure, it may be argued, he has had to acquire language, but this is a general precondition for all, or almost all, cases of propositional knowledge, and something more distinctive seems to be indicated, something that relates to the access of the particular competence and not to past occurrences thought necessary for all. Finally, this proposal lacks specificity: what sort of capacity is knowing, if it is a capacity? What does it enable the knower to do?

Ryle himself suggests an answer to this question which gives rise to a third proposal. For he speaks of knowing as a capacity to "bring things off, or get things right" (133). The element of achievement suggested here is that which forms, so to speak, the *object of* the capacity. It is what is *enabled by* acquiring the capacity, rather than located in the antecedent process through which the capacity may have been acquired. This process, where it has occurred, has been adjudged successful only because the capacity which is its outcome is thought to facilitate or enable an independent and distinctive form of achievement.

We have here the kernel of a third proposal which seems much closer to Ryle's intent. This proposal is to construe knowing not as itself an achievement but rather as a capacity *for* achievement. The proposal thus classifies knowing as dispositional and remains thus inconsistent with those passages which treat knowing as itself an achievement. But it is nevertheless suggested in certain sentences of Ryle's discussion. It is, further, a natural proposal in the sense that it enables us to understand the source of Ryle's difficulty: Starting with the idea of episodic achievements, he glides easily to the more abstract notion of the capacity for such achievements, without marking the transition. So in saying that knowing is an achievement, he is really to be interpreted more accurately as intending to say it is a capacity for achievement.

The suggestion receives further support from the following passages: (1) Just after describing achievement words as signifying that a performance has been carried through successfully, Ryle says, "Now successes are sometimes due to luck.... But when we say of a person that he can bring off things of a certain sort... we mean that he can be relied on to succeed reasonably often even without the aid of luck" (130). He thus shifts from considering the *single actual occurrence* of a successful performance to considering the *capacity* to carry out successful performances of analogous kind.

(2) On the same page he writes, "When we use, as we often do use, the phrase 'can tell' as a paraphrase of 'know', we mean by 'tell', 'tell correctly'. We do not say that a child can tell the time, when all that he does is deliver random time-of-day statements, but only when he regularly reports the time of day in conformity with the position of the hands of the clock, or with the position of the sun, whatever these positions may be" (130).

The latter passage suggests paraphrasing 'know' as 'can tell'. If we can assume that the suggestion applies to *knowing that* (rather than, or as well as to *knowing how*), we have a proposal to construe propositional knowing as a capacity to perform successfully in certain ways, namely, in telling correctly or truly. Each instance of correct telling is an episode classifiable as an achievement, strictly speaking. Knowing is not to be identified with any such instance, however. Rather, it is the *capacity* to generate instances of this sort. Does this idea, however, provide an adequate conception of propositional knowledge?

5. DISCOVERING AND SAYING

We must begin our examination by noting an ambiguity in the verb 'tell'. Sometimes it means "say" and sometimes "discover." We might be tempted to rule out the latter meaning immediately, for, while we can speak of "saying correctly," we cannot speak of "discovering correctly." Nevertheless, Ryle suggests treating inquiry procedures as forming a "subservient task activity" for knowing. Let us then first consider knowing as a capacity for discovery, in the sense of inquiring successfully. Each instance of finding out is an achievement, strictly speaking, whereas knowing is the general capacity to find out.

This proposal does not, upon reflection, seem to me to be tenable. We all know many things that we lack the capacity to have found out. A boy may know the Pythagorean theorem in the sense that he can understand it and grasp its proof; yet he may not have had the capacity to discover it, nor need he now have the capacity to discover analogous theorems. (And what is the relevant notion of a class of analogous theorems, anyway?) Conversely, a person may have the capacity to find out what time it is, without knowing what time it is. A person may not know the car is in the garage, but he certainly could not therefore be said to lack the requisite inquiry procedure for determining that it is or is not.

Let us then turn to the other meaning of 'tell', i.e., "say." John has the capacity to say truly that it is three o'clock if and only if he knows that it is three o'clock. This proposal also seems to me untenable. For what interpretation shall we put on the capacity to say truly? Suppose John believes it is two o'clock, because his watch has stopped, whereas it *really is* three o'clock. Does he lack the capacity to say "three o'clock" and to say it, therefore, truly? If he knows English and has no speech impediments, he surely can say "three o'clock." Yet he does not believe it is three o'clock and therefore clearly does not know this. If he is taught a theory or told a story that is in fact true, he may well have not only the capacity but even the tendency to repeat it under suitable classroom circumstances and thus to "tell correctly"; yet he may not accept the theory or story, and, not believing it, cannot be said to know it. This version of the present proposal thus seems to me also to fail.

I cannot here enter into a consideration of alternative views, beyond noting the position taken in *Conditions of Knowledge,* that "knowing appears to resemble rather those things that fit the categories of attainment, attitude, or, most broadly, *state*" (27-28). I do hold that propositional knowledge involves a state of belief as well as some independent factual condition truly described by the belief in question. Such knowledge is not, in my view, a performance, nor is it even a capacity, at least in any direct and straightforward sense of this term. It is independent of discovery and certainly of linguistic response. Yet it involves all of these, in complex ways. That is to say, it qualifies performance, presupposes and modifies capacity, is partly evidenced in discovery, and influences utterance. Like theoretical entities in science, it is deeply and widely linked in an explanatory way with observational phenomena and low-level dispositional traits. Yet, like such entities, it is not reducible to phenomena and dispositions, characterizing, in independent and complex ways that are as yet problematic, "the orientation of the person in the world."[4]

NOTES

1. Gilbert Ryle, *The Concept of Mind* (London: Hutchinson; New York: Barnes & Noble, 1949). Page references will be given in parentheses following passages quoted in the text.

2. Ryle's general contrast between tasks and achievements is criticized on independent grounds in my *Conditions of Knowledge* (Chicago: Scott, Foresman, 1965), 31-33.

3. For related points, see Jane Roland Martin, "On the Reduction of 'Knowing That' to 'Knowing How'," in B. Othanel Smith and Robert H. Ennis, eds., *Language and Concepts in Education* (Chicago: Rand McNally, 1961), 59-71. This is a revised version of her article which appeared first in *Philosophical Review* 67 (July 1958): 379-88.

4. *Conditions of Knowledge*, 90.

6. MORAL EDUCATION AND THE DEMOCRATIC IDEAL

WHAT SHOULD BE the purpose and content of an educational system in a democratic society, in so far as it relates to moral concerns? This is a very large question, with many and diverse ramifications. Only its broadest aspects can here be treated, but a broad treatment, though it must ignore detail, may still be useful in orienting our thought and highlighting fundamental distinctions and priorities.

Commitment to the ideal of democracy as an organizing principle of society has radical and far-reaching consequences, not only for basic political and legal institutions, but also for the educational conceptions that guide the development of our children. All institutions, indeed, operate through the instrumentality of persons; social arrangements are 'mechanisms' only in a misleading metaphorical sense. In so far as education is considered broadly, as embracing all those processes through which a society's persons are developed, it is thus of fundamental import for all the institutions of society, without exception. A society committed to the democratic ideal is one that makes peculiarly difficult and challenging demands of its members; it accordingly also makes stringent demands of those processes through which its members are educated.

What is the democratic ideal, then, as a principle of social organization? It aims so to structure the arrangements of society as to rest them ultimately upon the freely given consent of its members. Such an aim requires the institutionalization of reasoned procedures for the critical and public review of policy; it demands that judgments of policy be viewed not as the fixed privilege of any class or élite but as the common task of all, and it requires the supplanting of arbitrary and violent alteration of policy with institutionally channeled change ordered by reasoned persuasion and informed consent.

The democratic ideal is that of an open and dynamic society: open, in

---

This paper appeared in *Educational Research: Prospects and Priorities*, Appendix 1 to Hearings on H. R. 3606, 92d Congress, 2d Session, Committee on Education and Labor (U. S. Government Printing Office, 1972). It also appears in my *Reason and Teaching* (London: Routledge & Kegan Paul, 1973).

that there is no antecedent social blueprint which is itself to be taken as a dogma immune to critical evaluation in the public forum; dynamic, in that its fundamental institutions are not designed to arrest change but to order and channel it by exposing it to public scrutiny and resting it ultimately upon the choices of its members. The democratic ideal is antithetical to the notion of a fixed class of rulers, with privileges resting upon social myths which it is forbidden to question. It envisions rather a society that sustains itself not by the indoctrination of myth, but by the reasoned choices of its citizens, who continue to favor it in the light of a critical scrutiny both of it and its alternatives. Choice of the democratic ideal rests upon the hope that this ideal will be sustained and strengthened by critical and responsible inquiry into the truth about social matters. The democratic faith consists not in a dogma, but in a reasonable trust that unfettered inquiry and free choice will themselves be chosen, and chosen again, by free and informed men.

The demands made upon education in accord with the democratic ideal are stringent indeed; yet these demands are not ancillary but essential to it. As Ralph Barton Perry has said:

> Education is not merely a boon conferred by democracy, but a condition of its survival and of its becoming that which it undertakes to be. Democracy is that form of social organization which most depends on personal character and moral autonomy. The members of a democratic society cannot be the wards of their betters; for there is no class of betters . . . Democracy demands of every man what in other forms of social organization is demanded only of a segment of society . . . Democratic education is therefore a peculiarly ambitious education. It does not educate men for prescribed places in life, shaping them to fit the requirements of a preexisting and rigid division of labor. Its idea is that the social system itself, which determines what places there are to fill, shall be created by the men who fill them. It is true that in order to live and to live effectively men must be adapted to their social environment, but only in order that they may in the long run adapt that environment to themselves. Men are not building materials to be fitted to a preestablished order, but are themselves the architects of order. They are not forced into Procrustean beds, but themselves design the beds in which they lie. Such figures of speech symbolize the underlying moral goal of democracy as a society in which the social whole justifies itself to its personal members.[1]

To see how radical such a vision is in human history, we have only to reflect how differently education has been conceived. In traditional authoritarian societies education has typically been thought to be a process of perpetuating the received lore, considered to embody the central doctrines upon which human arrangements were based. These doctrines were to be inculcated through education; they were not to be questioned. Since, however, a division between the rulers and the ruled was fundamental in such societies, the education of governing élites was sharply differentiated from the training and opinion-formation reserved for the masses. Plato's *Republic,* the chief work of educational philosophy in our ancient literature, outlines an education for the rulers in a hierarchical utopia in which the rest of the members are to be deliberately nourished on myths. And an authoritative contemporary Soviet textbook on *Pedagogy* declares that, 'Education in the USSR is a weapon for strengthening the Soviet state and the building of a classless society ... the work of the school is carried on by specially trained people who are guided by the state.'[2] The school was indeed defined by the party program of March 1919 as 'an instrument of the class struggle. It was not only to teach the general principles of communism but "to transmit the spiritual, organizational, and educative influence of the proletariat to the half- and nonproletarian strata of the working masses." '[3] In nondemocratic societies, education is two-faced: it is a weapon or an instrument for shaping the minds of the ruled in accord with the favored and dogmatic myth of the rulers; it is, however, for the latter, an induction into the prerogatives and arts of rule, including the arts of manipulating the opinions of the masses.

To choose the democratic ideal for society is wholly to reject the conception of education as an *instrument* of rule; it is to surrender the idea of shaping or molding the mind of the pupil. The function of education in a democracy is rather to liberate the mind, strengthen its critical powers, inform it with knowledge and the capacity for independent inquiry, engage its human sympathies, and illuminate its moral and practical choices. This function is, further, not to be limited to any given subclass of members, but to be extended, in so far as possible, to all citizens, since all are called upon to take part in processes of debate, criticism, choice, and co-operative effort upon which the common social structure depends. 'A democracy which educates for democracy is bound to regard all of its members as heirs who must so far as possible be qualified to enter into their birthright.'[4]

Education, in its broad sense, is more comprehensive than schooling, since it encompasses all those processes through which a society's members are developed. Indeed, all institutions influence the development of persons working within, or affected by, them. Institutions are complex structures of actions and expectations, and to live within their scope is to order one's own actions and expectations in a manner that is modified, directly or subtly, by that fact. Democratic institutions, in particular, requiring as they do the engagement and active concern of all citizens, constitute profoundly educative resources. It is important to note this fact in connection with our theme, for it suggests that formal agencies of schooling do not, and cannot, carry the whole burden of education in a democratic society, in particular moral and character education. All institutions have an educational side, no matter what their primary functions may be. The question of moral education in a democracy must accordingly be raised not only within the scope of the classroom but also within the several realms of institutional conduct. Are political policies and arrangements genuinely open to rational scrutiny and public control? Do the courts and agencies of government operate fairly? What standards of service and integrity are prevalent in public offices? Does the level of political debate meet appropriate requirements of candor and logical argument? Do journalism and the mass media expose facts and alternatives, or appeal to fads and emotionalism? These and many other allied questions pertain to the status of moral education within a democratic society. To take them seriously is to recognize that moral education presents a challenge not only to the schools, but also to every other institution of society.

Yet the issue must certainly be raised specifically in connection with schools and schooling. What is the province of morality in the school, particularly the democratic school? Can morality conceivably be construed as a *subject,* consisting in a set of maxims of conduct, or an account of current mores, or a list of rules derived from some authoritative source? Is the function of moral education rather to ensure conformity to a certain code of behavior regulating the school? Is it, perhaps, to involve pupils in the activities of student organizations or in discussion of 'the problems of democracy'? Or, since morality pertains to the whole of what transpires in school, is the very notion of specific moral schooling altogether misguided?

These questions are very difficult, not only as matters of implementation, but also in theory. For it can hardly be said that there is firm agreement among moralists and educators as to the content and scope of morality. Yet the tradition of moral philosophy reveals a sense of morality as a comprehensive institution over and beyond particular moral codes, which seems to me especially consonant with the democratic ideal, and can, at least in outline, be profitably explored in the context of schooling. What is this sense?

It may perhaps be initially perceived by attention to the language of moral judgment. To say that an action is 'right,' or that some course 'ought' to be followed, is not simply to express one's taste or preference; it is also to make a claim. It is to convey that the judgment is backed by reasons, and it is further to invite discussions of such reasons. It is, finally, to suggest that these reasons will be found compelling when looked at impartially and objectively, that is to say, taking all relevant facts and interests into account and judging the matter as fairly as possible. To make a moral claim is, typically, to rule out the simple expression of feelings, the mere giving of commands, or the mere citation of authorities. It is to commit oneself, at least in principle, to the 'moral point of view,' that is, to the claim that one's recommended course has a point which can be clearly seen if one takes the trouble to survey the situation comprehensively, with impartial and sympathetic consideration of the interests at stake, and with respect for the persons involved in the issue. The details vary in different philosophical accounts, but the broad outlines are generally acknowledged by contemporary moral theorists.[5]

If morality can be thus described, as an institution, then it is clear that we err if we confuse our allegiance to any particular code with our commitment to this institution; we err in mistaking our prevalent code for the *moral point of view* itself. Of course, we typically hold our code to be justifiable from the moral point of view. However, if we are truly committed to the latter, we must allow the possibility that further consideration or new information or emergent human conditions may require revision in our code. The situation is perfectly analogous to the case of science education; we err if we confuse our allegiance to the current corpus of scientific doctrines with our commitment to scientific method. Of course we hold our current science to be justifiable by scientific method, but that very method itself commits us to hold contemporary doctrines fallible and revisable in the light of new arguments or new evidence that the future may bring to light. For scientific doctrines are not held simply as a matter of arbitrary preference; they are held for reasons. To affirm them is to invite all who are competent to survey these reasons and to judge the issues comprehensively and fairly on their merits.

Neither in the case of morality nor in that of science is it possible to convey the underlying *point of view* in the abstract. It would make no sense to say, 'Since our presently held science is likely to be revised for cause in the future, let us just teach scientific method and give up the teaching of content.' The content is important in and of itself, and as a basis for further development in the future. Moreover, one who knew nothing about specific materials of science in the concrete could have no conception of the import of an abstract and second-order scientific method. Nevertheless, it certainly does not follow that the method is of no consequence. On the contrary, to teach current science without any sense of the reasons that underlie it, and of the logical criteria by which it may itself be altered in the future, is to prevent its further intelligent development. Analogously, it makes no sense to say that we ought to teach the moral point of view in the abstract since our given practices are likely to call for change in the future. Given practices are indispensable, not only in organizing present energies, but in making future refinements and revisions possible. Moreover, one who had no concrete awareness of a given tradition of practice, who had no conception of what rule-governed conduct is, could hardly be expected to comprehend what the moral point of view might be, as a second-order vantage point on practice. Nevertheless, it does not follow that the latter vantage point is insignificant. Indeed, it is fundamental in so far as we hold our given practices to be reasonable, that is, justifiable in principle upon fair and comprehensive survey of the facts and interests involved.

There is, then, a strong analogy between the moral and the scientific points of view, and it is no accident that we speak of reasons in both cases. We can be reasonable in matters of practice as well as in matters of theory. We can make a fair assessment of the evidence bearing on a hypothesis of fact, as we can make a fair disposition of interests in conflict. In either case, we are called upon to overcome our initial tendencies to self-assertiveness and partiality by a more fundamental allegiance to standards of reasonable judgment comprehensible to all who are competent to investigate the issues. In forming such an allegiance, we commit ourselves to the theoretical possibility that we may need to revise our current beliefs and practices as a consequence of 'listening to reason.' We reject arbitrariness in principle, and accept the responsibility of critical justification of our current doctrines and rules of conduct.

It is evident, moreover, that there is a close connection between the general concept of *reasonableness,* underlying the moral and the scientific points of view, and the democratic ideal. For the latter demands the institutionalization of 'appeals to reason' in the sphere of social conduct. In requiring that social policy be subject to open and public review, and institutionally revisable in the light of such review, the democratic ideal rejects the rule of dogma and of arbitrary authority as the ultimate arbiter of social conduct. In fundamental allegiance to channels of open debate, public review, rational persuasion and orderly change, a democratic society in effect holds its own current practices open to revision in the future. For it considers these practices to be not self-evident, or guaranteed by some fixed and higher authority, or decidable exclusively by some privileged élite, but subject to rational criticism, that is, purporting to sustain themselves in the process of free exchange of reasons in an attempt to reach a fair and comprehensive judgment.

Here, it seems to me, is the central connection between moral, scientific, and democratic education, and it is this central connection that provides, in my opinion, the basic clue for school practice. For what it suggests is that the fundamental trait to be encouraged is that of reasonableness. To cultivate this trait is to liberate the mind from dogmatic adherence to prevalent ideological fashions, as well as from the dictates of authority. For the rational mind is encouraged to go behind such fashions and dictates and to ask for their justifications, whether the issue be factual or practical. In training our students to reason we train them to be critical. We encourage them to ask questions, to look for evidence, to seek and scrutinize alternatives, to be critical of their own ideas as well as those of others. This educational course precludes taking schooling as an instrument for shaping their minds to a preconceived idea. For if they seek reasons, it is their evaluation of such reasons that will determine what ideas they eventually accept.

Such a direction in schooling is fraught with risk, for it means entrusting our current conceptions to the judgment of our pupils. In exposing these conceptions to their rational evaluation we are inviting them to see for themselves whether our conceptions are adequate, proper, fair. Such a risk is central to scientific education, where we deliberately subject our current theories to the test of continuous evaluation by future generations of our student-scientists. It is central also to our moral code, *in so far as* we ourselves take the moral point of view toward this code. And, finally, it is central to the democratic commitment which holds social policies to be continually open to free and public review. In sum, rationality liberates, but there is no liberty without risk.

Let no one, however, suppose that the liberating of minds is equivalent to freeing them from discipline. *Laissez-faire* is not the opposite of dogma. To be reasonable is a difficult achievement. The habit of reasonableness is not an airy abstract entity that can be skimmed off the concrete body of thought and practice. Consider again the case of science: scientific method can be learned only in and through its corpus of current materials. Reasonableness in science is an aspect or dimension of scientific tradition, and the body of the tradition is indispensable as a base for grasping this dimension. Science needs to be taught in such a way as to bring out this dimension as a consequence, but the consequence cannot be taken neat. Analogously for the art of moral choice: the moral point of view is attained, if at all, by acquiring a tradition of practice, embodied in rules and habits of conduct. Without a preliminary immersion in such a tradition – an appreciation of the import of its rules, obligations, rights, and demands – the concept of choice of actions and rules for oneself can hardly be achieved. Yet the prevalent tradition of practice can itself be taught in such a way as to encourage the ultimate attainment of a superordinate and comprehensive moral point of view.

The challenge of moral education is the challenge to develop critical thought in the sphere of practice and it is continuous with the challenge to develop critical thought in all aspects and phases of schooling. Moral schooling is not, therefore, a thing apart, something to be embodied in a list of maxims, something to be reckoned as simply another subject, or another activity, curricular or extracurricular. It does, indeed, have to pervade the *whole* of the school experience.

Nor is it thereby implied that moral education ought to concern itself solely with the general structure of this experience, or with the effectiveness of the total 'learning environment' in forming the child's habits. The critical questions concern the *quality* of the environment: what is the *nature* of the particular school experience, comprising content as well as structure? Does it liberate the child in the long run, as he grows to adulthood? Does it encourage respect for persons, and for the arguments and reasons offered in personal exchanges? Does it open itself to questioning and discussion? Does it provide the child with fundamental schooling in the traditions of reason, and the arts that are embodied therein? Does it, for example, encourage the development of linguistic and mathematical abilities, the capacity to read a page and follow an argument? Does it provide an exposure to the range of historical experience and the realms of personal and social life embodied in literature, the law, and the social sciences? Does it also provide an exposure to particular domains of scientific work in which the canons of logical reasoning and evidential deliberation may begin to be appreciated? Does it afford opportunity for individual initiative in reflective inquiry and practical projects? Does it provide a stable personal milieu in which the dignity of others and the variation of opinion may be appreciated, but in which a common and overriding love for truth and fairness may begin to be seen as binding oneself and one's fellows in a universal human community?

If the answer is negative, it matters not how effective the environment is in shaping concrete results in conduct. For the point of moral education in a democracy is antithetical to mere shaping. It is rather to liberate.

NOTES

1. Ralph Barton Perry, *Realms of Value* (Cambridge: Harvard University Press, 1954), 431-32. Excerpt reprinted in I. Scheffler, ed., *Philosophy and Education,* 2d ed. (Boston: Allyn & Bacon, 1966), 32 ff.

2. Cited in Introduction to George S. Counts and Nucia P. Lodge, eds and translators, *I Want To Be Like Stalin: From the Russian Text on Pedagogy* by B. P. Yesipov and N. K. Goncharov (New York: John Day, 1947), pp. 14, 18. (The materials cited are from the 3d ed. of the *Pedagogy,* published in 1946.)

3. Frederic Lilge, 'Lenin and the Politics of Education,' *Slavic Review* 27 (June 1968): 255.

4. Ralph Barton Perry, 432.

5. See, for example, Kurt Baier, *The Moral Point of View* (Ithaca: Cornell University Press, 1958); William K. Frankena, *Ethics,* (Englewood Cliffs, N.J.: Prentice Hall, 1963), and R. S. Peters, *Ethics and Education* (Glenview, Ill.: Scott Foresman, 1967). Additional articles of interest may be found in Sect. 5, 'Moral Education' and Sect. 6, 'Education, Religion, and Politics,' in I. Scheffler, ed., *Philosophy and Education.*

7. BASIC MATHEMATICAL SKILLS: SOME PHILOSOPHICAL AND PRACTICAL REMARKS

FAMILIARITY – in life – may breed contempt. In thought, it typically breeds complacency and misunderstanding. The very familiarity of educational concepts often masks critical features of the problems facing us, lulling us the while into a false sense of clarity and the fitness of things.

Thus, we divide the matter of education into familiar "subject" categories and think thereby to have simplified and clarified the tasks of teaching. The subjects are, after all, drawn directly from parent disciplines, each with its distinctive and authoritative purchase on the world, each with its characteristic methodology and set of truths. Every subject is intellectually homogeneous within, and separable from every other without. Subjects are for the knowing, and knowing them is a matter of mastering their respective stores of truth and acquiring the respective methodologies from which their truths have sprung. Mastery of truths has to do with getting the appropriate beliefs; acquisition of methods and operations involves getting the right skills. For each subject there are characteristic and peculiar truths as well as distinctive and appropriate skills. To find these and to state them is to produce a curriculum. What could be more familiar – or more misguided?

Subjects are not, in fact, drawn directly or readily from their parent studies, and parent studies are not all disciplines.[1] (Is social science a discipline? Is the study of English language and literature? Is history?) Neither adult studies nor school subjects are written in the sky. The former are arranged for the expedient advancement of investigations and researches, the latter for the facilitation of learning and teaching in particular contexts – purposes that generate independent and powerful constraints. Neither studies nor subjects are internally homogeneous, nor are they wholly discrete from one another. Their aims, structures, methods, and boundaries change

This paper appeared in *Teachers College Record* 78 (No. 2, 1976): 205-12.

over time, and there are overlappings and branchings of various sorts at any given time. The "foreign relations" of subject areas at a time (generally of little interest to the specialist) are, moreover, of particular concern to the educator, interested as he must be both in economizing educational effort and in broadening the student's intellectual and cultural perspectives.[2]

Nor is the concept of *knowing* – taken as comprising acquisition of the distinctive beliefs of a subject and its distinctive methods – anywhere near as complex or broad enough to capture the tasks of curriculum formation.[3] The aims of education must encompass also the formation of habits of judgment and the development of character, the elevation of standards, the facilitation of understanding, the development of taste and discrimination, the stimulation of curiosity and wonder, the fostering of style and a sense of beauty, the growth of a thirst for new ideas and visions of the yet unknown. The articulated truths and methods of a subject are raw materials; they have no fixed locus in the curriculum. They are given special forms of life by the curriculum design that puts them to use for educative purposes.

Finally, it is (to say the least) gratuitous to suppose that *methods* and *skills* are exactly correlated; the classification of methods is typically a "logical," epistemological, or normative matter, while the classification of skills is often a matter of the psychology of learning or cognition. There is no reason to suppose that *methods* or *operations,* as catalogued by disciplinary or subject specialists, are identifiable with *skills,* as conceived by psychologists of cognition. Nor is there any substance to the notion that there must be simple rules for translating methods or operations into underlying psychological processes. The word "skills," it is true, is ambiguous, often used interchangeably with "operations" or with "methods"; or referring to what is *classifiable* by appeal to the latter as, e.g., "capacities to perform *such and such operations* or to apply *such and such methods.*" The skills of a subject, in the latter sense, are trivially derivable from a knowledge of its ingredient methods and operations. It is, however, fallacious to pass from the latter sense of the word to the sense in which "skills" refers to the basic processes of learning or cognition involved in applying a method or performing an operation. For the educator, this fallacy is a significant hazard: He must be concerned both with the intellectual methods of the subjects and with the psychological processes engaged in learning and applying them.

THE MEANING OF "BASIC"

Let us direct some of the above reflections to the case of *basic mathematical skills*. The very phrase is difficult for a variety of reasons. I have suggested an ambiguity in the term "skills" as between "operations or methods" and "processes of learning or cognition." Consider now the force of the term "basic": Does the phrase refer to skills *important* (perhaps necessary) for mathematics, to skills *peculiar* to mathematics, or to both? These senses diverge. The ability to follow an argument, for example, is certainly necessary for mathematics. But surely it is not peculiar to mathematics.

Take, then, the specific ability to follow a *mathematical* argument: Is *it* not both important and peculiar to mathematics? The idea suggests, if it does not imply, that mathematics is internally homogeneous in respect of the argumentation it displays – a point to be considered in the next section. Moreover, the idea begs the critical question as to whether the ability to follow a mathematical argument – no matter how such argument may be characterized – is a separable skill from a psychological and educational point of view. Is it, indeed, optimally developed in isolation from non-mathematical materials, and is its sphere of exercise limited to the mathematical domain? If, in short, *mathematical argument* is peculiar to mathematics, it does not follow that the *ability to learn, follow, or apply mathematical argument* is peculiar to mathematics.

Further, recent neuropsychological studies have cast empirical doubt on the correlation of subjects with mental processes. Such studies have tended to destroy the *a priori* skill-clustering fostered by traditional subject divisions: Certain symbol-processing abilities appear to cut across these divisions in various ways, while other inherited rubrics appear to require new divisions.[4] It is conceivable to me that very many (possibly all) of the psychologically significant skills important for mathematics may be important outside mathematics, that the particular perceptual, symbolic, inferential, mnemonic, questioning,[5] strategic, and imaginative capacities exercised in mathematics are also exercised outside it. It seems to me, moreover, overwhelmingly likely that successful performance in mathematics rests not only on general skills but also on general attitudes and traits such as perseverance, self-confidence, willingness to try out a hunch, appreciation for exactness, and yet others.

THE SCOPE OF "MATHEMATICAL"

Consider now the adjective "mathematical" in the phrase "basic mathematical skills." I have suggested that fundamental psychological processes connect mathematics with other fields of study. Here I argue that the subject itself is not all of a piece. Further, I conjecture that the oversimplified educational concept of a "subject" merges with the false public image of mathematics to form a quite misleading conception for the purposes of education: Since it is a subject, runs the myth, it must be homogeneous, and in what way homogeneous? Exact mechanical, numerical, and precise – yielding for every question a decisive and unique answer in accordance with an effective routine. It is no wonder that this conception isolates mathematics from other subjects, since what is here described is not so much a form of thinking as a substitute for thinking. What is in point is the process of calculation or computation, the deployment of a set routine with no room for ingenuity or flair, no place for guesswork or surprise, no chance for discovery, no need for the human being in fact.

Now calculation is certainly indispensable to mathematics, but it is not *mathematics*. When children are for the first time brought beyond the sphere of elementary calculation to the stage of problem-solving (perhaps in geometry), they naturally bring with them the impression that they are still learning the same subject. Yet here there are no routines for getting the right answer, here trial and error reign and there is ample scope for invention and surprise. The great gulf between mere calculation and problem-solving occurs within the subject, not beyond it. The tasks set, the purposes envisaged, the rules and constraints of the game are of a fundamentally different quality and are likely to evoke different applications and combinations of mental capacity. The notion of skill is thus in general not self-sufficient; it cannot eliminate needed reference to the nature of the tasks in question, their governing purposes and expected styles of execution.

The division I have just discussed, between computation and problem-solving, is a division internal to the school subject of mathematics. To elaborate it in teaching would help (I believe) not only to improve the process, but also to break down the mechanical stereotype of mathematics and relate the subject to other areas of creative thought. Other internal divisions are suggested by recent psychological studies, but are not as yet well understood. Certain sorts of injury to the brain may, for example, destroy the capacity to read while leaving intact the ability to recognize pictures, though both capacities are visual.[6] To what extent does success in geometry depend upon pictorial, as distinct from linguistic, processing capacities? In what ways indeed is a diagram like, and unlike, a picture? A map? An equation? A word description?[7]

Both the division between computation and problem-solving, and the possible division between geometry and other branches of the subject, are threatened by the popular conception of deduction. This conception deserves comment, for it offers the public a seemingly general way to override both distinctions and assimilate mathematics once again as a species of mechanism. For consider an ordered list of statements comprising a deductive proof: Each statement is an axiom or follows by single application of a stated rule from earlier statements. The chain is held together by necessity, the strongest conceptual glue: If the premises be true, the conclusion cannot fail of truth. Where is there any looseness or leeway? The whole is tightly made, an army of statements marching in order by command, a machine whose gears mesh inexorably according to fixed structural patterns. How can there be any mention of trial and error, of discovery and surprise, in the same breath with deduction?

The flaw in this example is that the proof is *given* at the outset. Though each of its statements "follows" by rule, the proof is not generated by rule. The determination of its character *as* a proof may indeed be made by mechanical routine. It does not follow that such routines exist for the construction of proofs. Indeed, it is demonstrable that no such routines are generally available. To *find* a proof is no merely mechanical matter, but an open and creative challenge in which ingenuity and good fortune, trial and error, and – at best – heuristic maxims hold sway.[8]

Moreover, the use of any formalism requires intelligence in application. The problematic material to which it is applied must, ordinarily, undergo suitable preparation, a phase of "words into symbols" that is not itself governed by mechanical rules but rather by good sense and an intuitive grasp of the problem's context and description.[9] One can impeccably run through a formal routine stupidly applied; the cure for stupidity is, furthermore, not formal. The application of formalisms and problem-solving strategies takes place always in a context and for a purpose. Intelligent deployment in context requires not only proper management of the formalism but appropriate application; and the latter involves accurate observation, sensible reading, logical analysis of problematic statements, translation into appropriate symbolic form, and eventual translation back to suitable statements – all such tasks belonging to mathematical applications but also to virtually all other spheres of human thought.

COMPREHENSION AND SKILL

I have said much about skill, virtually nothing about comprehension. What sort of skill is *that?* Elsewhere I have argued that it is not a skill at all.[10] To approach education as if it were always a matter of equipping the pupil with skills distorts our thinking. The category of skills has special features; these cannot be transported just anywhere.

A skill is capable of repeated exercise in separate episodes of performance, whereas comprehension is not thus exercised in performance. One who knows how to swim may swim every Thursday at 4:00; can we say, comparably, that one who knows how to understand quantum theory understands it every Thursday at 4:00? A skilled person may decide not to exercise his skill; a man who can play tennis may choose not to. A person with an understanding of quantum theory cannot, however, choose not to understand it. Nor can one speak of practice in the realm of comprehension as one does in reference to skills. One cannot develop an understanding of the quantum theory by understanding it over and over again, nor deepen one's understanding by faithfully repeated performances of understanding. One can tell a pupil to practice writing out a proof; it makes no sense to tell him to practice understanding it.

If comprehension is not a skill, what is it then? Can understanding a proof be merely a matter of checking its demonstrative character? Such a check would yield the conclusion that what purports to be a proof really is one. Would it guarantee an understanding of the proof? If the answer is negative, the question is: What else could possibly be required as a condition of understanding?

The point is elusive, but I suggest it has to do with appreciating the generality of the reasons behind each step.[11] These reasons are, further, of roughly two sorts: (a) *Deductive:* Those that characterize a given line as axiomatic or else derivable, by a single application of a stated rule, from earlier lines. Comprehension here requires an appreciation of the generality of the rule, an ability to recognize analogous cases and to apply it elsewhere. (b) *Strategic:* Those that characterize a given line as a promising step, in virtue of a certain strategic principle, toward the desired theorem. Comprehension here requires an appreciation of the general strategic principle, and it is also evinced in the treatment of parallel cases.

It is perhaps not misleading to describe deductive reasons as looking backward while strategic reasons look forward. Strategic reasons enable us, not to judge the *validity of the product,* but rather the *rationale of its step-by-step construction,* thus to enter into the mind of the maker. They answer the questions that are often so puzzling to the student: "How did the author of the proof think to apply such and such rule to get the next line? Granted, the step is valid, but where in the world did he get the idea to take that step in the first place, and what made him think that it would bring him nearer the desired conclusion?" To such questions, it is no answer at all to be told that the step is valid. And without an appropriate answer, one can hardly be said to understand the proof in the full sense of the word. A student who blames himself for failure to understand may (I suggest) never have been helped to see the special character of his questions, and the special nature of *strategic* – as distinct from *deductive* – reasons. Poincaré speaks of the matter in terms of intuition and image, describing the needed insight as "seeing the end from afar."[12] I have elsewhere described it as a matter akin to grasping the author's *motive,* the embodiment of which is his strategy: What did he hope to achieve by this step and how is such achievement related to his final goal?[13]

Understanding is not a skill but rather a state – an attainment – in which are ingredient general capacities. It is a fundamental and important aim of education, because it places the particular item in a general framework of rules and principles. It not only gives evidence of the pupil's right to be sure of his particular items and, hence, of his knowledge of them. It reaches out from these items to whole infinities of parallel cases in which evaluations are to be made but, moreover, new efforts undertaken. Nor is it an all-or-nothing affair, since it may grow gradually with the attempt to see how to deal with new cases. The very process of testing for understanding tends to develop it by forcing accommodation of the particular case with general principles.[14] The more reflective a grasp the pupil has of such principles and the more adequate they are to available cases, the less arbitrary the cases look and the more reasonable the principles; the more adequate, moreover, the pupil's orientation to new problems he may confront.

PROBLEMS AND RESEARCH

As suggested throughout the previous discussion, there is, I believe, much basic research to be undertaken, both of an analytical and of an empirical sort. There are also important studies to be conducted of a clinical and a practical kind. I shall comment briefly on these varieties.

As I hope to have shown earlier, the familiar categories in which educational thinking is cast are replete with difficulties. These difficulties are conceptual, but they critically affect the organization of practice. Analytical study of educational concepts needs to be undertaken, with particular reference to their mathematical applications.

The materials of mathematics need to be studied, both in relation to the educational analyses just proposed, and independently. I hope the foregoing discussion has suggested the importance of the logical, or normative, analysis of mathematical operations and methods, as helping to set the aims of mathematical education. Such analysis should investigate the diversity of tasks and purposes embodied in the several areas of mathematics. The study of methods ought also to be brought into connection with other fields than mathematics to discover new relations of a logical and epistemological sort.

Neither form of analytical study proposed is, of course, self-sufficient. Both need to be brought into communication with empirical considerations and inquiries of various kinds. I have mentioned some recent neuropsychological investigations into cognitive processes and the new articulations of skills and capacities to which they lead. Such studies, turned to the special concerns of mathematics, may well result in new pedagogical ideas of importance.

The study of so called "disabilities" in reading and other language functions is one mode of access to an understanding of underlying processes. Analogously, the systematic study of "disabilities" and deficiencies in mathematical areas may reveal new insights into forms and limitations of comprehension, with pedagogical reverberations and suggestions for improvement. I suggest a large series of studies of deficiency of all sorts, including investigations into mathematical "trauma," illiteracy, and misunderstanding in children and adults.

The relation of psychological studies of mental process to normative studies of mathematical method needs to be systematically investigated. The relation should, moreover, also be set in the context of general aims of education. For mathematics, as I have earlier argued, is not an island: Its linkages with all other areas of education need to be taken seriously, and studied systematically.

I have stated some conjectures about popular conceptions of mathematics and mathematical operations. All education is affected by prevailing attitudes and images concerning the content taught.[15] Studies of public attitudes toward mathematics might, it seems to me, reveal the sources of many difficulties, and perhaps point the way to some remedies.

Finally, I urge the study of teaching practice. What are the successful practices of good teachers? Why do they work? What skills do they embody? Such study ought to have a historical and comparative side and not restrict itself to local current custom.[16] But many good teachers now at work are no doubt doing good things capable of generalization. They are, however, unknown and generally cannot inform others of their work through publication. They should be sought out and studied. Furthermore, teachers should be encouraged to develop new practices, and educators to design new patterns of teaching. I am persuaded that the intuitive practice of teachers is an important – perhaps the single most important – source of new notions for the improvement of practice and even for theoretical ideas.

The proposals for study here put forth require collaborative effort. Such collaboration is a difficult but crucially important element in any program for advancing our knowledge and practice in education. Mathematicians, teachers, psychologists, philosophers, social scientists, educational theorists, and others need to find appropriate channels for sharing ideas and learning from one another. To develop such channels would be a contribution of great importance.

NOTES

1. On the notion of "discipline," see Israel Scheffler. *Reason and Teaching* (Indianapolis: Bobbs-Merrill, 1973), ch 4., pp. 45-57.

2. Ibid., 89.

3. On this point, see Israel Scheffler. *Conditions of Knowledge* (Chicago, Ill.: Scott, Foresman, 1965.), 106-107.

4. See Howard Gardner, "A Psychological Investigation of Nelson Goodman's Theory of Symbols," *The Monist* 58 (No. 2, April 1974): 318-26; Nelson Goodman, "On Reconceiving Cognition," ibid., 339-42; and H. Gardner, V. Howard, and D. Perkins, "Symbol Systems: A Philosophical, Psychological, and Educational Investigation," in *Yearbook of the National Society for the Study of Education,* ed. D. Olson (Chicago: University of Chicago Press, 1974).

5. Cp. the emphasis on questions and the *posing* of problems, by Stephen Brown and Marion Walter. When referring to problem-solving in this paper, I mean to include also the phase of problem-formulation stressed by Brown and Walter.

6. Gardner, 324.

7. On these questions, see Nelson Goodman, *Languages of Art* (Indianapolis: Bobbs-Merrill, 1968).

8. See, for the demonstration, A. Church, "A Note on the Entscheidungsproblem," *Journal of Symbolic Logic* 1 (1936): 40-41, 101-102; and B. Rosser, "An Informal Exposition of Proofs of Gödel's Theorems and Church's Theorem," ibid., 4 (1939): 53-60. For general remarks, see W. V. O. Quine, *Methods of Logic* (New York, N.Y.: Henry Holt, 1950), 190-91; and the entry by Alonzo Church, "Logic, Formal," in D. D. Runes, *Dictionary of Philosophy* (New York, N.Y.: Philosophical Library, 1942), 170ff., esp. 172 and 175.

9. See the section "Words into Symbols," in Quine, 39-46.

10. See Scheffler, *Conditions of Knowledge,* 17-21.

11. For a related discussion, see Scheffler, *Conditions of Knowledge,* 70ff.

12. Henri Poincaré, "Mathematical Definitions and Education," in Henri Poincaré. *Science and Method,* trans. Francis Maitland (New York, N.Y.: Dover Publications, 1952), 117-42. The relevant passage is on pp. 129-30.

13. Scheffler, *Conditions of Knowledge,* 73.

14. Ibid.

15. See L. J. Cronbach and P. Suppes, eds. *Research for Tomorrow's Schools* (London: Collier, 1969), 125ff.

16. Why, e.g., did Poland produce such a dazzling array of logicians between the World Wars?

8. IN PRAISE OF THE COGNITIVE EMOTIONS

THE MENTION of cognitive emotions may well evoke emotions of perplexity or incredulity. For cognition and emotion, as everyone knows, are hostile worlds apart. Cognition is sober inspection; it is the scientist's calm apprehension of fact after fact in his relentless pursuit of Truth. Emotion, on the other hand, is commotion – an unruly inner turbulence fatal to such pursuit but finding its own constructive outlets in aesthetic experience and moral or religious commitment.

Strongly entrenched, this opposition of cognition and emotion must nevertheless be challenged for it distorts everything it touches: Mechanizing science, it sentimentalizes art, while portraying ethics and religion as twin swamps of feeling and unreasoned commitment. Education, meanwhile – that is to say, the development of mind and attitudes in the young – is split into two grotesque parts – unfeeling knowledge and mindless arousal. My purpose here is to help overcome the breach by outlining basic aspects of emotion in the cognitive process.

Some misgivings about this purpose will, I hope, be allayed by a preliminary word. My aim, to begin with, is not reductive; I am concerned neither to reduce emotion to cognition nor cognition to emotion – only to show how cognitive functioning employs and incorporates diverse emotional elements – these elements themselves acquiring cognitive significance thereby. I am emphatically not suggesting that cognitions are essentially emotions, or that emotions are, in reality, only cognitions. Nevertheless, I hold that cognition cannot be cleanly sundered from emotion and assigned to science, while emotion is ceded to the arts, ethics, and religion. All these spheres of life involve both fact and feeling; they relate to sense as well as sensibility.

Secondly, though applauding the cognitive import of emotions, I do not propose to surrender intellectual controls to wishful thinking, nor shall I portray the heart as giving special access to a higher truth.[1] Control of wishful thinking is utterly essential in cognition; it operates, however, not through an unfeeling faculty of Reason but through the organization of

This paper appeared in *Teachers College Record* 79 (1977): 171-86.

countervailing critical interests in the process of inquiry. These interests of a critical intellect are, in principle, no less emotive in their bearing than those of wayward wish. The heart, in sum, provides no substitute for critical inquiry; it beats in the service of science as well as of private desire.

Finally, I concede it to be undeniable that certain emotional states may be at odds with sound processes of judgment and decision making. Overpowering agitations may derail the course of reasoning; greed, jealousy, or lust may misdirect it; depression or terror may bring it to a total halt. Conversely, the effect of rational judgment may well be to moderate, even wholly to dissipate, certain emotions by falsifying their factual presuppositions: Anger fades, for example, when it turns out the injury was accidental or caused by someone other than first supposed; fear evaporates when the menacing figure becomes the tree's dancing shadow. It does not follow from these cases, however, that *emotion* as such is uniformly hostile to cognitive endeavors, nor may we properly conclude that *cognition* is, in general, free of emotional engagement. Indeed, emotion without cognition is blind and, as I shall hope particularly to show in the sequel, cognition without emotion is vacuous.

EMOTIONS IN THE SERVICE OF COGNITION

Considering now the various roles of emotion in cognition, I divide the field, for convenience, into two main parts, the first having to do with the organization of *emotions generally* in the service of critical inquiry, and the second having to do with *specifically cognitive emotions.* Under the first rubric I shall treat: (a) rational passions,[2] (b) perceptive feelings, and (c) theoretical imagination, and I turn first to the rational passions, that is to say, the emotions undergirding the life of reason.

Rational passions

The life of reason is one in which cognitive processes are organized in accord with controlling rational ideals and norms. Such organization involves characteristic patterns of thought, action, and evaluation comprising what may be called rational character. It also thus requires suitable emotional dispositions. It demands, for example, a love of truth and a contempt of lying, a concern for accuracy in observation and inference, and a corresponding repugnance of error in logic or fact. It demands revulsion at distortion, disgust at evasion, admiration of theoretical achievement, respect for the considered arguments of others. Failing such demands, we incur rational shame; fulfilling them makes for rational self-respect.

Like moral character, rational character requires that the right acts and judgments be habitual; it also requires that the right emotions be attached to the right acts and judgments.[3] "A rational man," says R. S. Peters, "cannot, without some special explanation, slap his sides and roar with laughter or shrug his shoulders with indifference if he is told that what he says is irrelevant, that his thinking is confused and inconsistent or that it flies in the face of the evidence."[4] The suitable deployment in conduct of emotional dispositions such as love and hate, contempt and disgust, shame and self-esteem, respect and admiration indeed defines what is meant, quite generally, by the internalization of ideals and principles in character. The wonder is not that *rational* character is thus related to the emotions but that anyone should ever have supposed it to be an exception to the general rule.

Rational character constitutes an intellectual conscience; it monitors and curbs evasions and distortions; it combats inconsistency, unfairness to the facts, and wishful thinking. In thus exercising control over undesirable impulses, it works for a balance in thought, an epistemic justice, which requires its own special renunciations and develops a characteristic cognitive discipline. There is, however, no question here of the control of impulses through a "bloodless reason,"[5] as control is exercised through the structuring of emotions themselves. Rationality, as John Dewey put it,

> is not a force to evoke against impulse and habit. It is the attainment of a working harmony among diverse desires.... The elaborate systems of science are born not of reason but of impulses at first slight and flickering; impulses to handle, move about, to hunt, to uncover, to mix things separated and divide things combined, to talk and to listen. Method is their effectual organization into continuous dispositions of inquiry, development and testing. It occurs after these acts and because of their consequences. Reason, the rational attitude, is the resulting disposition.... The man who would intelligently cultivate intelligence will widen, not narrow, his life of strong impulses while aiming at their happy coincidence in operation.[6]

This coincidence, I emphasize, requires appropriate organization of feelings and sentiments in the interests of intelligent control.

Perceptive feelings

Having seen the role of emotions in the internalization of rational norms, let us consider now their employment in perception. For they are not only interwoven with our cognitive ideals and evaluative principles; they are also intimately tied to our vision of the external world. Indeed they help to construct that vision and to define the critical features of that world.

These critical features – however specified – are the objects of our evaluative attitudes, the foci of our appraisals of the environment. Our habits and judgments are keyed in to these appraisals; we define ourselves and orient our action in the light of our situation as appraised. Characteristic orientations are associated with distinctive emotional dispositions, and both involve seeing the environment in a certain light: Is it, for example, beneficial or harmful, promising or threatening, fulfilling or thwarting?[7] The subtle and intricate web relating adult feeling and orientation to adult perception of the environment is a product of evolutionary development, to be sure, but also of the special circumstances of individual biography. Acquiring human significance through biographical linkage with critical features of the environment, our feelings come indeed to *signify* – to serve as available cues for interpreting the situation.

Fear of a particular person, for example, presupposes that that person is regarded as dangerous – danger being a critical feature of the environment calling for a special orientation in response. There need, however, be no *independent* evidence, in every case, of the threat we sense: The characteristic feeling that has become associated for us with past dangers itself serves us as a cue. Interpreting that feeling *as* fear, we at once characterize our own state and ascribe danger to the environment. Indeed, we may thence proceed to an explicit attribution of danger, prompted by cues of feeling. Pursuing a more abstract direction in forming our cognitive concepts, we may, further, come to describe a certain situation as *terrifying,* ascribing to it, *independently of our own state,* the capacity to arouse fear. Thus employing the emotions as parameters, we gain enormous new powers of fundamental description, while abstracting from actual conditions of feeling.

The notion that aesthetic experience, for example, is peculiarly and purely a matter of emotion ignores such manifold connections of feeling and fact – both fact as *embodied* in the art work and fact as *represented* therein. Relative to the latter, H. D. Aiken writes,

> Just as in ordinary circumstances an emotional response is the product of a perceived situation which is apprehended by the individual as promising or threatening, so the expressiveness of an imaginative work arises, at least in part, from the fact that it provides a dramatic representation of an action of which the evoked emotion is the expressive counterpart. And such a representation must be understood as such if the expressive values of the work are to become actual; without it such emotion as the observer might experience would have no ground, and if, by a miracle, it could be sustained, it would still remain the private, dumb, inexpressive importation of the observer himself. As such it would be nothing more than an accidental, adventitious subjective coloring which, having no artistic basis in the thing perceived, would be devoid of aesthetic relevance to it. Aesthetically relevant

emotion in art is something which is expressed to us by the action or gesture of the work itself; it is something aroused and sustained by the work as an object for contemplation, and it is found there as a projected quality of the action.[8]

That the emotion is thus tied to a representational understanding of the work of art does not imply, however, that this understanding must be antecedently fashioned, in complete isolation from the feelings. This point must be especially emphasized since the familiar notion of the work of art as "an object for contemplation" may carry contrary, and therefore misleading, connotations. In fact, I believe, the very feelings through which we respond to the content of a work serve us also in interpreting this content. Reading our feelings and reading the work are, in general, virtually inseparable processes.

The cognitive role of the emotions in aesthetic contexts has been emphasized by Nelson Goodman in a recent discussion. He writes,

> The work of art is apprehended through the feelings as well as through the senses. Emotional numbness disables here as definitely if not as completely as blindness or deafness. Nor are the feelings used exclusively for exploring the emotional content of a work. To some extent, we may feel how a painting looks as we may see how it feels. The actor or dancer – or the spectator – sometimes notes and remembers the feeling of a movement rather than its pattern, insofar as the two can be distinguished at all. Emotion in aesthetic experience is a means of discerning what properties a work has and expresses.[9]

The general point is, of course, not limited to the aesthetic realm since, as I have earlier emphasized, the emotions intimately mesh with all critical appraisals of the environment: The flow of feeling thus provides us with a continuous stream of cues significant for orientation to our changing contexts. Indeed, as Goodman remarks,

> In daily life, classification of things by feeling is often more vital than classification by other properties: we are likely to be better off if we are skilled in fearing, wanting, braving, or distrusting the right things, animate or inanimate, than if we perceive only their shapes, sizes, weights, etc. And the importance of discernment by feeling does not vanish when the motivation becomes theoretic rather than practical. . . . Indeed, in any science, while the requisite objectivity forbids wishful thinking, prejudicial reading of evidence, rejection of unwanted results, avoidance of ominous lines of inquiry, it does not forbid use of feeling in exploration and discovery, the impetus of inspiration and curiosity,

or the cues given by excitement over intriguing problems and promising hypotheses.[10]

Theoretical imagination

Mention of the context of theory brings us to the third role of emotions in the service of cognition, that of stimulus to the scientific imagination. This role is virtually annihilated by the stereotyped emotion-cognition dichotomy. For this dichotomy assigns all feeling and flair, all fantasy and fun, to the arts and humanities, conceiving the sciences as grim and humorless grind. The method of science is miserly caution – to gather the facts and guard the hoard. Imagination is a seductive distraction – a hindrance to serious scientific business.

This doctrine is, in fact, the death of theory. Theory is not reducible to mere fact-gathering, and theoretical creation is beyond the reach of any mechanical routine. Science controls theory by credibility, logic, and simplicity; it does not provide rules for the creation of theoretical ideas. Scientific objectivity demands allegiance to fair controls over theory, but fair controls cannot substitute for ideas. "All our thinking," said Albert Einstein, "is of this nature of a free play with concepts; the justification for this play lying in the measure of survey over the experience of the senses which we are able to achieve with its aid."[11]

The ideal theorist, loyal to the demands of rational character and the institutions of scientific objectivity, is not therefore passionless and prim. Theoretical inventiveness requires not caution but boldness, verve, speculative daring. Imagination is no hindrance but the very life of theory, without which there *is* no science.

Now the emotions relate to imaginative theorizing in a variety of ways. The emotional life, to begin with, is a rich source of substantive ideas. Drawing from the obscure wellsprings of this life, the mind's free play casts up novel patterns and images, exotic figures and analogies that, in an investigative context, may serve to place old facts in a new light. The dream of the nineteenth-century chemist F. A. Kekulé will provide a striking illustration. He had been trying for a long time to find a structural formula for the benzene molecule. Dozing in front of his fireplace one evening in 1865, he seemed, as he looked into the flames, "to see atoms dancing in snakelike arrays. Suddenly, one of the snakes formed a ring by seizing hold of its own tail and then whirled mockingly before him. Kekulé awoke in a flash: he had hit upon the now famous and familiar idea of representing the molecular structure of benzene by a hexagonal ring. He spent the rest of the night working out the consequences of this hypothesis."[12]

The emotions serve not merely as a *source* of imaginative patterns; they fulfill also a *selective* function, facilitating choice among these patterns, defining their salient features, focusing attention accordingly. The patterns developed in imagination, that is, carry their own emotive values; these values guide selection and emphasis. They help imagined patterns to structure the phenomena, highlighting factual features of interest to further inquiry. "Passions," as Michael Polanyi has said, "charge objects with emotions, making them repulsive or attractive; . . . Only a tiny fraction of all knowable facts are of interest to scientists, and scientific passion serves . . . as a guide in the assessment of what is of higher and what of lesser interest."[13]

Finally, the emotions play a directive role in the process of *applying* the fruits of imagination to the solution of problems. The course of problem-solving, as has already been intimated, is continually monitored by the theorist's cues of feeling, his sense of excitement or anticipation, his elation or suspicion or gloom. Moreover, imagined objects encountered in thought by the problem-solver affect his deliberation emotively, as real objects do, and influence his decisions in analogous ways. "In thought as well as in overt action," says Dewey, "the objects experienced in following out a course of action attract, repel, satisfy, annoy, promote and retard. Thus deliberation proceeds."[14] There is, no doubt, much yet to be learned about the interaction of emotions and imagination in all the ways I have sketched, and in others as well. It should, however, even now, be evident that creation is fed by the emotional life in the sphere of science no less than in the spheres of poetry and the arts.

COGNITIVE EMOTIONS

We have, until now, concerned ourselves with the organization of emotions generally in the service of cognition. I want now to deal with two emotions that are, in a sense to be explained, *specifically cognitive* in their bearing – the *joy of verification* and the *feeling of surprise*.

In what sense do I speak of an emotion as specifically cognitive? Consider first the notion of *moral emotions,* conceived as those resting upon suppositions of a moral sort: Thus, indignation, for example, rests upon the supposition of a moral grievance – a piece of injustice, and remorse presumes that one has in fact done something wrong. If the relevant moral suppositions are false or lack evidential foundation, the respective emotions may be thought unreasonable, but if these suppositions are not made at all, that is to say, if the suppositions do not exist, the emotions in question can hardly, in normal circumstances, be said to have occurred. Now I propose, analogously, to consider an emotion specifically *cognitive* if it rests upon a supposition of a cognitive sort – that is to say, a supposition relating to the content of the subject's cognitions (beliefs, predictions, expectations) and, in cases of special interest to us, bearing upon their epistemological status.

It is important to avoid misunderstanding of the terminology I have chosen here. It is indeed true that *all* suppositions may be considered cognitive in a broad sense, inasmuch as they make factual claims expressible in propositional form; moreover, emotions *generally*, as I have maintained, presuppose the existence of such claims concerning critical features of the environment. However, when I characterize an emotion as *specifically cognitive*, I mean more than this. In particular, I mean not simply that it presupposes the existence of a factual claim but that the claim in question specifically concerns the nature of the subject's cognitions (and, in cases of interest, is epistemologically relevant to them). A cognitive emotion, I should further emphasize, is thus decidedly an *emotion*, but an emotion of a certain *kind*, specifiable by its cognitive reference as just explained.[15]

The joy of verification

In his well-known paper of 1934 on "The Foundation of Knowledge,"[16] Moritz Schlick provides an example of such a cognitive emotion in outlining his theory of science, giving primary place in his theory to the joy that accompanies the fulfillment of an expectation. Cognition, in Schlick's view, has, from the earliest times, always been predictive but the value of reliable prediction lay originally in its practical service to life. "Now in science," he writes, "... cognition ... is not sought because of its utility. With the confirmation of prediction the scientific goal is achieved: the joy in cognition is the joy of verification, the triumphant feeling of having guessed correctly."[17] Such moments of joy are, in Schlick's opinion, of central importance in understanding scientific purpose. "They do not in any way," he says, "lie at the base of science; but like a flame, cognition, as it were, licks out to them, reaching each but for a moment and then at once consuming it. And newly fed and strengthened, it flames onward to the next. These moments of fulfillment and combustion are what is essential. All the light of knowledge comes from them."[18]

Now one need not agree with Schlick's general view of science in order to acknowledge that the satisfaction of a theoretical forecast may indeed occasion joy. Nor is it required that we concur with the extravagant suggestion that *all* predictive success brings elation. It may, for example, be countered that routine successes based on theory frequently, perhaps typically, go unnoticed, while soberly predicted events may be so dreadful as to occasion not joy but sorrow or despair. Nevertheless, we can agree that the fulfillment of a prediction may indeed crown an investigative achievement in science, producing in its wake what Schlick calls a "triumphant feeling of having guessed correctly."

This joyful feeling I consider a cognitive emotion, because it rests on a supposition (with epistemological relevance) as to the content of the guess in question: It presumes that what has happened is what had, in fact, been predicted. Without such presumption, this joy of verification cannot be said to occur. Whether the presumption is true, or is based on adequate grounds, is another story. Certainly one assumes that the emotion in question may be criticized as unreasonable if it can be shown that what has in fact happened had not, in fact, been predicted.

Can such a criticism, however, actually be entertained as a matter of psychological fact? Is it not, rather, true that our expectations are so powerful as consistently to warp our observations to fit? A whole library of psychological writings testifies to the powerful tendency of expectation to create its own confirmations in experience. The general theme of this testimony may be indicated by the role of normal cues in perception: The perceptual identification of objects "proceeds on the basis of cues normally sufficient to select the objects in question; when these cues fail in fact, we tend anyway to see them as having succeeded."[19] Bruner, Goodnow, and Austin comment on this point:

> If [a bird] has wings and feathers, the bill and legs are highly predictable. In coding or categorizing the environment, one builds up an expectancy of all of these features being present together. It is this unitary conception that has the configurational or Gestalt property of "birdness". Indeed, once a configuration has been established and the object is being identified in terms of configurational attributes, the perceiver will tend to "rectify" or "normalize" any of the original defining attributes that deviate from expectancy. Missing attributes are "filled in" ..., reversals righted ..., colors assimilated to expectancy.[20]

Some philosophers have further maintained that scientific observation itself is *systematically* theory-laden – presupposing the very theories it is naively thought to test.[21] If indeed we are, as suggested, so blinded by our own theoretical beliefs as to be incapable of acknowledging anything that might contradict them, we can hardly take the joy of verification to represent a cognitive triumph of science. Rather, we must count it an unearned and deluded joy, resulting not from a happy match between theory and experience but solely from our desperate rigging of experience to make it fit.

This conclusion, as I have elsewhere argued, seems to me too extreme for the facts.[22] It is undeniable that our beliefs greatly influence our perceptions, but neither psychology nor philosophy offers any proof of a preestablished harmony between what we believe and what we see. Expectations have the function of orienting us selectively toward the future, but this function does not require that they blind us to the unforeseen. Indeed, the presumption of mismatch between experience and expectation underlies another cognitive emotion: *surprise*. The existence of this emotion testifies that we are not, in principle, beyond acknowledging the predictive failures of our own theories, that we are not debarred by nature from capitalizing upon such failures in order to learn from experience. The genius of science is, in fact, to institutionalize such learning by wedding the free theoretical imagination to the rigorous probing for predictive failures.

The Significance of Surprise

Surprise is a cognitive emotion, resting on the (epistemologically relevant) supposition that what has happened conflicts with prior expectation. Without such presumption, surprise cannot be supposed to occur, although the truth of the presumption may, of course, be questioned in particular cases. Surprise must, in any event, not be confused with mere novelty. A novel – that is to say, a hitherto unencountered – contingency may well be anticipated in thought, while a familiar phenomenon, juxtaposed with available theory, may profoundly surprise. Thwarting expectation, the surprising element may indeed provoke the revision of theory, even the reorganization of categories, thus *producing* novelty as a result. It is itself, however, never a mere matter of novelty, but always of conflict with prior belief. The concept of *unexpectedness,* it should be noted, is too weak to make this critical distinction, for it covers both the case of a feature that has simply not been anticipated, and that of a feature that has been positively ruled out by anticipation.[23]

To the extent that we are capable of surprise, the possibility that our expectations are wrong is alive for us and thus our joy in verification, if it occurs, is not utterly deluded. Receptive to surprise, we are capable of learning from experience – capable, that is, of acknowledging the inadequacies of our initial beliefs, and recognizing the need for their improvement. It is thus that the testing of theories, no less than their generation, calls upon appropriate emotional dispositions.

Receptivity to surprise involves, however, a certain vulnerability; it means accepting the risk of a possibly painful unsettlement of one's beliefs, with the attendant need to rework one's expectations and redirect one's conduct. To be sure, where the relevant beliefs are weakly held, or relatively segregated, or of peripheral significance for one's basic orientation – or where the required alterations are likely to be readily effected – the risk may be easily borne, even by the cautious. Surprise may, in such circumstances, not in fact distress but amuse – even enchant, as will be evident from even brief reflection on the role of surprise in humor, in music, in literature, and generally in the arts. Moreover, there are, in all realms of life, pleasant surprises, where the value of the unexpected event, or even of the unsettlement itself, outweighs the stress of disorientation and the concomitant costs of revision in belief.

One cannot, however, reasonably count on all – even most – surprises in life to be thus amusing or pleasant; it must be conceded that a general openness to surprise involves a real risk of epistemic distress. This risk may to varying degrees become palatable, even exciting; certainly accepting it is one of the normal requirements of rational character. Yet it *is* a risk of possibly painful disorientation, and it requires emotional strength to face and to master. To commit oneself to learning from experience is, in short, a significant attitude – supported by mature reflection, to be sure – but exacting a price in return for the prospect of improvement in one's system of beliefs.

Three alternative attitudes promise an avoidance of the price by erecting wholesale defenses against surprise. Since surprise presumes prior expectation, a defense may be sought, to begin with, in the *rejection* of all expectation – in effect, the denial of all belief. This is the attitude of the radical skeptic, who hopes to make himself immune to surprise by any contingency through renouncing all anticipations to the contrary, that is to say, all anticipations without exception. Of whatever happens, he says, in effect, "It doesn't surprise me since I never expected it not to happen!" A second – apparently opposite – attitude is that of utter credulity or gullibility – the *acceptance* of all beliefs or expectations without distinction. Here the formula in response to every contingency is "I'm not surprised! I expected that too!" Both radical skepticism and radical credulity are, however, alike forms of epistemic apathy: To reject all expectations is to be indifferent to each, while to accept all as equally good is in actuality to choose none, having no reason to expect anything at all rather than something else. It is no wonder that these seemingly contrary attitudes have so often been remarked to be psychologically akin, and together opposed to the selective hypothesis-formation characteristic of scientific thought. "Complete doubt," as Peirce noted, is "a mere self-deception" and no one who follows the method of radical skepticism "will ever be satisfied until he has formally recovered all those beliefs which in form he has given up."[24]

Moreover, each of these two alternative attitudes exacts its own heavy price. Neither can in fact be realized as a genuine option over a significant area of conduct. The skeptic, despite himself, forms positive expectations in executing his actions, while the radically credulous person, generally hospitable to inconsistencies, perforce rules out certain contingencies in carrying through the activities of daily life. Only in a local and intermittent way can these attitudes be attempted. They are perhaps more accurately described as poses or pretenses, the effect of which is, however, perfectly real – to aid the denial of responsibility for one's beliefs and so to block the possibility of their improvement through the educative medium of surprise.

The third attitude promising a defense against surprise is that of dogmatism. Unlike the radical skeptic and the total believer, the dogmatist is perfectly firm about the beliefs he espouses and the beliefs he rejects. He blocks surprise not by disclaiming responsibility for his doctrines, but rather by denying all experience that purports to contradict them. He not only avoids the systematic testing of his beliefs; he closes off the very possibility of recognizing negative evidence, by early and stout denial of its existence. Theory-laden to the point of blindness, his observations are predictably positive, the joy he takes in verification thus unearned and hollow. Dogmatism is also a difficult attitude to maintain if only because (as Peirce saw)[25] it is impossible to filter all negative indications in advance by a systematic method. Yet it can be carried a long way, preventing the acknowledgment of surprise, and, hence, the application of new and surprising experience to the improvement of initial beliefs and orientations. Dogmatism, no less than skepticism and gullibility, conflicts with the effort at such improvement. To accept this effort, with its associated vulnerability to the unsettlement of surprise, is to choose a distinctive emotional as well as cognitive path.

But how, it may be asked, is receptivity to surprise possible? Surprise is, after all, unsettling; it risks the distress of disorientation and the potential pain of relearning. In similar vein, Schlick contrasts the joy of verification with the disappointment of falsification[26] – the disappointment following upon the violation of beliefs in which we had put our trust. How can one counsel receptivity to surprise: Is this not an impossibly mixed emotion, like elation at despair, or happiness at depression?

We must, first of all, reject the suggestion that surprise is always painful. If Schlick's notion of uniformly joyful verification is to be rejected as extravagant, in line with our earlier remarks, his parallel notion of uniformly disappointing falsification must equally be criticized. Some falsifications are, as we have previously intimated, delightful, some disruptions of expectation pleasantly exciting, some occasions of relearning fraught with engaging challenge. Schlick, I suggest, confuses expectation with hope, but the two are clearly separable, and only the former is necessary for surprise.

Yet if surprise is not always painful, surely it sometimes is: It must therefore be conceded, at the very least, to be *uncertain* in its quality. The question then recurs in a new version: How can one counsel receptivity to uncertainty? Here, however, an immediate reply is forthcoming. The original version of the question raised the issue of impossibly mixed emotions, whereas the present version no longer does so. For uncertainty is not an emotion; it is rather a prospect or condition, while the *feeling* of uncertainty mixes readily with receptive and aversive attitudes. Uncertainty is indeed consistently faced in varying ways: Some persons tend to shrink from, while others tend to welcome the prospect. The receptivity to surprise that is implicated in the capacity to learn from experience is, in any case, perfectly coherent in its emotional composition.

Moreover, *receptivity* to surprise is not to be confused with elation or happiness. It is rather the capability of acknowledging surprise than joy in its occurrence that is here in point. Analogously, to acknowledge one's grief does not entail being elated by it. Acknowledgment itself is a possible and a significant attitude, opening the way beyond the acknowledged circumstance.

Yet receptivity is, of course, not enough to characterize the testing phase of inquiry. How we cope with surprise, once it is acknowledged, is of critical importance. Surprise may be dissipated and evaporate into lethargy. It may culminate in confusion or panic. It may be swiftly overcome by a redoubled dogmatism. Or it may be transformed into wonder or curiosity, and so become an educative occasion. Curiosity replaces the impact of surprise with the demand for explanation;[27] it turns confusion into question. To answer the question is to reconstruct initial beliefs so that they may consistently incorporate what had earlier been unassimilable. It is to provide an improved framework of premises by which the surprising event might have been anticipated and for which parallel events will no longer surprise.

Critical inquiry in pursuit of explanation is a constructive outcome of surprise, transforming initial disorientation into motivated search. There is, as we have seen, no mechanical routine that guarantees success in the search for explanatory theory. Yet, an emotional value of such search is to offer mature consolation for the stress of surprise and the renunciation of inadequate beliefs.[28] Achieving superordinate status in the economy of science, the value of inquiry becomes, indeed, autonomous, pressing new explanations deliberately into situations of risk, testing their vulnerability in novel ways, exposing their implicit predictions systematically to the chance of new surprise.

The constructive conquest of surprise is registered in the achievement of new explanatory structures, while cognitive application of these structures provokes surprise once more. Surprise is vanquished by theory, and theory is, in turn, overcome by surprise. Cognition is thus two-sided and has its own rhythm; it stabilizes and coordinates; it also unsettles and divides. It is responsible for shaping our patterned orientations to the future, but it must also be responsive to the insistent need to learn from the future. Establishing habits, it must stand ready to break them. Unlearning old ways of thought, it must also power the quest for new, and greater, expectations.[29] These stringent demands upon our cognitive processes also constitute stringent demands upon our emotional capacities. The growth of cognition is thus, in fact, inseparable from the education of the emotions.

NOTES

1. For a discussion of this theme in the context of the history of American thought, see Morton White, *Science and Sentiment in America* (New York: Oxford University Press, 1972).

2. On this topic see R. S. Peters, "Reason and Passion," in *Education and the Development of Reason*, eds. R. F. Dearden, P.H. Hirst, and R.S. Peters (London: Routledge & Kegan Paul, 1972), especially the section on the rational passions, pp. 225-27. See also John Rawls, *A Theory of Justice* (Cambridge: Harvard University Press, 1971), especially sections 67, 73-75; P. Foot, "Moral Beliefs," *Proceedings of the Aristotelian Society* 59 (1958-59); and B. A. O. Williams, "Morality and the Emotions," in *Problems of the Self*, B. A. O. Williams (Cambridge: Cambridge University Press, 1973). A significant recent book, dealing with a wide range of related topics, is Robert C. Solomon, *The Passions* (New York: Doubleday, Anchor Press, 1976).

3. Cp. Aristotle, *Nicomachean Ethics*, Book II, 3.

4. Peters, "Reason and Passion," 226.

5. John Dewey, *Human Nature and Conduct* (New York: Henry Holt, 1922, 1930), 196.

6. Ibid.

7. Related points are discussed in Peters, "Reason and Passion"; R. S. Peters, "The Education of the Emotions," Dearden, Hirst, and Peters, *Education and the Development of Reason;* G. Pitcher, "Emotion," Dearden, Hirst, and Peters, *Education and the Development of Reason;* and R. W. Hepburn, "The Arts and the Education of Feeling and Emotion," Dearden, Hirst, and Peters, *Education and the Development of Reason.* See also the article W. P. Alston, "Emotion and Feeling," in *The Encyclopedia of Philosophy* (New York: Macmillan, 1967), vol. 2, pp. 479-86.

8. Henry David Aiken, "Some Notes Concerning the Aesthetic and the Cognitive," *Journal of Aesthetics and Art Criticism* 13 (1955): 390-91.

9. Nelson Goodman, *Languages of Art* (Indianapolis: Bobbs-Merrill, 1968), 248.

10. Ibid., 251.

11. Albert Einstein, "Autobiographical Notes," tr. Paul Arthur Schilpp, in *Albert Einstein: Philosopher-Scientist*, ed. P. A. Schilpp (New York: Tudor Publishing, 1949), p. 7. (Now published by the Open Court Publishing Company, La Salle, Illinois.) The passage is quoted in a discussion of these and related points in I. Scheffler, *Science and Subjectivity* (Indianapolis: Bobbs-Merrill, 1967), esp. ch. 4.

12. Carl G. Hempel, *Philosophy of Natural Science* (Englewood Cliffs, N.J.: Prentice-Hall, 1966), 16.

13. Michael Polanyi, *Personal Knowledge* (New York: Harper & Row, 1958, 1962), 134-35.

14. Dewey, *Human Nature and Conduct*, 192.

15. For discussion helpful in clarifying certain points in this section, I am grateful to Professors Eli Hirsch and Jonas Soltis.

16. Moritz Schlick, "Uber das Fundament der Erkenntnis," *Erkenntnis* 4 (1934), trans. as David Rynin, "The Foundations of Knowledge," in *Logical Positivism*, ed. A.J. Ayer (New York: Free Press, 1959).

17. Schlick, "Uber das Fundament der Erkenntnis," in Ayer, *Logical Positivism*, 222-23. There is a general discussion of Schlick's paper in Scheffler, *Science and Subjectivity*, ch. 5.

18. Schlick, "Uber das Fundament der Erkenntnis," in Ayer, *Logical Positivism*, 227.

19. Scheffler, *Science and Subjectivity*, 30.

20. Jerome S. Bruner, J. Goodnow, and G.A. Austin, *A Study of Thinking* (New York: John Wiley, 1956), 47.

21. See N.R. Hanson, *Patterns of Discovery* (Cambridge: Cambridge University Press, 1958), 18-19, and elsewhere. For a general discussion see Scheffler, *Science and Subjectivity*, especially chs. 1 and 2.

22. Ibid., ch. 2.

23. On these points, there is disagreement among previous writers. For informative historical, as well as other material see M. M. Desai, "Surprise: A Historical and Experimental Study," *British Journal of Psychology Monograph Supplements*, No. 22, 1939; D. Berlyne, "Emotional Aspects of Learning," *Annual Review of Psychology* 15 (1964): 115-42; and W. R. Charlesworth, "The Role of Surprise in Cognitive Development," in *Studies in Cognitive Development*, D. Elkind and J.H. Flavell (New York: Oxford University Press, 1969). Although I agree with various points in these psychological papers (e.g., Charlesworth, pp. 270, 276), they tend to focus on individual behavior in a relatively local situation, whereas I tend to link surprise with failed prediction in the context of discussions in philosophy of science.

24. C. S. Peirce, "Some Consequences of Four Incapacities," in *Collected Papers of Charles Sanders Peirce*, ed. Charles Hartshorne and Paul Weiss, vol. 5 (Cambridge: Harvard University Press, 1934), 264-65. See also Israel Scheffler, *Four Pragmatists* (London: Routledge & Kegan Paul, 1974), 52-53, 69-70.

25. C. S. Peirce, "The Fixation of Belief," Hartshorne and Weiss, *Collected Papers of Charles Sanders Peirce*, 5:382. See also a general discussion in Scheffler, *Four Pragmatists*, 60ff.

26. Schlick, "Uber das Fundament der Erkenntnis," in Ayer, *Logical Positivism*, 223.

27. On explanation generally, see Israel Scheffler, *Anatomy of Inquiry* (New York: Alfred A. Knopf, 1963), Part I; also Bobbs-Merrill edition, 1971. I here use the term in a very broad sense.

28. Interesting discussion of this sort of point and related psychological issues is contained in Fay H. Sawyier, "About Surprise" (Paper read to the annual meeting of the Western Division of the American Philosophical Association, St. Louis, Missouri, 1974). For a discussion of the pedagogical use of surprise in the teaching of mathematics see Stephen I. Brown, "Rationality, Irrationality and Surprise," *Mathematics Teaching: The Bulletin of The Association of Teachers of Mathematics* 55 (Summer 1971).

29. On related points see the papers by H. Gardner, M. W. Wartofsky, and N. Goodman in *The Monist* 58 (1974): 319-42.

9. DEWEY'S SOCIAL AND EDUCATIONAL THEORY

THE KEY TO Dewey's social and educational theory is his emphasis on wholeness. He urges an increasing awareness of the infinite context of our action, a continual growth in meaning through an expansion of intelligent activity. Thus, he views education as continuous growth. Thus, he judges social and political institutions by their capacity to enable individual persons to develop in power and awareness. Thus, he demands of schools that they present the studies in relation to one another and link available knowledge with the live context beyond the classroom.

Dewey is not a religious thinker, but, in *Human Nature and Conduct*, he offers an interpretation of the theme of wholeness in religious terms:

> Infinite relationships of man with his fellows and with nature already exist. The ideal means, as we have seen, a sense of these encompassing continuities with their infinite reach. This meaning even now attaches to present activities because they are set in a whole to which they belong and which belongs to them. Even in the midst of conflict, struggle and defeat a consciousness is possible of the enduring and comprehending whole.[1]

Such a consciousness requires symbols, but the symbols of the past no longer serve.[2] 'Religion has lost itself in cults, dogmas, and myths.'

> Religion as a sense of the whole is the most individualized of all things, the most spontaneous, undefinable and varied. For individuality signifies unique connections in the whole. Yet it has been perverted into something uniform and immutable.... Instead of marking the freedom and peace of the individual as a member of an infinite whole, it has been petrified into a slavery of thought and sentiment, an intolerant superiority on the part of the few and an intolerable burden on the part of the many.[3]

Although he is critical of actual religions, Dewey identifies the religious sense as a sense of the whole, and holds that 'every act may carry within

---

This selection is an extract from my *Four Pragmatists* (London: Routledge & Kegan Paul, 1974), 240-55.

itself a consoling and supporting consciousness of the whole to which it belongs and which in some sense belongs to it'.[4] In a vein reminiscent of James's discussion of the moral holidays afforded by belief in the Absolute, Dewey suggests that a consciousness of the whole allows an emancipation from its burdens, which is yet consistent with responsible action.

> There is a conceit fostered by perversion of religion which assimilates the universe to our personal desires; but there is also a conceit of carrying the load of the universe from which religion liberates us. Within the flickering inconsequential acts of separate selves dwells a sense of the whole which claims and dignifies them. In its presence we put off mortality and live in the universal. The life of the community in which we live and have our being is the fit symbol of this relationship. The acts in which we express our perception of the ties which bind us to others are its only rites and ceremonies.[5]

The community is, thus, a symbol of the whole, consciousness of which offers the only religious consolation Dewey acknowledges as significant. The whole is, however, infinite and so cannot be grasped as complete. We may approximate it in our experience through the conception of growth, growth without end – in awareness, sensitivity, and meaning. The community ideally fit to symbolize the whole is one that frees itself and its members to grow.

An ideal society, for Dewey, is an association that allows for maximum growth of each person, through his own activity and self-development. Such an association aims to institutionalize intelligence in matters of conduct, as natural science institutionalizes intelligence in investigations of nature. It is free of artificial barriers dividing its members from one another, it fosters the free exchange of ideas, and it treats the ideas underlying its common activities as hypotheses – open to the test of experience, criticizable by all whom such activities affect, and revisable by procedures enlisting their common consent.

This is the ideal of democracy. The machinery of democracy is not an end in itself but a means directed toward such an ideal. The justification of democracy is not to be sought in some mythical infallibility of democratic procedures. Rather it is to be sought in the *quality* of human action promoted by institutions that acknowledge each person's dignity and judgment in forms of public exchange and participation in the public life.

> The keynote of democracy as a way of life may be expressed, it seems to me, as the necessity for the participation of every mature human being in formation of the values that regulate the living of men together:

which is necessary from the standpoint of both the general social welfare and the full development of human beings as individuals.[6]

Democratic political forms are means to an end. They rest

> upon the idea that no man or limited set of men is wise enough or good enough to rule others without their consent; the positive meaning of this statement is that all those who are affected by social institutions must have a share in producing and managing them. The two facts that each one is influenced in what he does and enjoys and in what he becomes by the institutions under which he lives, and that therefore he shall have, in a democracy, a voice in shaping them, are the passive and active sides of the same fact.[7]

When social institutions exclude certain persons from the development of their own powers, it is not only they as individuals who suffer, but 'the whole social body' that is deprived of their intelligence, judgment, and contribution. And there is one thing in particular that excluded persons are 'wiser about than anybody else can be, and that is where the shoe pinches, the troubles they suffer from'.[8] Authoritarian schemes assume that the value of a person's contribution may be judged

> by some *prior* principle, if not of family and birth or race and color or possession of material wealth, then by the position and rank a person occupies in the existing social scheme. The democratic faith in equality is the faith that each individual shall have the chance and opportunity to contribute whatever he is capable of contributing and that the value of his contribution be decided by its place and function in the organized total of similar contributions, not on the basis of prior status of any kind whatever.[9]

Of all the freedoms required by the democratic outlook, freedom of *mind* is basic, for without it, individuals are not genuinely free to develop. 'Freed intelligence... is necessary to direct and to warrant freedom of action.'[10] Cultivation of intelligence under conditions of freedom is thus at once, for Dewey, the fundamental imperative of democracy and the main task of education.

Institutionally, education may, in fact, be viewed as the formal agency for fostering intelligence. Its primary aim is to develop the habits and mentality of critical thinking in application to all spheres of life. As an institution, it may itself be operated more or less intelligently. For it to be run intelligently, it should incorporate the general values of the democratic ideal, and its procedures should be developed and reviewed in a critical and scientific manner. This implies, fundamentally, that educational policy is to be stated and criticized in the public forum and that all concerned in education ought to be heard. It implies also that educational procedures are to be judged by their fruits rather than their origins; curriculum and teaching methods, school organization and grouping, grading and testing – all are open to critical review in the light of empirical consequences scientifically assessed. It emphatically does not follow from Dewey's view that the educational past is to be rejected in a wholesale manner. On the contrary, the funded wisdom of the past provides a valuable guide to present activity. But guiding ideas are not dogmas; they are tested by the very activities they help to organize. They must continue to prove themselves by their consequences. If we are alert to this fact, we will profit from their guidance and we will also learn from critical experience how to improve upon them.

The aim of education, according to Dewey, is first and foremost to develop critical methods of thought. Its task is not to indoctrinate a particular point of view, but rather to help generate those powers of assessment and criticism by which diverse points of view may themselves be responsibly judged. In pursuit of this task, the school ought to *exemplify* the application of critical method to all the domains of human life. This implies the need to present these domains with an emphasis upon their *meaning,* that is, in their relatedness to one another but, most particularly, in their bearing upon the realm of purposive activity. For the more meaning we grasp, the greater the context we can take into account and the more we are able to evaluate critically. This is the central idea of Dewey's theory of education, which he develops into a notion of proper method and curriculum.

> Study is effectual in the degree in which the pupil realizes the place of the ... truth he is dealing with in carrying to fruition activities in which he is concerned. This connection of an object and a topic with the promotion of an activity having a purpose is the first and the last word of a genuine theory of interest in education.[11]

Proper method requires that the subject be placed in a broad, and growing, context – a context that embraces the student's own purposes and potential activities as well as the urgent problems confronting the human community of which he is a part.

In the matter of curriculum, Dewey's emphasis is on continuity and meaning. Selection and specialization are, of course, necessary in modern schooling, but we must take care not to erect practical separations into hard and fast divisions among the studies. For every such division disrupts an array of real connections and so impoverishes the meaning of the subjects taught.

> The subject matter of education consists primarily of the meanings which supply content to existing social life. The continuity of social

life means that many of these meanings are contributed to present activity by past collective experience. As social life grows more complex, these factors increase in number and import. There is need of special selection, formulation, and organization in order that they may be adequately transmitted to the new generation. But this very process tends to set up subject matter as something of value just by itself, apart from its function in promoting the realization of the meanings implied in the present experience of the immature.[12]

Divisions between higher and lower studies, between theoretical and applied, between scientific and humanistic, between literary and technological studies – all are devices of convenience at best. Taken in any more serious fashion, they are all mischievous. We may need to use them, but we need also to help the student to see through them. For the full meaning of technology cannot be appreciated unless it is put into connection with its theoretical base and its human import; humanistic studies are, likewise, impoverished if they are isolated from contact with the social world and the science which is transforming it. A fundamental continuity of intimation, development, and human significance, moreover, unites studies that are elementary and advanced, basic and applied.

Integration of the curriculum is primarily a matter of recognizing that 'all studies grow out of relations in the one great common world'.[13] Studies are to be interrelated as varied avenues of access to this world and as valuable resources for the solution of common problems. In *Democracy and Education* Dewey argues that the school is not simply a mirror image of society; it represents a simplified, idealized, and balanced environment with its own long-range goals of cultivating intelligent habits of mind.[14] Yet the impact of such cultivation is ultimately social, and the process of cultivation of such habits is one that requires genuine reference to environing social conditions within which the school has its being and role. It is this long-range social role that provides integration and coherence to the varied specialized activities of schooling.

Dewey envisages a society that allows the maximum growth of each person through an expansion of his own capabilities for intelligent and effective action. Such a society, although it is based on the concept of indefinite and varied growth in its individual members, is far from indefinite and indeterminate itself. It requires institutions that foster free expression of ideas, toleration of diversity, and the participation of its members in 'formation of the values that regulate the living of men together'.[15] Moreover, such institutions themselves require supporting habits of mind, in particular, habits that are consonant with critical and scientific thought. Dewey has been criticized for taking growth as his basic value, without specifying the direction or ultimate goal of growth. But, as he might reply, the outcomes of growth cannot thus be restricted without substituting an uncritical constraint in advance for the operation of intelligence, thereby placing an unwarranted limitation upon the freedom of activity. Furthermore, as we have seen, the ideals of intelligence, growth, and freedom, open-ended as they are, are not amorphous or directionless; indeed, they make the most stringent of demands upon those who would embody them in human institutions and strive to rear their young by their light.

The schooling Dewey advocates is analogous in its rationale. It is designed to cultivate critical habits of mind. This aim, according to Dewey's general analysis of thought, requires the involvement of the pupil's own purposes in the learning process and the relating of the school's studies to problems of the environing society. But schooling, in Dewey's scheme, is not therefore amorphous or undemanding. It is organized around problems and it is directed toward internalizing in the pupil the discipline of critical and responsible thinking.

Dewey's educational views have been subjected to contrary criticisms. On the one hand, he has been charged with counselling extreme permissiveness – a kind of anarchy in the classroom. On the other hand, he has been accused of exalting the social role of the school to the detriment of individuality – urging a society of conformists. He has been held to be radically disruptive of the ordered ways and traditions of the past and also to be a conservative, stamping in an inflexible belief in the values of the surrounding society.

These opposed forms of criticism cannot be simultaneously held, for they cancel one another. Moreover, each overshoots the mark, exaggerating some element of Dewey's thought to the point of distortion and ignoring the fundamental allegiance to critical, experimental thought that lends balance to his educational doctrines.

Dewey wants to enlist the student's purposes in learning, so that the relation of the various studies to his own choices may become evident. Such a procedure requires a problem-organization of materials; it is calculated to enhance the meaning of the studies while increasing the student's sense of effectiveness as a purposive and intelligent agent. It is a crude mistake to take Dewey as advocating activity for its own sake. The whole point of activity in his scheme is that it should, in so far as possible, be made educative through the guiding power of ideas and the critical assessment of consequences. This undertaking may be very difficult to execute, or otherwise inadequate; it surely cannot be characterized as simply permissive. Not only does it retain the materials of adult studies under a new educational organization, but it imposes the structure inherent in definite purposes, requiring special instrumentalities for their realization. Dewey is not opposed to discipline; he wants the school, so far as possible, to strive for an internalization of intellectual discipline rather than construing discipline as primarily a matter of enforcement from without. As with scientific research or the practice of the arts, crafts and professions, discipline ought, he believes, to grow from the firsthand struggle to solve specific problems, encountering the resistance of available resources, experimenting with alternative ideas, putting them to the hard test of experience, evaluating them in the light of guiding purposes. To acquire the discipline born of struggle with problems is to incorporate habits of critical method; such incorporation is not a *laissez faire* or easy task – it both enlists and helps to foster dedicated effort, care, responsibility, and self-control. It is this conception of discipline as growing out of purposive problem-solving activity that ought, in Dewey's view, to predominate in education.

Nor is Dewey eager to foster a society of conformists, as should be evident from even a cursory reading of his work. Certainly, he wants the work of the school to be related to basic problems of the environing society. In this way, he believes, the meaning of school work may be enlarged and the effectiveness of knowledge made increasingly apparent; at the same time, the urgent difficulties of the common life may be illuminated and the moral habit of coping with them reliably may be encouraged in the young. To take social problems as a significant focus for schooling is, however, by no means to advocate inculcation of a social dogma. Dewey explicitly warns against social indoctrination; the fundamental allegiance of the school ought to be an allegiance to critical methods of analysis. Such methods are inimical to conformity: they demand responsible and independent judgment of social issues by canons of scientific reasoning and assessment of data.

Dewey emphasizes the social climate of the school, to be sure. His point, however, is not to foster conformity, but to alert teachers to the learning that goes on outside the formal lessons of the classroom. In so far as critical habits of mind and character are related to the human arrangements within which academic lessons are set, teachers need to be aware of the influence of such arrangements and to take responsibility for the moral and intellectual habits they foster. These habits ought, ideally, to be consonant with scientific and responsible thought, they ought to foster the independence of mind, the respect for others, and the capability of adjudicating differences by orderly methods that are characteristic both of democratic and of scientific attitudes. Allegiance to method rather than conformity to creed is the keynote.

Dewey's educational vision is designed neither to uphold the past, as such, nor to disrupt it, as such. Continuity with the past is, in any case, inevitable, as are departures from it, in one or another respect. The point is to strive to develop habits of intelligence that may be applied to life's problems. The deliverances of intelligent analysis may, in certain cases, counsel revision or rejection of inherited ways; in other cases, they may counsel retention. The widespread development of intelligent and responsible habits of mind would, however, in itself, constitute a large change in society, capable of bringing critical evaluation to bear on the problems and practices of our common life. It is this long-range development of intelligent habits of mind which is the school's role, in Dewey's view. To succeed in this development, the school must take society's problems seriously, but this by no means requires it to stamp prevalent social values into the minds of its children. The school must also strive to deepen reflection, strengthen independence, and develop critical skills in application to social issues, to a degree that has not yet been achieved. But this is by no means tantamount to counselling a wholesale disruption of the past. It is, rather, to conceive the school's task as enabling society to cope with its problems more intelligently, more effectively, more imaginatively, and more responsibly than it has so far done.

I should, myself, however, wish to offer certain criticisms of Dewey's views, despite my agreement with much of what he has to say on educational matters. Since I have elsewhere elaborated some of these criticisms,[16] I will here note them only briefly, and try to connect them with earlier critical points related to pragmatic themes.

First, I suggest that the notion of continuity is exaggerated in Dewey's treatment. Continuities are certainly important in education, and Dewey's emphasis on bringing together the humanistic and the technical, the elementary and the advanced, the disciplinary and the problematic elements of the educational process is a salutary one. He is, I believe, certainly right in attacking the idea that the studies are independent and external entities, self-enclosed and somehow rooted in nature. He is right in advocating that technical matters be illuminated by theoretical knowledge and seen from the perspective of their human significance. He is right, too, in demanding that an appreciation of values be supplemented by a realistic understanding of their natural conditions and vicissitudes. Nevertheless, discontinuities and distinctions are as natural as continuities, and they need also to be acknowledged where they exist.

Theory is, for example, surely connected with observation and with practice, but it is also autonomous; it has its own career and life. The general tendency of pragmatism, as we have seen, is to interpret theory as intermediary between practical problem and practical resolution, and to construe its content wholly in terms of observable transformations of the world through practical effort. As indicated in earlier discussions of both Peirce and Dewey, this characterization of theory does not seem to me tenable. Neither the content nor the function of theory can be fully understood by reference to the resolution of practical problems through transformations of the world. One must, to appreciate the force of a theory, grasp more than just its practical ramifications; theories serve not simply to guide practice, but to afford us an intelligible and coherent representation of fundamental natural processes.

The process of theorizing is a creative process. It is not just a matter of cataloguing the functional relations among phenomenal changes, nor is it, in any plausible sense, generated out of experience. The theorist is free to invent, simplify, postulate, categorize, extrapolate, idealize – he may need to back away from the detail of phenomenal change and practical urgency in order to strive to 'see through' to underlying elements and patterns. Distance, in other words, is functional for the theorist, who strives for ever deeper insights and broader perspectives on nature. The value of theoretical distance must be acknowledged in education, and distinguished from mere remoteness and pedantry. In opposing the latter, we must avoid destruction of the former. Education ought, indeed, to encourage the theoretical motive, which, whether or not it promises to relate to practical solutions of social problems, aims to achieve a penetrating vision of natural structures.

Second, the problem-theory of thinking, as developed by Peirce and Dewey, seems to me inadequate. Problems cannot be identified with difficulties of practice; the problems that organize scientific research arise in a context of prior theory and experimentation. Moreover, the life of science is not exhausted in resolving problems that arise without effort; the scientist's thought does not subside when his questions have been answered. Problem-finding is as important as problem-solving, and scientific thought of the greatest significance is expended in seeking, formulating and elaborating questions that have not yet intruded on practice.

We have noted the difficulties in Peirce's conception of real doubt as the origin of inquiry, and his own recognition of the importance of feigned, hypothetical, and speculative questions. We have also seen the shortcomings of the problem-theory of reflection as support for Dewey's notion of continuous reconstruction and critical testing of ideas in action. If we reflect upon the import of these criticisms for education, we should need to acknowledge both the possibility and the importance of encouraging the pupil to seek problems, of helping him to a wider perception and a richer sensitivity, a more insistent curiosity and a more active imagination. We should continue to value problem-solving as a method of educational organization, but we should place it within the context of a growing awareness that reaches out steadily to problems unperceived before.

The problem-theory of thinking seems to me also difficult in attempting to give a uniform analysis of thought in all realms. I have argued that it does not even give an adequate picture of scientific thought, the preferred domain of pragmatic interpretations. Much less does it provide an analysis that can smoothly contain not only the scientific imagination, but the work of the artist, the historian, the poet, the translator, the inventor, the novelist, the mathematician. Much of our thinking is problem-oriented, but much is not, growing out of free speculation, playfulness, curiosity, or out of the need to express, describe, or create. One can attempt to force all these varieties into a common abstract framework, but the advantages of doing so would need to be so evident as to override the cost in artificiality and generality. I do not think the problem-theory is adequate in this respect; even if it were, we should need to provide, in education, a concrete and realistic awareness of the special features differentiating science from history, art from mathematics, poetry from legal reasoning, philology from philosophy. We should, in other words, need to transmit the several traditions of thought as we now possess them, rather than simply filtering them through an abstract philosophical schema of thinking as problem-solving.

Third, while I applaud Dewey's emphasis on meaning in his account of education, I suggest that it is of variable relevance in the educational process. That is to say, the meaning of subjects is indeed enhanced through their mutual connectedness as well as their incorporation into the student's context of purpose and potential action. Moreover, the teacher himself ought to have as clear a conception as possible of the aims, values, and criteria that animate his educational choices of curriculum as well as methods and organization. Nevertheless, learning does not need to proceed, at every point, through linkages to prior purpose; the pupil does not, I suggest, need to learn everything in the context of its meanings and uses. Children learn many things as they do language, or games – through participation, curiosity, or the sheer joy of the activity. The teacher ought to be clear as to the meaning and value of introducing a particular subject or educational activity and enlisting participation as if it were a game. But the child may well learn it, efficiently and joyfully, as a game. It is, in principle, an empirical question whether mathematics is better taught through elaborate reference to the meaning of fundamental concepts and operations than through intuitive and gamelike methods. This question ought, at any rate, not be begged by a philosophical predilection for meaning in education.

What is certainly of basic importance is that, no matter how a child has been introduced to a subject, his capacity for meaningful action, intelligent criticism, and growth in understanding should not be destroyed or stunted through his education. Whatever hospitals do, as Florence Nightingale is reputed to have said, they at least ought not to spread disease. No matter what schools do, they should not cripple educational growth. The child's questions, his attempts to understand, should always be respected and sincerely met. His efforts to criticize, to relate, to utilize, and to elaborate ought always to be strengthened and encouraged, so that he becomes ever more aware of his intellectual agency, its powers and responsibilities. This does not imply, however, that he needs to have all things explained to him before questions arise, that he is incapable of stepping out of the familiar circle of his projects and purposes to explore the world beyond without prior assurances of meaning.

Finally, I believe that Dewey's view of the school underestimates its autonomy, for it emphasizes as the primary role of the school its long-range transformation of society through its ultimate impact on problems of the larger culture. Dewey's view of the school is by no means a simple-minded one; he does not see the school as a mere reflection of society, nor does he suppose it to be an instrument for accomplishing social purposes set in advance. The social end that is served by the school is represented, for him, not by society as it happens to be but by a reformed society, illuminated by critical intelligence. Nevertheless, he emphasizes the intermediary role of the school, as an agency capable of transforming problematic conditions, through the cultivation of intelligence, into more harmonious and satisfactory arrangements. To this end, he stresses the continuity between school and society, placing social problems at the centre of the school's focus.

As is already apparent from my earlier remarks on the autonomy of theory, I believe the work of the school is not adequately represented by Dewey's account. In fostering theory, the school ought, in a basic sense, to stand apart from life, not by propagating pedantry and myth, but by encouraging the theoretical illumination of a world that is wider than the school and wider even than the society in which the school is placed. The school requires sufficient distance from society to enable it to develop intellectual concerns and cultural standards that have their own worth, quite apart from the resolution of social problems, and that may, moreover, place those very problems in a new perspective. The school may be viewed as an intermediary agency helping to improve society in the long run, but society may equally be viewed as an intermediary agency to be judged by its dedication to the autonomous values of intelligence, criticism, knowledge, and art, of which the school is the guardian. The school, in my view, ought to see itself *not simply* as instrumental to an improved society, although it ought to see itself in that way, surely. Its job is not only to serve but also to enlighten, create, understand, and illuminate, efforts which have intrinsic value and dignity, efforts which are themselves to be served by the society of men.

NOTES

1. *Human Nature and Conduct* (New York: Henry Holt, 1922, 1930), 330.

2. Ibid.

3. Ibid., 331.

4. Ibid.

5. Ibid., 331-32.

6. John Dewey, 'Democracy and Educational Administration' (an address to the National Education Association, 1937), *School and Society*, 45 (3 April 1937), 457-62. Cited passage reprinted in Joseph Ratner, ed., *Intelligence in the Modern World: John Dewey's Philosophy* (New York: Modern Library, 1939), 400.

7. Ibid., 401.

8. Ibid., 402.

9. Ibid., 403-404.

10. Ibid., 404.

11. John Dewey, *Democracy and Education* (New York: Macmillan (original date 1916), 1961), 135.

12. Ibid., 192.

13. John Dewey, *The School and Society* (University of Chicago Press, 1899), 103.

14. See Dewey, *Democracy and Education*, 20-22.

15. Dewey, in Joseph Ratner, ed., *Intelligence in the Modern World: John Dewey's Philosophy* (New York: Modern Library, 1939), 400.

16. I. Scheffler, 'Educational Liberalism and Dewey's Philosophy', *Harvard Educational Review* 26 (1956): 190-98; and Scheffler, 'Reflections on Educational Relevance,' *Journal of Philosophy* 66 (1969): 764-73. Both papers are reprinted in my *Reason and Teaching* (London: Routledge & Kegan Paul, 1973).

10. PRAGMATISM AS A PHILOSOPHY

PRAGMATISM IS popularly taken simply as an attitude or a style, an emphasis on the practical or the social to the detriment of theoretical reflection and individual values. There are natural causes for such a construal, aside from the mere connotation of the word itself in everyday use. For pragmatic thinkers do in fact lay great stress upon practice, emphasizing the role of action in human thought, from the humblest bit of learning by a child exploring its room, to the most refined learning of the scientist manipulating the environment experimentally in order to explore the universe. Pragmatists, moreover, stress the social import of thinking – the structure of science as a community of investigators, the influence of historical contexts on the course of philosophy, and the relevance of philosophical inquiry to the problems of men.

Nevertheless, the popular conception of pragmatism is extremely misleading, untrue to the movement as a whole and to the works of its individual thinkers. For, while it accurately reflects pragmatic emphases on action and society, it neglects to represent such emphases as arising out of philosophical inquiry, failing utterly to register the source of such inquiry in a struggle with abstract questions and issues. Seeing pragmatism simply as an emphasis, it ignores pragmatic efforts to formulate philosophical accounts of meaning and thought, truth and knowledge, reality and value.

No investigation of such efforts can avoid the conclusion that pragmatism is a serious philosophy, with deep roots in the philosophical tradition. With the exception of William James, whose thought derived from British empiricism, the Cambridge pragmatists were indeed, as Murphey has called them, Kant's children,[1] and the movement as a whole may be largely interpreted as an extension of Kantian themes into the scientific and social worlds of the nineteenth century. Charles Sanders Peirce, who took the very name "pragmatism" from Kant, was steeped in the history of philosophy and developed his thinking through interaction with the great philosophical masters of the past as well as the science and mathematics of his own day. Rejecting Cartesianism, he formulated challenging new conceptions of meaning and reality through reflection on the methods of the natural sciences,

---

This paper appeared in *Zeichen und Realität* (Strauffenburg Verlag, 1984), ed. Klaus Oehler.

and strove all his life to develop a complete system of philosophy, in the spirit of the Kantian architectonic.

William James, sensitive psychologist as well as bold philosophical thinker, sought not only to advance the empirical study of mental phenomena but also to develop an adequate metaphysical interpretation of such phenomena and of the main features of human life as known to us by whatever source. He sought a view that would, in particular, do justice to individual freedom, to the reality of human choice, and to the stubbornness and disjointedness of particular facts as against what he called the "block universe" of Idealism. George Herbert Mead, philosopher and social psychologist, worked out a developmental approach to the mind-body problem, in which the central role was played by symbolism. Starting from Wilhelm Wundt's theory of gesture as communication of meaning, Mead elaborated a radical view of mind and self as well as language, as emerging out of society, rather than the reverse. His study of symbolism yielded rich and suggestive interpretations of communication, conscience, and human community. John Dewey, finally, who began his thinking career as a Hegelian, retained forever after characteristic marks of his Hegelian beginnings: a pronounced developmentalism, a respect for the force of ideas and, above all, an urge to unify opposites – to achieve a vision of the inclusive whole within which particular doctrines in conflict may be seen to represent but partial accounts of reality.

These pragmatist thinkers were intellectually and spiritually very different. Peirce's inspiration was the logic of science; his vision was of the ideal and the general: the ideal community of investigators, the general purport of ideas, the long run approximation to reality through the self-corrective method of science. James' inspiration was rather the individual life of the individual creature, the predicaments of personal choice, the open options defining the particular act, the flow of time and mind, the religious perspective of the single human agent. Mead's vision was, further, social and evolutionary, his effort to understand the distinctive features of human community, as made possible through the growth of symbolic function. And Dewey's aim was in the broadest sense practical and moral, to reconstruct human arrangements through fostering the habit of intelligence: making reflection practical and practice reflective by relating both thought and action to their anticipated meanings in experience.

Though different in their several primary concerns and emphases, these pragmatic thinkers nevertheless joined in producing a distinctive philosophical orientation, rooted in the intellectual past but responsive to the intellectual challenges of the present. They sought indeed to bring philosophical conceptions up-to-date through analysis of the new science and its methods, while placing science itself within a philosophical framework featuring a new approach to meaning. And they sought further to apply their philosophical conceptions in understanding the unprecedented circumstances and the challenges facing human society in the modern world.

The nineteenth century world in which pragmatism developed was a world in which important oppositions were at work: science versus religion, positivism versus romanticism, intuition versus sense experience. The characteristic posture of pragmatism in response to such oppositions was that of a mediating philosophy, attempting to bridge science and religion, theory and practice, fact and value, speculative thought and analysis, tender-minded and tough-minded temperaments (as James put it), and (with Dewey) school and life. This mediating posture differentiates pragmatism from other philosophical tendencies inspired by science. In particular, pragmatism contrasts with positivism in refusing to assimilate intellectual interests generally to some simplified model of positive science. Responsive in particular to evolutionary thought and the new statistical modes of inference, pragmatism was indeed led to revise inherited conceptions of science itself. And, rather than using science as a device for dismissing or downgrading other modes of experience, such as art, history, morality, religion, philosophy, and social practice, pragmatism has taken science primarily as exemplifying general concepts of critical thought, in terms of which important continuities among all the modes might be revealed, and in light of which they might all be refined and brought to bear intelligently upon human problems.

PROBLEMS OF PRAGMATISM AND PRAGMATIC RESPONSES

Pragmatism developed in a period of enormous social and intellectual change – a time that Max Fisch has named the "classical period" in America, from the end of the Civil War to the eve of the Second World War.[2] Dewey's life spanned this period, and Gail Kennedy has described its course as follows:

> He was born on the eve of the great war that was to ensure the triumph in America of industrialism and economic enterprise, in the year that Darwin published his "Origin of Species", the book which marked the coming of age of modern science. He grew up in the environment of the older America, in the Vermont town of Burlington. Here life was still largely unaffected by the newer science and by modern industrialism. From this small community with its simple and intimate round of handicraft and agricultural occupations, the form of society that Jefferson knew, he was to go out into the complex world

created by modern science and mass-production industries, to the first American university, the newly founded Johns Hopkins, to the fermenting democracy of the Middle West, in his years of teaching at the Universities of Michigan and Minnesota, then to the great industrial and commercial cities of Chicago and New York. Dewey has said in an autobiographical essay that the forces which influenced him came 'from persons and from situations' rather than from books. It was the transition from the America of his boyhood to the new America of his maturity that created the basic problems and formed the central theme of his philosophy.[3]

This period of transition brought with it major intellectual, and not merely social, changes. Not only was there a challenge in science to traditional religion and morality; there was also a challenge to inherited conceptions of science and classical views of knowledge. The most important influence was that of evolution, promoting the biologizing of man's intelligence and the continuity between mankind's capacities and those of the lower animals. The rise of experimental and historical sciences of man, as well, reinforced evolutionary ideas of process and continuity, and also brought out the adaptive variability in human custom. The new prominence of probabilistic and statistical concepts both in physics and biology required a revision of older conceptions of logic and science. Finally, while great social changes were complicating life, making liberty and choice more precarious, the new human sciences painted a more flexible picture to replace older notions of cultural fixity; the idea of a *social* science indeed held out the prospect of bringing tradition itself under a measure of control. Knowledge, it seemed, had now to be reconceived: arising out of a biological matrix, continuous with adaptive action addressed to environmental problems, it yielded provisional solutions rather than necessary truths, promising increased social control but therefore imposing increased moral responsibility.[4]

The changes here noted were taken by pragmatism to pose the following broad philosophical problems: First, how is our contemporary theory of knowledge to assimilate the new scientific understanding of change, of process, of biological and social factors, of probable reasoning? Classical rationalism and classical empiricism seemed both inadequate to the task, the one taking knowing as the work of the individual mind drawing up eternal truths from within, the other as the mind's passively registering ideas stamped on it from without.

Second, and more generally, how are we to articulate the new emphasis on *continuity,* connecting man's life with the world of nature, relating his knowledge and his values, his cognitions with his feelings and actions, his life as an intellect with his career as an organism in a particular biological and evolutionary setting? The basic problem, for pragmatism, was to find a way of overcoming inherited dualisms of knower and known, mind and body, fact and value, theory and practice, ends and means.

Third, how are we to find new sources of stability in the face of radical changes in scientific belief and – more urgently still – how to find sources of stability consonant with the experimental habit of mind underlying science? Rejecting what Dewey called "the quest for certainty" and adopting instead Peirce's attitude of "fallibilism", how could sufficient stability yet be found to sustain the arts of inquiry, of education, of culture, and the public life?

Fourth, how are we to conceive the prospects of individual selfhood and democratic community under the new conditions of industrial society? How can policy formation be institutionalized in today's circumstances so as to be responsive to those whom policy affects? How can technological advance be reconciled with humane purpose, with the values of the arts and of associated life, and with the primacy of critical intelligence as the chief ideal of education?

The pragmatists' response to these problems cannot be recounted here in any detail, but its main features may be related, in outline, to a single starting point, namely, the rejection of Cartesian philosophy. Rejecting the mind-body dualism of Descartes, pragmatism is, first of all, led to develop a *functional view of thought,* relating cognition to the purposive life of the organism, responding to problems set by its environment. Secondly, giving up Cartesian certainty, pragmatism proposes instead a *fallibilistic view of knowledge* as a provisional scheme of hypotheses, resting upon probable reasoning and pointed toward the future, remaining ever subject to the test of further experience. Thirdly, surrendering Cartesian individualism, pragmatists offer in its place a *social conception of science* as the effort, not of single inquirers, but of an open-ended community of investigators to learn from experience in a systematic way. Finally, giving up Cartesian intuition, pragmatists alternatively propose the *representative character of thinking,* holding thought to be always and throughout symbolic, channeled through networks of interdependent sign processes, thus incapable of ever yielding either fixity or certainty. Symbolism enables thought to frame ends-in-view independent of actual outcomes, and so to anticipate and regulate conduct. Thus, writes Dewey, "The invention or discovery of symbols is doubtless by far the single greatest event in the history of man". Signs or symbols themselves are not *images* or *pictures* of reality; they are rather to be interpreted as devices of the purposeful life. "What now is a *conception?*" asks William James, and he answers, "It is a *teleological instrument.* It is a partial aspect of a thing which *for our purpose* we regard as its essential aspect, as the representative of the entire thing." "Wherever intelligence operates", writes Dewey, "things are judged in their capacity of signs of other things. If scientific knowledge enables us to estimate more accurately the worth of things as signs, we can afford to exchange a loss of theoretical certitude for a gain in practical judgment. For if we can judge events as indications of other events, we can prepare in all cases for the coming of what is anticipated", and take part as knowers in the purposive "direction of change".[5]

The consequences of these anti-Cartesian positions are far-reaching. Once certainty and individualism are surrendered as epistemic ideals and science reinterpreted as the continuous learning effort of an ideal community, stability is to be sought in the *intellectual method* defining such community. The conclusions of particular inquiries are indeed all provisional; they are probable at best and subject to revision by further investigation. But despite recurrent revisions of scientific doctrine, the community dedicated to systematic learning from experience is itself a continuous entity, unified by its allegiance to critical method. It is such allegiance that gives us ground to stand on even as we alter particular items of belief. The very self-correctiveness of science which forces the revision of its theories when they fall out of accord with the evidence constitutes a steady ideal standing firm throughout change. And while we cannot hope to be sure of any of our particular theories, we can be sure that the method of science will yield increasingly adequate theories through continued inquiry by the investigative community of mankind.

This investigative community offers, finally, a suggestive model for conceiving democratic society generally. As science institutionalizes procedures for investigating hypotheses about nature, so democratic society institutionalizes procedures for the critical testing of social ideas, plans, and policies – all to be conceived as hypothetical. Provisional agreements on particulars, whether in scientific research or in social action, are necessary and, indeed, sufficient to organize further collaborative efforts but all particular ideas remain subject to the continuing test of experience, to be revised when necessary, in accord with the underlying unity of method.

In science, the open communication of ideas and their availability for testing by rival theorists is an essential point of method. In democratic society, too, an essential need is to ensure free communication among persons, so that their special perspectives may be appreciated by others and made available, moreover, for the general testing of social arrangements. Dewey's notion of *shared experience* does not refer to the having of the same experiences by all, but rather to the communication of diverse experiences by means of shared symbolic structures. The social problem is to develop and sustain such structures and, moreover, to facilitate their proper use. This requires breaking down artificial barriers to sympathetic communication and educating individuals for those skills and traits of character peculiarly consonant with democratic institutions. Of all the freedoms required by such institutions, freedom of mind is basic for, without it, individuals are not genuinely free to develop. "Freed intelligence", says Dewey, "is necessary to direct and to warrant freedom of action."[6] It is the cultivation of free, sympathetic, and critical intelligence that constitutes at once the fundamental imperative of democracy and the main task of its education.

The foregoing sketch of pragmatic ideas provides at best a general and composite picture rather than an individual portrait of any single pragmatist. I shall therefore devote the remainder of my remarks to the pioneering work of Charles Sanders Peirce, the founder of the movement, focussing in particular, upon his rejection of Descartes and the development of his influential alternative theory of belief, doubt, and inquiry.

PEIRCE'S THEORY OF BELIEF, DOUBT, AND INQUIRY

In place of Descartes' emphasis on *radical doubt* and Locke's emphasis on *sensations,* Peirce emphasizes *belief,* indeed placing the notion of belief at the center of his theory of inquiry. Thought, or inquiry, arises always in a context of belief and it is precipitated by doubt. Doubt is, however, not the mere lack of belief; it is an active state of irritation, focussed and specific rather than wholesale and diffuse like the Cartesian variety. Provoked by such irritation, inquiry arises, inquiry being the active process of passing from doubt to belief. Unlike doubt, belief is itself a calm, settled state of readiness, in the nature of a habit; it is not an episode or occurrence but more like a disposition or set. In Alexander Bain's words, it is that upon which we are prepared to act. Orienting us thus to future experience, belief is always, in consequence, open to upset by experience. Certainty of belief is therefore precluded. Indeed since belief is expressible only in signs, with implicit reference to other signs, it is always mediated and never direct, hence in principle incapable of certainty. "From the proposition that every thought is a sign," says Peirce, "it follows that every thought must address itself to some other, must determine some other, since that is the essence of a sign."[7] All thoughts are thus in the same boat, all fallible, all interdependent. In place of Cartesian certainty, Peirce in fact espouses rather what he calls *fallibilism.*

The underpinnings of these ideas are developed in two papers of 1868, both devoted to criticism of Descartes. In the first of these, "Questions Concerning Certain Faculties Claimed for Man", Peirce criticizes the doctrine of intuition or immediate knowledge. In the second, "Some Consequences of Four Incapacities", he concentrates on issues of logic and methodology. "We cannot", he says in the latter paper, 'begin with complete doubt," since doubt requires a positive reason. Even in philosophy, this principle ought to hold. "Let us not pretend to doubt in philosophy what we do not doubt in our hearts."[8] Radical Cartesian doubt, since it is in fact impossible, must be empty and self-deceptive. Its impotence is revealed by the fact that the Cartesian method that begins with radical doubt ends by recovering all the beliefs with which the doubter began. Real doubt, on the other hand, as illustrated by a typical research question in the sciences, is focussed and motivated, framed by a variety of assumptions meanwhile taken for granted. Such doubt is not impotent; it has, indeed, the power to stimulate inquiry that may in the end alter the beliefs of the doubter. Each provisional assumption may, furthermore, be doubted in its turn, but at no time are *all* assumptions thrown into doubt at once.

Peirce considered Bain's above-mentioned definition of belief as "that upon which a man is prepared to act" as the basis of his pragmatism.[9] Bain's view, as Murphey suggests, supplied the "psychological foundation for Peirce's denial of Cartesian doubt, for Bain holds that men are naturally believers and that doubt is produced only by events which disrupt our beliefs – not by pretense."[10] Rather than supposing that the "natural" state is utter lack of belief, i.e. *radical doubt,* so that every belief we have requires justification from scratch, Bain offers Peirce a theory that reverses the order of naturalness, as the modern concept of inertia reversed the natural state from rest to motion. The natural psychological state is now held to be that of belief, with no possibility of wholesale and radical justification. Rather, doubt arising in the body of our beliefs now wants a positive reason, and it finds resolution in the recapture of belief. Using Bain's idea, Peirce can, as Murphey points out,

> fit his whole theory of inquiry into an evolutionary frame of reference. Beliefs may be regarded as adjustive habits while failure of adjustment leads to doubt [...] this biological perspective [...] provides him with a new definition of the nature of a problem – a definition subsequently developed by Dewey. A problem situation exists whenever we find our established habits of conduct inadequate to attain a desired end, [...] and the effect of a problem situation upon us is the production of doubt. This being the case, Cartesian doubt is nonsense, for there is no problem situation. But secondly, the theory provides a clarification of the nature of an answer. An answer is any rule of action which enables us to attain our desired ends. Accordingly, our objective is to find a rule which will always lead us to that which we desire. So in the investigation of a real object, our objective is a knowledge of how to act respecting that object so as to attain our desired ends. Thus, as

pragmatism asserts, the concept of the object can mean nothing to us but all the habits it involves. The attainment of a stable belief – belief that will stand in the long run – is thus the goal of inquiry. Such belief we define as true, and its objects as reality."[11]

In his 1877 paper "The Fixation of Belief", Peirce presents a full statement of his theory. Doubt differs from belief in three respects, he says. First, "there is a dissimilarity between the sensation of doubting and that of believing;" secondly, "the feeling of believing is a more or less sure indication of there being established in our nature some habit which will determine our actions. Doubt never has such an effect;" and thirdly, "doubt is an uneasy and dissatisfied state from which we struggle to free ourselves and pass into the state of belief, while the latter is a calm and satisfactory state which we do not wish to avoid, or to change to a belief in anything else. On the contrary, we cling tenaciously, not merely to believing, but to believing just what we do believe."[12]

Inquiry, now, is the struggle to overcome doubt and attain belief, and it has in doubt its "only immediate motive". It begins with doubt and ends only with the cessation of doubt. Therefore, says Peirce, "The sole object of inquiry is the settlement of opinion." When opinion is settled and "real and living doubt" overcome, genuine inquiry cannot arise. When no *actual* doubt affects any given proposition, it does not matter that it might *possibly* be thrown into doubt by hypothetical considerations.

PEIRCE'S COMPARISON OF METHODS

If the function of inquiry is indeed the settlement of opinion, or the fixation of belief, the question may be raised as to the relative effectiveness of alternative methods by which this function may be carried out. Considering this question, Peirce proceeds to a comparison of four such methods: the method of tenacity, the method of authority, the *a priori* method, and the method of science. Simple tenacity, i.e. a reiteration of the belief, "dwelling on all which may conduce to that belief, and learning to turn with contempt and hatred from anything which might disturb it" is a method "really pursued by many men" and offering "great peace of mind" despite some inconveniences. But it is ineffective for, as Peirce says, "the social impulse is against it". Finding oneself confronted with the differing opinions of others, one's confidence in one's own tenaciously held beliefs is shaken. Nor can we shield ourselves from contacts with others unless we become hermits. Tenacity thus leaves us vulnerable to continual unsettlement of our beliefs.

The method of authority, transferring tenacity to the group, utilizes social or political institutions to inculcate preferred doctrines and to stamp out contrary views. A whole array of repressive measures is available, e.g. censorship, indoctrination, terror, with occasional massacres as needed for, as Peirce remarks, these have proved "very effective means of settling opinion in a country". The method of authority is capable of atrocity for, says Peirce, "the officer of a society does not feel justified in surrendering the interests of that society for the sake of mercy, as he might his own private interests". As to effectiveness, it is superior to tenacity, shielding the individual, by and large, from encounters with differing opinions. But it has its own sources of inefficiency nevertheless: Social regulation cannot extend to all opinions whatever, and unregulated opinion always poses a potential threat to settled belief. Individuals may reflect that other societies and other ages have held quite different beliefs, and conclude that it is mere historical accident that has led them to the official doctrines they have. Such doubts must, says Peirce, affect "every belief which seems to be determined by the caprice either of themselves or of those who originated the popular opinions".

The *a priori* method rejects tenacity as well as the effort to force one's beliefs on others. Rather, it follows the natural preferences of "men conversing together and regarding matters in different lights". The chief example of the operation of this method is to be found in "the history of metaphysical philosophy", where beliefs have been formed not in the effort to account for observed facts but rather in the effort to formulate what seemed "agreeable to reason". Enjoying greater intellectual respectability than either of the others already considered, this method nevertheless fails equally. It is ineffective since it assimilates inquiry to the development of taste, always a matter of fashion, and thus never culminates in agreement but remains always subject to pendulum swings over time. When we reflect on the diversity of fashion, we recognize our own beliefs to have been formed by such "accidental causes", and new doubts arise again to unsettle these beliefs.

The method of science, finally, is one which purports to form beliefs by reference to external permanencies rather than human causes. It supposes real things with properties "entirely independent of our opinions about them". It is true, says Peirce, that the supposition of realities cannot be proved by science, since it *underlies* science, but practice of the scientific method never leads us to doubt this supposition, whereas practice of the other methods does lead us to doubt them. To question the existence of real things *in general* is idle: "If there be anybody with a living doubt upon the subject," says Peirce, "let him consider it." But the fundamental contrast between the method of science and all the others is that it is the only one that presents "any distinction of a right and a wrong way". The method, that is to say, is self-corrective, acknowledging the possibility of errors in application corrigible by further use of the method itself. By contrast, the result of applying any of the other methods is necessarily correct according to the method in question, so that no errors can be admitted, much less corrected, by the method itself. While the other methods have their virtues, a man should reflect, says Peirce, that "after all, he wishes his opinions to coincide with the fact, and that there is no reason why the results of those first three methods should do so. To bring about this effect is the prerogative of the method of science."

DIFFICULTIES IN PEIRCE'S TREATMENT

The doubt-belief theory of inquiry is central to Peirce's general conceptions of mind, meaning, truth, and reality. It appealed to him in the first instance as providing a psychological foundation for his epistemological critique of Descartes. Read purely as psychology, it seems to me, however, obscure on several points, which I have elaborated more fully in my book "Four Pragmatists". For example, is doubt always conscious or may it simply be inferred from its characteristic disruptive effects on conduct? Though described as an occurrent state of irritation, it might still theoretically be construed as not implying consciousness in every case; yet Peirce affirms a "sensation of doubting". Does he then call the characteristic disruptive effects minus the characteristic sensation "doubt" or not? The answer is unclear.

To take another example, how, if belief is a habit or set, can Peirce speak of a "sensation of believing" and describe it (in the companion essay "How to Make Our Ideas Clear") as "something we are aware of'? More importantly, how can he say of belief (loc. cit.) that it "appeases the irritation of doubt", having earlier argued that belief is prior to doubt, constituting indeed the natural state of the mind before inquiry. Finally, what does Peirce intend in saying (loc. cit.) that belief involves the establishment of a rule of action or a habit? Since not all habits are associated with beliefs, does he have any way of specifying which subclass of habits is peculiarly belief-related? Does he, similarly, have any way of indicating which disruptions of conduct are constitutive of doubt? To me, at any rate, these unanswered questions indicate basic obscurities in the theory, under a psychological interpretation.

However, the theory may be given an *epistemological* rather than a *psychological* interpretation; taken thus, it is not necessarily vulnerable to the difficulties just outlined, and it requires a fresh evaluation. Interpreted epistemologically, the theory purports not to *describe* but rather to *prescribe* the course of thought, construed as a critical or scientific effort. Properly, the theory declares, such thought always addresses specific questions arising from real doubt, proceeding in every case by taking a variety of assumptions for granted throughout the inquiry, and subject to evaluation by seeing how well it turns out to resolve the questions from which it arose. The theory thus rejects the idea that there can be scientific investigations without assumptions altogether, and it equally rejects the idea that assumptions actually adopted must be absolutely indubitable.

Although this epistemological version of the theory does indeed escape the problems of the psychological reading, Peirce's insistence on "real and living doubt" as the proper origin of inquiry still poses a difficulty. For there is, in fact, much thinking of a significant kind that does not originate in doubt. Imagination, recollection, perception, translation, composition – all seem to provide counter-instances. In reply, it will be said that Peirce is concerned, not with thinking in general, but with *inquiry* specifically, in particular as exemplified in scientific research. Is it then the case that all such research originates, or should originate, in real and living doubt? Does there really need to be an active irritation, a breakdown in earlier habits before scientific research can be initiated? Does not theoretical curiosity have a role in the stimulation of inquiry? Peirce insists, to the contrary, that genuine thought arises from real and active irritation rather than from theoretical or speculative motives.

Nevertheless, he displays increasing ambivalence on this central point. As early as 1878, in "How to Make Our Ideas Clear", he speaks of "feigned hesitancy", saying that "whether feigned for mere amusement or with a lofty purpose", it "plays a great part in the production of scientific inquiry".[13] Then, in a note to "The Fixation of Belief" added in 1893, he says that doubt is typically "anticipated hesitancy about what I shall do hereafter, or a feigned hesitancy about a fictitious state of things. It is the power of making believe we hesitate, together with the pregnant fact that the decision upon the merely make-believe dilemma goes toward forming a bona fide habit that will be operative in a real emergency."[14] Finally, in a note of 1903, he says that "for the sake of the pleasures of inquiry, men may like to seek out doubts".[15] The net effect of these qualifications is surely to deny that research must always spring from actual difficulties, real irritations, or living doubts; it is also to acknowledge that the researcher is motivated not only to solve problems but to seek them. Research activity, in sum, does not subside when real doubts are dispelled, real problems resolved – for the generation of feigned doubts and hypothetical problems continues unabated.

Peirce's theory of inquiry, even under epistemological interpretation, thus remains difficult. In its unqualified form, it clashes with the fact of theoretical motivation in research. Taken together with its supplementary qualifications, it appears inconsistent. The qualified theory, moreover, undercuts Peirce's earlier criticism of Descartes. For, having himself insisted on the role of feigned or hypothetical doubt in science, how can Peirce dismiss Descartes' radical doubt as mere idle pretense?

AN EPISTEMOLOGICAL INTERPRETATION

My own view is that Peirce's formulation: *that all inquiry must begin in real and living doubt,* is indeed untenable. It implies that without real and living doubt there can be no inquiry; yet Peirce admits the importance of inquiries springing from doubts that are merely feigned. Once feigned inquiries are admitted, however, what differentiates Peirce's notion of doubt from the radical doubt of Descartes? What is the distinctive import of Peirce's theory?

I suggest the answer lies in the *role* ascribed to feigned doubt. For Descartes' method, feigned doubt disqualifies a proposition from serving as an assumption, since what he seeks are assumptions not only undoubted but indubitable. For Peirce, on the other hand, a proposition that is as a matter of fact undoubted, i.e. free of real and living doubt, still qualifies as an assumption, even though subject to feigned doubt. The mere fact that one might *hypothetically* doubt such a proposition does not *require* us to reject it as an assumption and try to replace it, or perhaps reinstate it through additional argumentation. We are *required* to disqualify assumptions only if they are subject to real and living doubts; that is, doubts that are specific to the propositions in question and that rest on positive reasons. But a proposition we are not *required* to reject as an assumption may be rejected anyhow for the space of a given hypothetical inquiry, during which other undoubted assumptions are meanwhile retained. Inquiries may, in other words, indeed originate in feigned doubt; such feigning is consistent with use of the proposition in question as an assumption in other inquiries; the mere possibility of such feigning does not render assumptions generally useless.

Peirce, according to this interpretation, is here rejecting the unconditional or wholesale doubt of radical scepticism, insisting that inquiry may stand on undoubted though dubitable assumptions, even as it proceeds to investigate others taken as problematic. The sceptic, doubting all assumptions short of indubitability, leaves himself no room to stand and allows himself no resources for dealing with the problems he raises. He errs, not in his feigning as such, but in his demand that all hypothetically dubitable propositions be feigned useless as assumptions simultaneously. By contrast, scientific doubt, whether it is real or feigned, is always specific, resting on provisional assumptions that serve usefully as premises of the inquiry even though they fall short of absolute certainty. Nor is the scientific researcher's work done when his problem is solved, for he will then try to find, imagine, or construct new problems specific enough to be formulable as testable questions. The answer to the sceptical yearning for certainty at the outset thus lies in the continuity of fallible inquiries tending toward the fixation of beliefs in the future.

THE PRIMACY OF METHOD

Consider now Peirce's comparison of methods in "The Fixation of Belief", a comparison that is very puzzling indeed. For Peirce promises to compare his four methods solely by reference to their relative effectiveness in stabilizing belief since, as he says, "the settlement of opinion is the sole object of inquiry". Yet, in defending the method of science, he does not even mention its superior effectiveness, but invokes instead a variety of new considerations, some metaphysical (relating to the supposition of real things), some methodological (relating to self-correctiveness), some epistemological (relating to the need for opinions to aspire to coincidence with fact), and some even moral ("to avoid looking into the support of any belief from a fear that it may turn out rotten is quite as immoral as it is disadvantageous").

Moreover, to defend science as more successful than the other methods in settling belief seems doomed to failure anyhow. Science does not, like the method of tenacity, yield "great peace of mind". Indeed the rate of change of scientific opinions would seem to be higher than that associated with any of the other methods. What is characteristic of science is that it places all its claims in perpetual jeopardy, making them forever vulnerable to unsettlement. It might perhaps be thought plausible to defend scientific method through appeal to the restless spirit of man, relishing the prospect of a continued unfixing of beliefs and recasting of received doctrines. How could Peirce have hoped to succeed in the exactly opposite course?

Perhaps, it might be said, the mere *change* of scientific opinion is not fatal to the notion of science as stabilizing belief. For such change may be construed as uniformly *progressive,* i.e. as adding to reliable information without disturbing the already available stock, as sharpening the vaguer formulations of the past, or as steadily narrowing the range of opinion in a process of approximation to an ideal limit. Such a conception is reflected in the important passage in "How to Make Our Ideas Clear" in which Peirce says that truth is "the opinion which is fated to be ultimately agreed to by all who investigate [...] and the object represented in this opinion is the real". Scientists, he says "may at first obtain different results, but, as each perfects his method and his processes, the results will move steadily together toward a destined center".[16]

This idea is, however, vulnerable to two criticisms: First, the concept of approximation may be suitable for measurements, but it does not fit theories. As Quine has remarked, "the notion of limit depends on that of 'nearer than', which is defined for numbers and not for theories".[17] Secondly, science does not simply add information or sharpen vague formulations or steadily converge in opinion; it often changes theoretical direction and rejects previous beliefs. Even if experimental and technological knowledge *does* tend to accumulate, even if later theories, in accounting for a wider range of such knowledge, are considered not merely *different from* but *superior to* earlier theories, still theoretical change must be recognized to be non-progressive: The theoretical agreement of a given period is often, that is to say, uprooted and superseded by a conflicting agreement in a later period. And this is incompatible with the project of showing science to be maximally effective in *fixing beliefs in general*. It is perhaps, I conjecture, some such train of thought that accounts for the first difficulty with the essay, i.e. that the defence of the method of science shifts ground, moving from a consideration of effectiveness to other considerations of various sorts.

These latter considerations have this in common: they transfer attention from the stability of *particular beliefs* to that of *methods,* arguing that the method of science is *itself* firmer than the other methods discussed. Because it rests on the undoubted supposition of real things, because it is self-corrective, because it tests beliefs not by reference to human attitudes, intuitions, or institutions but rather by reference to those facts to which the beliefs purport to refer, scientific method is itself capable of standing firm through the change of specific beliefs. To challenge a particular belief sanctioned by any of the other methods calls the method itself into question because none of these methods is capable of allowing consistent correction of its own pronouncements. These methods are *brittle,* incapable of absorbing change without fracture. The method of science, by contrast, achieves stability through flexibility. Rejecting pretensions to certainty, opening wide the testing process to all members of the ideal community of investigators, requiring continual correction to account for all available facts, the method is itself capable of absorbing change without upset.

Since, however, the essay thus shifts the question of stability from the level of belief to that of method, it does not in fact fulfill its promise. Yet the defence of science it offers is of interest in its own right, exemplifying, moreover, that emphasis on the primacy of method which is characteristic of pragmatic philosophy in all its variants. It is method rather than doctrine that defines the community of investigation, and it is the stability of method in pursuit of the truth that holds this community together throughout doctrinal change. Similarly, for pragmatic social theory, it is the method embodied in democratic institutions that defines the community dedicated to the qualities of human freedom and dignity, and it is the stability of democratic methods that may hold the community together through changes of policy and belief.

The task of education, finally, is to teach proper method. Its special function, as Dewey emphasized, is not to indoctrinate a particular point of view but rather to develop those powers of logical assessment and criticism by which diverse points of view may themselves be evaluated. Method, moreover, is the key to intellectual advance, and education therefore has another powerful motivation to stress method as primary. It is this aspect of education to which Peirce attaches the greatest importance. Construing logic as a methodical study of methods, and associating it with the general theory of signs, he writes: "When new paths have to be struck out, a spinal cord is not enough; a brain is needed, and that brain an organ of mind, and that mind perfected by a liberal education. And a liberal education – so far as its relation to the understanding goes – means *logic*".[18]

NOTES

I have drawn upon the treatment in my "Four Pragmatists" (London: Routledge & Kegan Paul, 1974), for various aspects of the present paper.

1. Murray G. Murphey, "Kant's Children: The Cambridge Pragmatists," "Transactions of the Chares S. Peirce Society," 4, 3-33.

2. Max H. Fisch, "Classic American Philosophers" (New York: Appleton-Century-Crofts, 1951, 1966), preface and general introduction.

3. Gail Kennedy, "Introduction to John Dewey," in Fisch, 328-29.

4. See Fisch, "The Classic Period in American Philosophy," in Fisch, 10-12.

5. John Dewey, "The Quest for Certainty" (New York: Minton, Balch & Co., 1929), 151; William James, "Collected Essays and Reviews" (New York: Longmans, Green, 1920), 86-7; quoted in Fisch, 26; and Dewey, 213.

6. John Dewey, "Democracy and Educational Administration," "School and Society" (1937), 457-62; reprinted in J. Ratner, "Intelligence in the Modern World" (New York: Modern Library, 1939), 404.

7. Peirce, "Questions Concerning Certain Faculties Claimed for Man," "Collected Papers," 5.253.

8. CP, 5.265.

9. CP, 5.12-13.

10. Murray G. Murphey, "The Development of Peirce's Philosophy" (Cambridge: Harvard University Press, 1961), 161.

11. Ibid., 163.

12. CP, 5.370ff.
13. CP, 5.394.
14. CP, 5.373, n. 1.
15. CP, 5.372, n. 2.
16. CP, 5.407.

17. W. V. Quine, "Word and Object" (New York: Technology Press of M.I.T. and John Wiley, 1960), 23.

18. From the Johns Hopkins University Circulars (Nov. 1882) as excerpted by M. H. Fisch and J. I. Cope, "Peirce at the Johns Hopkins University" in "Studies in the Philosophy of C. S. Peirce," ed. Philip P. Wiener and Frederic H. Young (Cambridge: Harvard University Press, 1952), 289-90. reprinted in Wiener, "C. S. Peirce Selected Writings" (New York: Dover, 1958), 336-37.

WRITINGS OF ISRAEL SCHEFFLER

BOOKS

The Language of Education. Springfield: Charles C Thomas, 1960.
The Anatomy of Inquiry: Philosophical Studies in the Theory of Science. New York: Alfred A. Knopf, 1963. New printing, Indianapolis: Hackett, 1981.
Conditions of Knowledge: An Introduction to Epistemology and Education. Chicago: Scott, Foresman, 1965.
Science and Subjectivity. Indianapolis: Bobbs-Merrill, 1967; 2d edition, Indianapolis: Hackett, 1982.
Reason and Teaching. London: Routledge & Kegan Paul, 1973. New printing, Indianapolis: Hackett.
Four Pragmatists. London: Routledge & Kegan Paul, 1974.
Beyond the Letter: A Philosophical Inquiry into Ambiguity, Vagueness, and Metaphor in Language. London: Routledge & Kegan Paul, 1979.
Of Human Potential. London: Routledge & Kegan Paul, 1985.

COLLECTIONS

Editor, *Philosophy and Education.* Boston: Allyn & Bacon, 1958; 2d edition, 1966.
Editor, with Richard Rudner. *Logic and Art.* Indianapolis: Bobbs-Merrill, 1972; new printing, Indianapolis: Hackett.

ARTICLES

Verifiability in History: A Reply to Miss Masi. *Journal of Philosophy* 47 (1950): 158-66.
The New Dualism: Psychological and Physical Terms. *Journal of Philosophy* 47 (1950): 737-52.
Anti-Naturalist Restrictions in Ethics. *Journal of Philosophy* 50 (1953): 457-66.

On Justification and Commitment. *Journal of Philosophy* 51 (1954): 180-90.
Is the Dewey-like Notion of Desirability Absurd? *Journal of Philosophy* 51 (1954): 577-82.
An Inscriptional Approach to Indirect Quotation. *Analysis,* n.s. 14 (1954): 83-90.
Toward an Analytic Philosophy of Education. *Harvard Educational Review* 24 (1954): 223-30.
On Synonymy and Indirect Discourse. *Philosophy of Science* 22 (1955): 39-44.
Science, Morals, and Educational Policy. *Harvard Educational Review* 26 (1956): 1-16.
Educational Liberalism and Dewey's Philosophy. *Harvard Educational Review* 26 (1956): 190-98.
Explanation, Prediction, and Abstraction. *British Journal for the Philosophy of Science* 7 (1957): 293-309.
Prospects of a Modest Empiricism, I. *Review of Metaphysics,* 10 (1957): 383-400.
Prospects of a Modest Empiricism, II. *Review of Metaphysics* 10 (1957): 602-25.
Inductive Inference: A New Approach. *Science* 127 (1958): 177-81.
Inscriptionalism and Indirect Quotation. *Analysis* 19 (1958): 12-18.
Justifying Curriculum Decisions. *The School Review* 66, (1958): 461-72.
What is Said to Be, with N. Chomsky. *Proceedings,* Aristotelian Society, London, 1958-59.
Thoughts on Teleology. *British Journal for the Philosophy of Science* 9 (1959): 265-84.
Comment on Geiger. In *Education in Transition.* Ed. Gruber. Philadelphia: University of Pennsylvania Press, 1960.
A Note on Confirmation. *Philosophical Studies* 11 (1960): 21-23.
A Rejoinder on Confirmation. *Philosophical Studies* 12 (1961): 19-20.
A Note on Behaviorism as Educational Theory. *Harvard Educational Review* 32 (1962): 210-13.
Is Education a Discipline? In *Education as a Discipline.* Ed. John Walton and James Kuethe. Madison: University of Wisconsin Press, 1963.
Concepts of Education: Some Philosophical Reflections on the Current Scene. In *Guidance in American Education.* Ed. E. Landy and P. Perry. Cambridge: Harvard University Press, 1964.
Postscript on Inscriptionalism. *Journal of Philosophy* 62 (1965): 158-60.
Comments on Sellars. In *Boston Studies in the Philosophy of Science,* vol. 2. New York: Humanities Press, 1965.

Philosophical Models of Teaching. *Harvard Educational Review* 35 (1965): 131-43.
Reply to Elizabeth Flower. *Studies in Philosophy and Education* 4 (1965): 133-36.
Reply to Professor Robert D. Heslep. *Harvard Educational Review* 35 (1965): 365-67.
Reply to George F. Kneller. *Studies in Philosophy & Education* 5 (1966-67): 136-38.
On Ryle's Theory of Propositional Knowledge. *Journal of Philosophy* 65 (1968): 725-32.
Reflections on the Ramsey Method. *Journal of Philosophy* 65 (1968): 269-74.
University Scholarship and the Education of Teachers. *Teachers College Record* 70 (1968): 1-12.
Reflections on Educational Relevance. *Journal of Philosophy* 66 (1969): 764-73.
An Improvement in the Theory of Projectibility, with Schwartz and Goodman. *Journal of Philosophy* 67 (1970): 605-609.
Explanations, Desires, and Inscriptions. *British Journal for the Philosophy of Science* 22 (1971): 362-69.
Philosophy and the Curriculum. *Philosophic Exchange* 2 (1971): 59-66.
The Moral Content of American Public Education. *Educational Research: Prospects and Priorities*. 92d Congress, 2d Session, Committee on Education and Labor. Appendix 1 to Hearings on H.R. 3606. Washington, D.C.: U.S. Government Printing Office, 1972.
Supercalifragilistic Reduction: A Reply to Jan Berg. *Philosophy of Science,* 1971, p. 121.
Ambiguity: An Inscriptional Approach. In *Logic and Art,* pp. 251-72. Ed. Rudner and I. Scheffler. Indianapolis: Bobbs-Merrill, 1972; new printing, Indianapolis: Hackett Publishing.
Selective Confirmation and the Ravens: A Reply to Foster, with N. Goodman. *Journal of Philosophy* 69 (1972): 78-83.
Vision and Revolution: A Postscript on Kuhn. *Philosophy of Science* 39 (1972): 336-74.
Philosophy of Education at Harvard. *HGSE Association Bulletin* 18 (1974): 11-16.
Basic Mathematical Skills: Some Philosophical and Practical Remarks. Proceedings of N.I.E. Conference on Mathematical Skills and Learning. October 1975. *Teachers College Record* 78 (1976): 205-12.
In Praise of the Cognitive Emotions. *Teachers College Record* 79 (1977): 171-86.

Philosophy of Education: Some Recent Contributions. *Harvard Educational Review* 50 (1980): 402-406.
The Wonderful Worlds of Goodman. *Synthèse* 45 (1980): 201-209.
Ritual and Reference. *Synthèse* 46, No. 3 (1981), pp. 421-437.
Reply to Gareth Matthews. *Synthèse,* Vol. 46 (No. 3, 1981): 445-48.
Pragmatism as a Philosophy. In *Zeichen Und Realität,* pp. 29-43. Ed. K. Oehler. Proceedings of 3rd Semiotics Colloquium, sponsored by German Semiotic Society. Tübingen: Stauffenburg Verlag, 1984.
Projectibility: A Postscript. *Journal of Philosophy* 79 (1982): 334-36.
Four Questions of Fiction. *Poetics* 11 (1982): 279-84.
Human Nature and Potential. *Educational Studies* 14 (1983): 211-24.
On the Education of Policy-Makers. *Harvard Educational Review* 54 (1984): 152-65.
Computers at School? *Teachers College Record,* forthcoming.

REVIEWS

Psychoanalysis and Religion. *The Reconstructionist* 17 (1951): 26-29.
Civilization and Value. *Harvard Educational Review* 23 (1953): 110-16.
Moral Principles of Action. *The Reconstructionist* 19 (1953): 24-29.
Philosophy of Science, by Gustav Bergmann. *Science* 125 (1957): 822.
Toward Reunion in Philosophy, by Morton White. *Harvard Educational Review* 27 (1957): 156-58.
An Introduction to Philosophy of Education, by O'Connor. *Journal of Philosophy* 56 (1959): 766-70.

INDEX

A

Adams, John Couch, 84
Aiken, H. D., 350-51
Alexander, H. G., 199n
Alston, W. P., 360n
Anderson, Alan Ross, 223n
Aristotle, 24, 360n
Austin, G. A., 355
Austin, J. L., 58
Ayer, A. J., 123, 124, 126, 361n

B

Bain, Alexander, 381
Baier, Kurt, 337n
Bergson, Henri, 48-49, 66
Berlin, I., 126
Berlyne, D., 361n
Bigelow, J., 104, 105, 106, 107, 109
Bochenski, I. M., 26n
Bohnert, Herbert, 229-30n
Braithwaite, R. B., 93, 107, 108, 109, 229n
Bridgman, P. W., 171
Brown, Stephen, 346n, 362n
Bruner, Jerome S., 355

C

Carnap, R., 2, 8, 9, 11, 12, 14, 15-18, 21, 24, 30n, 135, 153n, 163-64n, 287
Cartwright, Richard L., 223n
Cassirer, Ernst, xiv, 41, 42, 43-45, 46, 47, 48-49, 50, 52, 53, 66-67
Charlesworth, W. R., 361n
Chomsky, A. N., ix, 18n, 19n, 26n, 81, 99n, 119n, 151n, 215-24, 291n, 302n

Church, Alonzo, 2-3, 8, 9, 10-12, 18n, 21, 22, 23-24, 25, 26n, 28-29, 126, 346n
Colleran, J. M., 321n
Cooley, J. C., 164n
Cope, J. I., 391n
Craig, W., 147, 148, 149, 150, 151, 225
Cronbach, L. J., 346n

D

Daitz, Edna, 257n
Davidson, Donald, 3-5, 210n
Desai, M. M., 361n
Descartes, R., xii, 375, 379, 380, 381, 382, 385, 387
Dewey, John, 18n, 153n, 279, 280-81, 303-7, 310-11, 349, 353, 363-73, 376, 377, 378, 379-81
Ducasse, C. J., 112, 116
Dunes, D. D., 346n
Durkheim, E., 45

E

Einstein, Albert, 352
Elgin, Catherine, ix, 3, 79n

F

Feigl, Herbert, 19n, 120n, 230n
Fermat, P., 18n
Fisch, Max, 377, 378, 390n
Flew, A. G. N., 257n
Foot, P., 360n
Foster, Lawrence, 203-6
Frankena, William K., 337n
Frankfort, H., 68n, 62
Frankfort, H. A., 68n, 62
Frege, G., 19n

Frye, Marilyn P., 28-29

G

Gaifman, Haim, 214n
Gardner, Howard, 346n, 362n.
Gaster, Theodor H., 68n
Gombrich, E., 172
Goncharov, N. K., 337n
Goodman, Nelson, ix, 1, 2, 7, 12n, 18-19n, 22, 24, 26n, 32, 42, 52-57, 58-60, 76, 77, 81, 82-85, 99n, 119, 126, 129, 136, 139-40, 145, 148, 151-52n, 153n, 154, 155-56, 157, 159, 161-63, 166-68, 172-73, 195, 196, 201-5, 207-10, 226, 251-52, 257n, 271-78, 289, 291n, 294, 302n, 346n, 351-52, 362n
Goodnow, J., 355
Gorovitz, S., 31-35, 37, 38-39
Graves, Robert, 77

H

Haack, R. J., 3
Hanson, N. R., 172, 361n
Hempel, C. G., 18n, 21-22, 25, 88, 91, 92, 102n, 125, 127-28, 132, 135, 147, 148-49, 151n, 176-79, 181-90, 192-95, 198, 201, 202, 203, 205, 226, 227, 230n, 352n
Hepburn, R. W., 360n
Hill, Christopher S., 6n
Hirsch, Eli, 361n
Hook, Sidney, 307n
Hosiasson-Lindenbaum, J., 190, 192
Howard, V., 346n
Hume, David, 125, 154-55, 166

J

Jacobsen, Thorkild, 61-62
James, William, 42, 364, 375, 376, 377, 379, 380

K

Kant, Immanuel, 125, 317-18, 375
Kaufmann, Y., 68n
Kekulé, F. A., 352
Kelvin, Lord, 170
Kennedy, Gail, 377-78
Kuhn, Thomas, 82, 259-69

L

Ladd, John, 308n
Lakatos, I., 269n
Langer, Susanne K., xiv, 41, 45-48, 50, 52, 60-61, 63-64
Langford, C. H., 13n
Leverrier, U. J. J., 84
Lévy-Bruhl, L., 43, 45, 49
Lewis, C. I., 94n, 137, 294, 307n
Lilge, Frederic, 337n
Linsky, L., 151n, 153n
Locke, John, 310, 381

M

Mandelbaum, M., 119n
Martin, Jane Roland, 329n
Martin, R. M., 25n, 227-28
Mates, Benson, 18-19n
Matthew, Murdoch, 66-67n
Matthews, Gareth, 67n, 56, 70-73
Maxwell, Grover, 171, 226, 227, 230n
Mead, George Herbert, 376
Meckler, L., 24
Mills, J. S., 137, 156-57, 307
Moore, G. E., 287
Morgenbesser, S., 199n, 302n
Müller, Max, 43
Murphey, Murray G., 375
Musgrave, A., 269n

N

Nagel, E., 101n, 153n, 170-71, 229n

Index

Neurath, Otto, xi, 82, 233-41, 243, 244, 249-50, 254-55, 256
Nicod, Jean, 175, 179, 180, 194, 198

O

Oppenheim, P., 88, 91, 92, 99n

P

Pasteur, Louis, 84
Passmore, John, 257n
Pears, David, 190-91, 192, 194-95
Peirce, C. S., xi, xii, 280, 357-58, 371, 375, 376, 379, 381-87, 388, 390
Perkins, D., 346n
Perry, Ralph Barton, 331, 337n
Peters, R. S., 67, 120n, 320-21, 337n, 349, 360n
Pitcher, G., 360n
Planck, Max, 264
Plato, 266, 313, 332
Plumpe, J. C., 321n
Poincaré, Henri, 343
Polanyi, Michael, 353
Popper, K. R., xi, 99n, 183-84, 185, 186, 187, 270n
Price, Kingsley, 321n
Putnam, H., 19n

Q

Quasten, J., 321n
Quine, W. V., 2, 3, 12n, 18n, 22, 32, 36, 37, 38, 40n, 74-75, 96, 99n, 114, 120n, 145, 153n, 168-70, 215-18, 220-21, 223, 230n, 253, 255, 289, 295, 346n, 389

R

Ramsey, Frank Plumpton, xiv, 225-29
Rawls, John, 360n
Reichenbach, Hans, 100n

Rosenblueth, A., 104, 105, 106, 107, 109
Rosser, B., 346n
Russell, Bertrand, 74-75, 78, 251, 255
Russell, E. S., 107
Ryle, Gilbert, 279, 280-81, 322-29
Rynin, David, 257n

S

St. Augustine, 313-15
Sawyier, Fay H., 362n
Scheffler, I., 1, 2, 3, 14, 21, 29n, 31, 32, 33, 37, 39, 51n, 78, 79n, 120n, 153n, 203, 229n, 346n, 360-61n
Scheffler, Samuel, ix, 278n
Schilpp, Paul Arthur, 230n
Schlick, Moritz, 82, 233, 240-41, 242-51, 256, 354, 358
Schwartz, Robert, ix, 3, 40n, 82, 207-10
Scriven, Michael, 230n
Sellars, W., 19n, 120n
Shapere, Dudley, 269n
Solomon, Robert C., 360n
Soltis, Jonas, 361n
Sorabella, JoAnne, ix
Spencer, Herbert, 43
Stevenson, C. L., 291
Suppes, P., 346n

T

Tarski, Alfred, 3, 27n, 255
Taylor, R., 105, 107
Toulmin, S. E., 93, 153n

U

Ullian, Joseph, 214n

V

Velikovsky, I., 24

W

Walter, Marion, 346n
Wartofsky, M. W., 362n
Watkins, J. W. N., 183-85, 186, 187, 189-90, 200n
Weinstein, Scott, 4
White, Morton, 18n, 133, 287, 289, 306-7, 360n
Wiener, N., 104, 105, 106, 107, 109
Wiener, Philip P., 391n
Williams, B. A. O., 360n

Wilson, John A., 68n
Wittgenstein, Ludwig, 257n
Woodger, J. H., 27n
Wundt, Wilhelm, 376

Y

Yesipov, B. P., 337n
Young, Frederic H., 391n

Z

Zabludowski, Andréj, 214n

www.ingramcontent.com/pod-product-compliance
Lightning Source LLC
Chambersburg PA
CBHW071229290426
44108CB00013B/1337